Brunei and the Future Transformation of Southeast Asia

Gabriel Yit Vui Yong · A. K. M. Ahsan Ullah
Editors

Brunei and the Future Transformation of Southeast Asia

Springer

Editors
Gabriel Yit Vui Yong
Faculty of Arts and Social Sciences
Universiti Brunei Darussalam
Gadong, Brunei Darussalam

A. K. M. Ahsan Ullah
Faculty of Arts and Social Sciences
Universiti Brunei Darussalam
Gadong, Brunei Darussalam

ISBN 978-981-96-7900-3 ISBN 978-981-96-7901-0 (eBook)
https://doi.org/10.1007/978-981-96-7901-0

This Springer imprint is published by the registered company Springer Nature Singapore Pte Ltd.
The registered company address is: 152 Beach Road, #21-01/04 Gateway East, Singapore 189721, Singapore

Foreword

A country with less than half a million people, Brunei Darussalam may appear to most outsiders as a backwater or an end-of-career destination. There are no bustling towns, no towering skyscrapers, no trains, and no traffic jams that last over thirty minutes. Restaurants are rarely packed; shopping centers are seldom crowded. Thick forests, meandering rivers, and wide beaches envelop much of the landscape. Fully air-conditioned mosques are visible in almost every district. It is uncommon to find four to five cars in a single home, bought on festive occasions, or during unplanned showroom trips, or perhaps because petrol is cheaper than a can of Coca-Cola. A typical weekend is spent tracking in canopy walks, failing that, in wet markets, culminating in gatherings at the homes of relatives and friends, where laughter, gossip, and food are plenty. When asked about the day's highlight, many locals would happily respond: "I met the King!" As they do on many other days.

While such a characterization of the few surviving Asian monarchical states may be true, it belies the wider significance of this enigmatic domain. Brunei may be small when viewed from the perspective of numbers. It is boring if seen from the viewpoint of contentious politics. And yet, it is a social scientist's paradise, a philosopher's delight, and an analyst's puzzle. For geographers, conservationists, and experts in development studies, Brunei presents some conundrums to wrestle with: How has such a tiny oil-rich state sustained its operations with such a small citizenry? What measures have been taken to enhance its international standing? Why is the country redefining its image from a secularized territory to a Shari'a-oriented kingdom? Who are the key players shaping its destiny? The ruler, the elites, or the people? Where does the country situated itself in light of the rise of a highly conflictual, technologically driven, ecologically challenged, and increasingly mobile world? Does Brunei have an optimistic trajectory?

This collection of essays answers such questions. Together, the contributors explore Brunei's journeys through the prism of spaces and interfaces. They uncover the strategic spaces that Brunei occupies in an evolving world order and the ASEAN region in which it is firmly ensconced. This much-need text explains how the country has morphed into a halal hub. It puts into sharp relief the effects of climate change and global connectivity on Brunei's natural environment. To be sure, Brunei is a terrain exposed to a host of vulnerabilities, from energy to food and everyday governance. Its strategic position situated in-between two major sea routes works in its favour, allowing the country to capitalize on the revived Maritime Silk Road project with China as the most influential force. As most essays in this book clearly explain, Brunei has grown to become an interface of such initiatives and offers much more. Post-COVID-19 crisis, the government revitalized its age-old reputation as a magnet for economic migrants. Education and consumption, especially among the young, have risen exponentially because of such policies. Demographic changes, the entry of more women into the workforce, as well as transformations in global trade and capital markets, are encouraging the government to hasten the process of economic diversification.

Whether Brunei will build more stable and sustainable institutions to shed its old skin is a matter of speculation. Whether she is willing to reinvent fast enough and realize its spatial and interfacial capacities remains to be seen. I have had the benefit of working at Brunei's premium university. I have spoken at length with many high-ranking officials there. I have witnessed firsthand the everyday travails of ordinary Bruneians. The country is blessed with highly educated youth and committed individuals all too willing to see their homeland prosper. It is a realm saturated with a million potentials, endless resources, and extensive opportunities. Whichever direction the monarchical state is headed, it must begin its venture with ideas. This book is a minefield towards that end. It deserves to be trawled and utilized for practical solutions to common local and global concerns. It is an indispensable companion for those invested in the country's future. Indeed, *Brunei and the Future Transformation in Southeast Asia* is a manifesto for a nation that is constantly in motion.

Khairudin Aljunied
National University of Singapore
Singapore, Singapore

Preface

The future of Southeast Asia (SEA) is an unfolding narrative shaped by the aspirations of the diverse nations that have emerged from a few centuries of Western colonialisation and the Second World War (WW2), and having to negotiate the challenges of the global order established. Among them, Brunei Darussalam, a small oil-rich country, is often overlooked due to its size and insignificance in regional and global affairs. However, its storied history and strategic location could offer valuable insights to navigating contemporary geopolitical and economic dynamics.

Prior to the arrival of the Europeans, it was a major thalassocracy and regional power. Its fortunes and drastic transformation are an outcome of interaction with external forces that it never adequately grasped until it was too late. Present-day Brunei is a modern nation state occupying only 5,765 km^2 split into two parts by Limbang (Malaysia). Yet, although its past glory was lost to the British, it was Britain that made it a wealthy modern state.

Brunei's story however would resonate with many in the region, and indeed across the globe. Many were once proud sovereign regions with rich cultural heritage that were drastically altered by colonialisation. They are also grappling with their identity, trying to reconcile traditions and heritage with the modern world, which is irreversibly transformed and governed by global Capitalism and technological advancement.

Although SEA nations have evolved differently, each at their own pace and having to overcome unique challenges, countries do not develop uniformly. There are always regions within that lags, because of circumstances. A Bruneian perspective could offer insights into impacts of larger regional-global dynamics on local conditions. We decided to write Brunei and the Future Transformation of SEA, because of the tendency to in regional analysis to take the macro view, which renders local-micro issues inconspicuous.

The significance of this book lies in its focus on global-regional change. It addresses (a) the changing global political-economic order and (b) the Climate Crisis. The former refers to the inevitable transition from a unipolar world led by the USA to an emerging multipolar one being developed by the Global South. The latter, widely regarded as an existential crisis by its advocates, affects different parts of the planet in different ways. Its impact on SEA could be anticipated with a good degree of certainty, which allows for appropriate adaptation.

This edited volume examines the implication of a changing global order on the region, from the vantage point of Brunei and other similarly small regions, such as parts of larger countries. It hopes to shed light on the repercussion of regional-global developments on small geographical regions, which are not adequately prepared to absorb the shocks. It takes a geographical perspective because regions have distinctive characteristics that ought to be considered in analysis and regional dynamics are not given due consideration.

The book contains a mix of empirical case studies and theoretical analyses pertaining to the themes highlighted. It addresses questions on sustainability, production networks and food security, migration and trade, and climate change. The contributors offer their thoughts based on extensive experience living and working in Brunei and in the region. As such, it offers a perspective that elucidates the nuances of the local and periphery as part of the larger in responding and adapting to the wider dynamics.

As a microcosm of regional dynamics, the case studies presented a lens through which broader transformations in SEA can be understood. This volume therefore fills an important gap in the scholarly literature, which addresses the third pillar of the Brundtland's Commission's definition of Sustainable Development, i.e., social justice and equity, which has not received adequate attention. We believe that this book will advance SEA studies and enrich debates on globalisation, regionalism and cultural identity in a time of profound change.

For Brunei, the insights offered in this book can serve as a resource for reflection and planning that better fits its context. In a world increasingly defined by technological innovation, environmental challenges and shifting geopolitical landscapes, the country needs to capitalise on its unique strengths while managing its vulnerabilities. This book aims to provide a platform for critical discussions on how Brunei and other small regions can adapt to and realise their unique potential to benefit from, and participate meaningfully in, regional and global progress.

For scholars and researchers, this book offers important insights from empirical evidence, theoretical frameworks and policy analyses that can inspire further research. It also highlights the importance of geographical considerations, particularly of regions and their distinctive characteristics, whether of nature, culture or human dynamics. Finally, it is hoped the book will make both a reflective and forward-looking contribution, and highlights the unlikely role of the less significant in elucidating the complexity of the challenges ahead.

The book will appeal to a wide audience, including academics, policy makers and professionals interested in SEA studies, international relations and socio-economic development. It offers a different perspective on the region and suggests how, despite its diversity, SEA countries can address common challenges such as climate change, economic shocks and ride the digitalisation wave to achieve a sustainable future.

Gadong, Brunei Darussalam
2025

Gabriel Yit Vui Yong
A. K. M. Ahsan Ullah

Contents

Editors and Contributors

About the Editors

Gabriel Yit Vui Yong is a lecturer in the Geography, Environment, and Development Programme at Universiti Brunei Darussalam. With a B.Sc. in Geological Geophysics and an MSc in Environmental Analysis, he has spent three decades examining the complexities of sustainable development in Brunei and the region. Actively involved in fieldwork and national-regional projects, his extensive experience has fostered a view of the world as an intricate, evolving system influenced by conceptual processes of individuals and groups.

A. K. M. Ahsan Ullah is Professor of Geography, Environment and Development at the University of Brunei Darussalam (UBD). Dr. Ullah has worked with institutions such as the City University of Hong Kong, IPH at the University of Ottawa, McMaster University, Saint Mary's University, Dalhousie University in Canada, the American University in Cairo (AUC), Osnabruck University, Germany, and the Asian Institute of Technology (AIT), Thailand. His research areas include migration and mobilities, intercultural encounters and development, with a geographic focus on the Asia-Pacific, Africa, and Middle East, and theoretical focus on globalization and neoliberalism, development and human rights, transnationalism, gender, intersectionality and the everyday life.

Contributors

Izni Azrein Noor Azalie Faculty of Arts and Social Science, Universiti Brunei Darussalam, Gadong, Brunei Darussalam

Diotima Chattoraj Nanyang Technological University, Singapore, Singapore

Noor Hasharina Hassan Faculty of Arts and Social Science, Universiti Brunei Darussalam, Gadong, Brunei Darussalam

Mary Dorothy dL. Jose College of Arts & Sciences, University of the Philippines Manila, Manila, Philippines

Norzurianie Kamarulzaman Faculty of Arts and Social Science, Universiti Brunei Darussalam, Gadong, Brunei Darussalam

Anabelle B. Ragsag School of Social Work, McMaster University, Ontario, Hamilton, Canada

Fredolin Tangang Geography, Environment and Development (GED) Programme, Faculty of Arts and Social Sciences, Universiti Brunei Darussalam, Gadong, Brunei Darussalam

A. K. M. Ahsan Ullah Faculty of Arts and Social Sciences, Universiti Brunei Darussalam, Gadong, Brunei Darussalam

Gabriel Yit Vui Yong Faculty of Arts and Social Sciences, Universiti Brunei Darussalam, Gadong, Brunei Darussalam

Abbreviations

3Rs	Reduce, Reuse, Recycle
AEC	ASEAN Economic Community
AFSIS	ASEAN Food Security Information System
AFTA	ASEAN Free Trade Area
AICHR	ASEAN Intergovernmental Commission on Human Rights
AIIB	Asian Infrastructure Investment Bank
APTERR	ASEAN Plus Three Emergency Rice Reserve
ASEAN	Association of Southeast Asian Nations
BIRC	Brunei Islamic Religious Council
BRI	Belt and Road Initiative
CEDAW	Convention on the Elimination of All Forms of Discrimination Against Women
CO_2	Carbon Dioxide
COP	Conference of the Parties
CRMP	Coastal Resource Management Project
DAA	Department of Agriculture and Agrifood
DRRM	Disaster Risk Reduction and Management
FAO	Food and Agriculture Organization
FoMO	Fear of Missing Out
GDP	Gross Domestic Product
GMSL	Global Mean Sea Level
GPN	Global Production Networks
GVC	Global Value Chains
HACCP	Hazard Analysis and Critical Control Points
ICNS	Information-Communication Network and System
ICZM	Integrated Coastal Zone Management
IEA	International Energy Agency
ILO	International Labour Organization
IOM	International Organization for Migration
IPCC	Intergovernmental Panel on Climate Change
ISO	International Standards Organization

MABIMS	Ministers of Religious Affairs of Brunei, Indonesia, Malaysia, and Singapore
MSR	Maritime Silk Road
NDP	National Development Plan
NDRRMP	National Disaster Risk Reduction and Management Plan
NELM	New Economics of Labour Migration
NLD	National League for Democracy
PMB	Pulau Muara Besar
POEA	Philippine Overseas Employment Administration
RCEP	Regional Comprehensive Economic Partnership
RCP	Representative Concentration Pathways
SDG	Sustainable Development Goals
SEA	Southeast Asia
SLR	Sea Level Rise
SPA-FS	Strategic Plan of Action on Food Security
SSP	Shared Socioeconomic Pathways
UN	United Nations
UNFCCC	United Nations Framework Convention on Climate Change
UNFPA	United Nations Population Fund
WCED	World Commission on Environment and Development
WFS	Women-Friendly Spaces
WW2	Second World War

List of Figures

List of Tables

Chapter 1
Introduction: Southeast Asia from a Brunei Perspective

Gabriel Yit Vui Yong and A. K. M. Ahsan Ullah

Abstract This introductory chapter presents the concept of the book, Brunei and the Future Transformation of Southeast Asia. It explains the choice of Brunei as a focus, not to serve as a model or exemplar, but as a vantage point to view the global changes that are unfolding before us, and consider our readiness to navigate the challenges anticipated. The main aim is to consider details to understanding realities of smaller areas, which are typically ignored because of the tendency to focus on more important and salient features. The rationale for adopting a geographical approach is then explained and elaborated. Geographical applications of 'regions' are illustrated, using the following themes: physical environment; ecosystems; premodern regions; colonial blocs; independent countries; and ASEAN. It concludes by providing brief overviews of the chapters: Changing Realities (Chap. 2); Climate Change and Sea Level Rise in Southeast Asia: Current Knowledge and Future Projections (Chap. 3): Changing Global Order and Climate: A Comparative Analysis of Cartesian and Organismic Perspectives (Chap. 4): Food Security Using Rice Production in Brunei Darussalam as a Case Study (Chap. 5); ASEAN's Halal Production Networks: Towards Regional Integration and Effective Industrial Governing (Chap. 6); Young Consumers' Understanding of Sustainable Consumption (Chap. 7), Politics and Security in Southeast Asia (Chap. 8); Unpacking Women's Participation in Disaster Management: Insights from the Philippines and Its Diaspora (Chap. 9); Brunei and the Maritime Silk Road (Chap. 10); Southeast Asia: Mobility, Development and Long-term Plan (Chap. 11); and Migration Routes in Southeast Asia: How Economy Shapes the Routes (Chap. 12).

Keywords Southeast Asia · Brunei · Vantage point · Geographical approach · Smaller areas

---

G. Y. V. Yong (✉) · A. K. M. A. Ullah
Faculty of Arts and Social Science, Universiti Brunei Darussalam, Gadong, Brunei Darussalam
e-mail: gabriel.yong@ubd.edu.bn

A. K. M. A. Ullah
e-mail: ahsan.ullah@ubd.edu.bn

G. Y. V. Yong and A.K.M. A. Ullah (eds.), *Brunei and the Future Transformation of Southeast Asia*, https://doi.org/10.1007/978-981-96-7901-0_1

Introduction

This volume examines the future transformation of Southeast Asia (SEA) using Brunei as a vantage point. It covers interrelated themes of strategic importance in the context of heightened geopolitical tensions, conflicts and trade wars, against a backdrop of unfolding climate change and advancing technology, particularly in the field of automation, new energy and artificial intelligence (AI), which are transforming social-cultural and economic landscapes. Of the eleven countries in the region, ten had suffered as well as benefited from Western colonialisation (Kratosa, 2021). Although Thailand was never colonised, it also underwent significant internal transformation as a consequence of Western pressure in the nineteenth–twentieth centuries, which forced it to align with the modernisation process, resulting in the loss of territory to Britain and France (Chakrabarty, 2010). In the period following the Second World War (WW2), SEA countries regained independence through the United Nations' (UN) decolonisation initiative under resolution 1514(XV), adopted on 14 December 1960: Declaration on the Granting of Independence to Colonial Countries and Peoples[1] (UN General Assembly, 1960); the UN was formed only 15 years earlier in 1945 in response to the two global wars, with a charter to maintain international peace and security, promote friendly relations among countries, encourage social progress and better living standards, and protect human rights (UN Charter, n.d.).

Brunei, a small oil state on the northwest coast of Borneo, is an unlikely focus, given its relative insignificance in the global arena and academic literature. It is often omitted in ASEAN reports,[2] due to lack of data and participation in the processes concerned. As a young nation-state, which has regained its independence from Great Britain just 40 years ago in 1984, Brunei is still in its nascent stage of nation building. Its modern façade and the extravagant lifestyle of the elite class have tended to draw attention away from the reality on the ground, where it shares a world of inequality and numerous development and sustainability issues with other countries in the world. Only locals and visitors who have spent time with communities outside the bubbles of expatriate professionals and politicians are aware and can relate to similar situations that exist throughout the region, even in countries that are advanced, economically robust, and highly developed. This book presents studies based on the experiences and thoughts by academics who have lived and worked in Brunei for extended periods. Some are locals, born and bred in various parts of the country. The volume is augmented by articles on other countries in the region, and the region overall. The purpose is to improve clarity of regional characteristics and issues to address the focus of the book, which is: the future transformation of SEA.

This introductory chapter will therefore begin with an overview of SEA countries, followed by an elaboration of the overall approach, i.e., employing a geographical perspective to elucidate the complex dynamics and unfolding transitions in the world

[1] https://legal.un.org/avl/ha/dicc/dicc.html.

[2] https://asean.org/library-of-asec/.

and region. It then filled in details of geographical regions, beginning with the physical environment, followed by the historical development of its people, culture and their diversity, to the range of issues and challenges confronting the region and members states at this juncture in its history. It will conclude by introducing the chapters and highlighting the salient points and relevance.

Brunei in Southeast Asia

In 1967, five SEA countries (Indonesia, Malaysia, Philippines, Singapore and Thailand) founded the Association of Southeast Asian Nations (ASEAN) to promote regional stability, economic growth, and cultural exchange.[3] Another five others joined later: Brunei (1984); Vietnam (1995); Lao PDR and Myanmar (1997); and Cambodia (1999). Timor-Leste, which regained independence in 2002 from Indonesia, after initially declaring the same from Portugal in 1975, has applied to joined ASEAN in 2002, but is still in the process of achieving full membership. Since its formation, ASEAN has made significant strides in regional cooperation and integration, including forming partnerships with Japan, South Korea, and China, and navigating numerous political, economic and global challenges, such as the 1997–1998 financial crisis (Kittiprapas, 2000) and Covid-19 pandemic, which lasted from 2019 until 2023 (Ullah & Chattoraj, 2022; Ullah et al., 2024). On 20 November 2007, ASEAN adopted a Charter to consolidate its gains and guide development (ASEAN, 2007), which came into force on December 15 of the following year.

According to its 2024 statistical highlights,[4] ASEAN has a population of 676.6 million in 2023, which accounted for 8.5% of the world's population. Indonesia is the largest country by land area and population with the largest economy in terms of nominal gross domestic product at current prices (see Table 1.1). The next most populous, with over 100 million, are the Philippines and Vietnam, each with similar land area of over 300 thousand km^2. Myanmar and Thailand, which are larger, at about 700 and 500 thousand km^2 respectively, have population of 56 million and 65 million, the latter having about 10 million more people. Lao PDR and Cambodia are quite large in area, but have relatively small population. At the bottom, the smallest country by area and population is Singapore and Brunei respectively. In terms of Life Expectancy (Expect. in Table 1.1), Singapore has the highest at 83.0 years while Myanmar has the lowest at 67.4 years. Apart from Lao PDR, the rest have Life Expectancy in the 72–76 range. The countries with the highest percentage of grey population (>65 years in Table 1.1) are Thailand (13.4%), Singapore (12.1%) and Vietnam (9.2%). Lao PDR has the lowest at 4.8%.

ASEAN is a demographically young region, where almost half of the people are below 30 years old. It has also urbanised rapidly, such that almost half its population are urban, compared to about 30% in 1967, when it was formed. Indonesia has

[3] https://asean.org/about-asean.

[4] https://www.aseanstats.org/publication/ash-2024/.

Table 1.1 Key statistics, ASEAN, 2022–2023

Country	Area (km^2)	Population (000)	% 65 yrs	Life expect. (Yrs)	GDP current prices (USD billions)	GDP grow. (%) Const. Prices	CO_2 (Mt)	% Reg.
Brunei	5765	450.5	6.9	76.2	15.1	1.4	9.1	0.5
Cambodia	181,035	17,091.5	6.6	73.3	42.0	5.0	14.7	0.9
Indonesia	1,892,410	278,696.2	7.0	73.9	1371.1	5.0	651.7	38.7
Lao PDR	236,800	7545.8	4.8	68.0	14.2	4.2	18.8	1.1
Malaysia	330,411	33,379.5	7.4	74.8	396.3	3.6	241.1	14.3
Myanmar	676,577	56,243.0	7.3	67.4	64.4	3.3	25.9	1.5
Philippines	300,000	111,912.4	5.9	72.7	437.2	5.5	137.5	8.2
Singapore	734	5917.6	12.1	83.0	501.3	1.1	46.6	2.8
Thailand	497,407	65,083.8	13.4	76.2	514.9	1.9	250.0	14.9
Vietnam	331,344	100,309.2	9.2	74.5	429.5	5.0	286.6	17.0

Sources ASEAN Statistical Highlights 2024; IEA Greenhouse Gas Emissions from Energy in 2022, https://www.iea.org/data-and-statistics/data-product/co2-emissions-from-fuel-combustion

the largest economy, with GDP in current prices of USD 1,371.1 billion in 2023. It leads a cluster of five countries by 2.7–3.5 times: Thailand, Singapore, Philippines, Vietnam and Malaysia, with GDP ranging from USD 515 to 396 billion. The second largest country by area, Myanmar, has the third smallest economy, while Brunei, often misperceived as a wealthy country has the second smallest economy at USD 15 billion. However, in terms of growth rate based on GDP at constant prices, i.e., excluding inflation and other extraneous factors, four countries (Philippines, Indonesia, Vietnam and Cambodia) recorded a rate of 5% or more, with the Philippines leading the pack. The country with the smallest economy, Lao PDR, has a growth rate of 4.2%, indicating upward mobility in economic development. Singapore, Brunei and Thailand appear to be experiencing considerable lag in growth.

With regards to CO_2 emission from fuel combustion (IEA, 2022), ASEAN contributes 4.5% of the global emission of 37,400 million tonnes (Mt) annually. Indonesia is the largest emitter, with 651.7 Mt, or 38.7% of regional (Reg.) contribution in 2022. It is followed by Vietnam, Thailand and Malaysia, which produced half as much as Indonesia individually. As expected, Brunei is the lowest emitter, despite being an oil producer, due to its small size, accounting for only 0.5% of ASEAN's total emission. In contrast, Singapore, nearly eight times smaller than Brunei, but with a population that is thirteen times larger, produced five times more CO_2. The less urbanised–industrialised member states (Cambodia, Lao PDR and Myanmar) accounted for 0.9–1.5% of regional emissions each. Modern development and population are therefore major factors.

It can be seen from Table 1.1 that Indonesia is the most sustainable country in ASEAN in terms of economic size, growth and Life Expectancy (a proxy for quality of life). Despite its large CO_2 emission, its per capita emission is only 2.3 tonnes (t), which is ranked 7th in the region. The small nation of Brunei, with its small sluggish economy and low emission in absolute terms, has a per capita CO_2 emission rate of 20.3 t, an order of magnitude greater than Indonesia. The Philippines and Cambodia appear to be on the right path, with low per capita CO_2 emissions (0.9 and 0.5, respectively), similar life expectancies of 73 years, and GDP growth of 5.5% and 5.0%, respectively. Vietnam is doing similarly well with GDP growth at 5.0%, life expectancy of 74.5 years and per capita CO_2 emission of 2.9 t. Singapore and Malaysia have relatively high emissions (7.9 and 7.7), but the former encountering slowing growth while the latter is growing by 3.6%. It is clear that Brunei is not a leader or model in ASEAN. It is interesting to note however that Singapore, the other small state, which is widely regarded as most advanced in the region and a leader in sustainable development, shares some unsustainable statistics with Brunei. There is therefore much to study in order to understand realities beyond superficial characterisation from cursory reference to published information.

Method: A Geographical Perspective

Geography seeks to understand patterns by synthesising knowledge from different perspectives. Geomorphology, ecology and other environmental sciences are necessary to understand how the physical geographic context responds to human activities and to anticipate the changes (Haggart, 2001). This applies to understanding environmental impacts and climate change, which are supported by a growing number of global monitoring systems that continuously provide important synoptic data for analysis and modelling. On the ground, geographers study the dynamic interplay between people, technology, development and the environment in conjunction with culture, economy and political relations (Purvis & Grainger, 2004; Sachs, 2020). Geospatial technology and systems, which have become indispensable in commerce, finance, economics and management, are also important geographical research tools. Maps, a means of visualising and presenting spatial information, are essential elements in the human sphere. Globalisation has led to the convergence of disciplines and enhanced the role of geography in research that involves a spatial dimension, from local to regional and global scale. It provides a suitable methodology for studying globalisation and geopolitics (Blouet & Blouet, 2001).

From a geographical perspective, the world is a mosaic of many distinct 'regions' defined by unique cultural and physical characteristics (Nijman et al., 2020). They are given meaning through human agency, identifying some as countries, states, private property and so on, including larger geographical extent, such as SEA. Regions are dynamic. They change in character as well as in extent primarily in response to the human stories that play out within and between them, such as: trade; wars; and conquests. Human mobility, interaction with environment, technology, and inter-regional relationships are key drivers of transformation. Regions in SEA have changed dramatically with the rise and fall of kingdoms. In the sixteenth century, European colonial powers began introducing and imposing systems and structures on various regions (polities, kingdoms, parts of kingdoms, etc.), effectively erasing the old and reconfigured them to align with colonial systems to facilitate exploitation and transferance of resources back to their home territories. These structures have been further developed into the current global systems governing financial transaction, trade and international relations. In SEA, structures implemented by Western powers continue to serve as guides as they develop and participate in the global systems established. However, historical structures, much of which were established by religion and past polities and cultures, co-exist with the modern systems, and their interference contributes to inefficiencies due to incongruence between the overlapping systems.

Regions are defined based on commonalities, such as, location, culture, history and political-economic alliance. Geographical analysis differentiates between distinct regions based on one or more aspects of interest. Further details could be provided by sub-disciplines such as geomorphology, climatology and biogeography for physical regions. Additionally, as regions are dynamic, i.e., they change in extent and characteristics over time, due to external forces, including planetary dynamics, and the developmental trajectories of the different parts of regions as well as their interactions, it is important to understand specific controlling factors in particular regions or sets of contiguous regions. This introductory chapter will therefore examine SEA from this perspective. ASEAN is important as a region. However, Table 1.1 suggests that differences and clusters exist.

Physical Regions

Gupta (2005) provides an excellent description of SEA's environmental regions. The north-west region is part of the Asian land mass, comprising present-day Myanmar, Lao PDR, Thailand, Vietnam, and the Malaysian peninsula. Eastwards, the region comprises mainly islands, most of which are volcanic, created by the convergence of tectonic plates in the western part of the Pacific Ring of Fire. The two archipelagic countries are Indonesia and the Philippines. Borneo is located off the active plate margins at the southern edge of the South China Sea (SCS). This large island is shared by Brunei, Malaysia (Sabah and Sarawak) and Indonesia (Kalimantan). Singapore is the small island off the southern tip of Peninsular Malaysia.

SEA is located within the Tropics of Cancer and Capricorn, east of the Greenwich meridian between the longitudes of 100–150°E. Several parts of Indonesia straddle the Equator, and Brunei, Singapore, Malaysia and Timor-Leste are close enough to it to be affected by the fluctuation of the Inter-Tropical Convergence Zone (ITCZ), which migrates to the north and south of the equator during the Boreal summer and winter respectively. This region, classified as "Humid Tropical" under the Köppen climate classification, is characteristically warm and wet throughout the year, with high humidity and precipitation. It coincides with the Rainforest in the World Wildlife

Fund (WWF) ecoregions classification. Due to differences in climate , there is a subdivision, with a wetter zone around the Equatorial belt (Tropical Moist Broadleaf Forests, or Indo-Malayan Rainforest), where there is no distinct dry season, and one with lower rainfall and a distinct dry season (Tropical Deciduous Forests) (Olson, 2001). The forests are also taller and evergreen in the equatorial region, while deciduous species are found in the drier region further north. Due to the high solar energy input and precipitation ranging from 1750 to over 6000 mm annually, natural vegetation grows rapidly, some developing into colossal trees. Wetlands are characterised by swamp forest (mangrove, peat swamps, freshwater swamps, and mixed) instead of marsh.

The SCS is the central physical feature. It is bounded by archipelagos, large islands and peninsulas, and the southwestern part of the Asian mainland in the northwest corner. The SCS is connected to China at its northern edge and to the Pacific, the Indian and Southern Oceans to the east, west and south respectively. The coastal regions around SCS forms a distinct ecological as well as cultural region, the latter a product of people interacting with the natural environment. This region includes the continental shelves created by sea level rise during the first 5,000–7,000 years of the current Holocene epoch, which began c.10,000 years ago, with the last retreat of Pleistocene glaciers. The SCS is a distinct marine region with respect to ecology and fisheries. The Sulu, Celebes and Banda seas to the southeast are also important but distinct marine regions in SEA. The coastal waters of Southeast Asia are characterised by rich biodiversity, including extensive coral reefs, mangroves, and seagrass beds. These ecosystems are among the most diverse in the world, supporting a wide variety of marine species and habitats. The region is particularly known for its high levels of endemism and the presence of numerous species that are not found elsewhere. The Indo-Pacific marine region is a biodiversity hotspot. It contains the Coral Triangle.[5]

The region's geology is the outcome of complex interaction of tectonic plates, volcanic activity, and sedimentary processes, because it lies at the convergence of several major tectonic plates (Indo-Australian, Eurasian, and Pacific). In the western region, the Himalayan orogeny created prominent features, e.g., the Thai highlands and mountain ranges of Myanmar, from folding and faulting. A line of active volcanoes, e.g. the Sunda Arc, marks the boundary between the Indo-Australian and Eurasian plates. Indonesia and the Philippines have numerous volcanoes, which produced fertile soils, unique ecosystems and contributed to productive agriculture. However, in contrast, much of coastal SEA consists of extensive lowland areas, composed of sedimentary deposits. River systems, e.g., the Mekong and the Irrawaddy, have shaped the physical landscape through erosion and sedimentation, creating vast deltas and floodplains that support rich agricultural land. Additionally, limestone formations are prevalent, particularly in regions in Thailand, Vietnam and Malaysia, producing spectacular karst landscapes with dramatic cliffs and caves. The physical landform is composed of a mosaic of regions of interest to geomorphologists, biogeographers and environmental scientists.

---

[5] https://www.coraltriangleinitiative.org/.

SEA has rich mineral resources due to its geology. They include tin, gold, coal and petroleum. Extensive gravel, sand and clay deposits are also valuable for economic development. It is the presence of natural resources—minerals, spices, and jungle products—that drew the European to the region in the sixteenth century.

Changes Through History

SEA is largely a maritime region with the SCS at its centre. As the name of the sea suggests, it lies south of China, and occupies a strategic position in maritime trade, including the ancient Chinese maritime 'silk road'. The people and cultures are therefore inextricably linked to maritime trade (Billé et al., 2022). It brought religion (Hinduism, Buddhism and Islam) to the region and established several thalassocracies: Srivijaya (c.700–c.1400), Majapahit (1293–c.1500) and Brunei (1368–1888). The Buddhist kingdom of Srivijaya on Sumatra controlled the Strait of Malacca and thus the trade between the Indian Ocean and the SCS. The Javanese kingdom of Majapahit ruled over the Indonesian archipelago. The Sultanate of Brunei was the bastion of Islam in the region in the fifteenth–sixteenth centuries. Its capital was "built entirely over the water" in the Inner Brunei Bay (Pigafetta, 1521, in Nicholl, 1975:11). Archaeological evidence shows that the region's coastal communities engaged in small-scale maritime trade as early as the fourth-century BCE via ports in present-day Malaysia, Cambodia, Thailand and Vietnam (UNESCO, n.d.). Maritime trade facilitated the spread of knowledge and skills as well as the establishment of local production and innovation centres in SEA (ibid.; Billé et al., 2022).

Historically, two distinct sub-regions, Nusantara and Indochina, co-existed in premodern SEA. Both played significant roles in the history, trade, and culture of the region. Nusantara refers to the Malay world, encompassing the Indonesian archipelago and surrounding islands. It also include the Malay Peninsula and parts of the Philippines. It is a region with a diverse range of ethnic groups, cultures, and polities that interacted with one another and with neighbouring regions. Indochina is a land region, comprising modern-day Vietnam, Laos, Cambodia, Thailand, and parts of Myanmar. It has strong historical ties to Indian and Chinese empires and cultures. The two regions were connected through maritime trade and cultural exchanges. The Malacca Strait was a vital artery linking the Nusantara islands with the Indochinese mainland. It facilitated the diffusion of religion, art, language, and political influence across both regions. In the sixteenth century, the established trade routes brought Western Europeans to its shores. Portugal, Spain, and later, Britain, the Netherlands and France, colonised much of the region, ruled over the populace, and exploited its resources.

Western colonisation profoundly altered the socio-economic, political, and cultural landscape of Nusantara, impacting its diverse societies (Kratosa, 2021). The Portuguese, Dutch, and British, in particular, introduced significant changes in the region. In the seventeenth century, the Dutch established the Dutch East India Company, which gained control over vast territories in present-day Indonesia to monopolise the valuable spice trade, significantly affecting local economies, shifting traditional subsistence farming to cash crop agriculture (e.g., coffee, sugar, etc.). The new governance structures imposed disregarded existing local systems, resulting in social stratification and fragmentation, and conflict among the indigenous populations. The introduction of Christianity further reshaped cultural dynamics, as missionary activities promoted new religious practices, replacing and altering traditional beliefs. Western education and legal systems led to changes in the social fabric and the emergence of a new educated elite. The colonisers exploited local resources for their benefit, causing drastic and deleterious environmental changes.

The British's impact on the Malay peninsula, Singapore and north Borneo (Brunei) was similar. At the time, Brunei was a sultanate with a territory that spanned the entire northern half of Borneo. The British carved up Brunei through two separate developments. James Brookes, a colonial adventurer and former Naval officer, who became involved in the regional politics and conflicts, managed to acquire a massive region in the western part of the sultanate in 1841, which his successor Charles Brooke later expanded at the expense of Brunei, including splitting the latter into two parts with the acquisition of the Limbang in 1890. The Brookes founded a dynasty of White rajahs which ruled Sarawak for over a century (Runciman, 1960). To the east (modern-day Sabah), Baron von Overbeck acquired the rights to exploit the region in 1877, but later sold it to the North Borneo Company (Seitelman, 1948). In 1888, Brunei accepted Britain's offer to become a British Protectorate, together with Sarawak and North Borneo (present-day Sabah), as other colonial powers were interested in their rich resources, notably rubber, rattan and coal. In 1906, Brunei accepted a British Resident to govern the remnant of its territory, until WW2. Britain's influence is deeply entrenched in modern Brunei's government, constitution, education, and even culture.

The French established control over Vietnam, Cambodia, and Laos, consolidating them into French Indochina by the late nineteenth century (Kratosa, 2021). It introduced a centralised governance structure, replacing traditional local authorities and imposing new laws that often marginalised indigenous customs and practices. The French exploited Indochina's rich natural resources, particularly rubber, rice, and minerals, to fuel economic development that primarily benefited France. They promoted infrastructure development, including railroads and roads, aimed at enhancing resource extraction and connecting urban centres, which also facilitated the movement of goods and people. However, this led to significant land dispossession, robbing local farmers of their means of livelihood. Culturally, the French introduced Western education, leading to a new educated elite that began to challenge colonial authority. Missionary activities promoted Christianity while also causing tensions with traditional beliefs. Nationalism emerged as a response to colonial oppression, with movements gaining momentum in the early twentieth century, such as the Viet Minh in Vietnam.

In summary, the impact of colonisation was a wholesale change, imposed without regard for existing people and culture. It also introduced roads, vehicles, and hydrocarbon-powered machinery, which paved the way to a host of environmental impacts, from deforestation, pollution, erosion and sedimentation, and climate change, which attained proportions that alarmed Western nations only in late twentieth century. Japan briefly occupied the region during the WW2 from 1942 to 1945, causing much devastation and suffering.

ASEAN and the Current West-Led World Order

Following the surrender of Japan in WW2, Indonesia took the opportunity presented by the power vacuum to declare independence in 1945. However, the Dutch returned, intending to resume control, resulting in a bloody struggle until 1949, when Indonesia finally gained its independence through the UN decolonisation process (Kratosa, 2021). Vietnam went through a similar process. It had declared independence from France in 1945, but the French returned to reoccupy it, culminating in the First Indochina War in 1954. North Vietnam finally achieved independence as the Democratic Republic of Vietnam at the end of this conflict. However, following the American War,[6] Vietnam became a unified Socialist Republic in 1976 (ibid.). The Philippines and Myanmar had achieved independence from USA (in 1946) and Britain (in 1948) respectively shortly after WW2. Laos and Cambodia achieved theirs in 1953 from the French; Malaysia in 1957 from Britain, with Sabah and Sarawak joining the federation in 1963; Singapore in 1965 from Britain; and Brunei in 1984 from Britain. In 2002, Timor-Leste finally achieved independence from Indonesia, which had occupied it after the withdrawal of Portugal in 1975. The SEA nations set up ASEAN to facilitate cooperation and enhance regional development. Timor-Leste applied to join in 2002 and was granted 'observer status in 2023 and will most likely become a full member in 2025 (Cardoso, 2024).

The largely British, Dutch and French colonial regions therefore gave way to regions of the modern nation states, federations and republics of today, which have banded together to form a larger bloc for regional cooperation. ASEAN's main objective is (a) peaceful economic growth and progress through mutual respect for national sovereignty, and (b) regional cooperation in the areas of economy, cultural exchange and security. The ASEAN Charter (ASEAN, 2007) aimed to strengthen co-operation and regional integration through three pillars: (i) Political-Security Community (APSC); (ii) Economic Community (AEC); and (iii) Socio-Cultural Community (ASCC). Each pillar is supported by roadmaps and blueprints. Despite a rather turbulent twentieth century, SEA as a region appears to be politically and economically stable at the turn of the century and is well on its way to becoming one of the four economic powers of the world due to its location and the institutional structures created (Laker, 2023).

---

[6] Or Vietnam War in western media and literature.

Confronting Geopolitics and Navigating Global Change

As the twenty-first century unfolds, how should the region navigate the geopolitical and economic currents (Islam, 2023, March 27), as well as confront the spectre of the climate emergency declared by the United Nations (UNEP, n.d.). A thought-provoking article in the Jakarta Post entitled, "ASEAN 2050: Towards a new ambition" (Galimberti, 2023), stated that, if ASEAN truly wants to become a relevant institution in the world, it needs to become relevant and noticed by its own people. The article lamented the painfully slow pace towards regional integration in the areas of sustainability, democracy and human rights, and shared prosperity. It identified the 'ASEAN way' as the main obstacle and called for a real change of heart and leadership from among the younger generations, individually and collectively. However, given the state and circumstances of/in SEA following WW2, independence struggles, and post-independence 'teething' problems, ASEAN's advancements have been impressive. Careful, measured and prudent decision-making, while slow, is evidently appropriate, given the lack of capabilities to navigate the modern, globalising world created by the collective 'West' (former colonial nations). The USA assumed the role of leader and defender of the 'free' world and its currency served as lynchpin of the global economic system. It is perhaps that ASEAN leaders not aligning with the agenda and rules set by the West that prompted Galimberti to question the ambition of ASEAN's leadership. However, it is apparent that the twenty-first century is turning out to be a period of unprecedented change, not dissimilar to the early decades of the previous century, and possibly, as tumultuous.

This book proposes that a geographical perspective of tensions, conflicts and changes unfolding before us from a Bruneian vantage point would be insightful to formulate appropriate strategies to navigate the potential global and regional shocks and their fallouts. As pointed out at the outset, Brunei is not proposed as a model or an exemplar of how to do it correctly. Rather, the rationale is quite the opposite. As one of the youngest nation-states, Brunei is still in the early stages of nation-building, trying to reconcile the modern political and state structures imposed on it by a foreign power with the traditional systems that has been in place for 600 years. The complexity of governance structure has been further complicated by the recent adoption of religious syariah system. The young nation has been highly reliant on foreign experts and migrant labour in the first decades of its development, even after approximately six decades of a local human resource scheme in which selected citizens were sent overseas for training. The small population and lack of appropriate supporting structures are major factors, and the system established by the British has become outmoded and require modernising. Brunei's development programme is fraught with problems and inefficiencies. Plans to diversify the economy have met with limited success. The domestic economy is sluggish and youth unemployment is one of the highest in the region despite the small population and oil wealth.

The purpose of using Brunei as a case study and measure is however not to highlight its problems or underachievements, which is often perceived as criticising the country and seen as unpatriotic, disrespectful, and ungrateful at home. On the contrary, an academic's contribution to national and regional development is to identify weaknesses, gaps, or entanglements in existing system so as to improve upon it. Brunei's situation is not unique. Rather, it is quite common across regions, if 'regions' are considered from a geographical perspective, as distinct areas across the gamut of scale. Such regions exist within every country, as lesser developing parts, or in the wider region, such as ASEAN, like the Brunei-Indonesia-Malaysia-Philippines East ASEAN Growth Area (BIMP-EAGA), which is less developed. Brunei's main impediments are (i) its size, which limits its capacity for ambitious projects, and (ii) its recent history, which means, it is still at the beginning part of the learning curve. In any case, disregarding smaller under-developed regions is antithetical to sustainable development (WCED, 1987). In the case of Brunei, its frequent dismissal due to lack of interest in the international arena, where its issues are not given due consideration, reflects a lack of regional commitment to sustainable development, which ought to be truly inclusive in order to achieve the triple bottomline of the concept.

The issues and problems highlighted are common to many geographical regions. In SEA, they share common features, from physical geography, environment, religion and culture, to people, migration history and colonial legacy. But they are also different and individually unique due to geography and history. Many are struggling to navigate the modern, globalised and increasingly digitalised world, often unaware of the geopolitical tensions or the economic and information war being waged beneath the surface, and of their influence. The goal of the book is to catalyse interest in the smaller regions, which are often overlooked because they are regarded as insignificant. In the context of sustainable development (WCED, 1987), they matter and their inclusion is needed to achieve the desired sustainability goals.

Presentation of the Book

In Chap. 2, Gabriel Yit Vui Yong sets the context of the book, highlighting the unfolding transition from the current unipolar global economic-political order led by the USA, to a multipolar world being developed by China and the Global South. It presents the consequential Changing Realities amidst a climate crisis, but also examines how reality is created in the mind of individuals and the collective. He addresses the role played by state agencies, private enterprises and media in shaping human reality to manufacture consent. The relentless advancement of the global information-communication network and system (ICNS) and its influence of humanity is then discussed in relation to Vernadsky's theory of the Biosphere and Noosphere. The chapter explored trawling the Internet as a methodology for accurate understanding of current geopolitical tensions, trade and proxy wars, and the impacts of US policies and actions on China and the Global South. It also discusses the tipping point and how a potential nuclear war poses an existential threat.

In Chap. 3, Fredolin Tanggang presents an overview of Climate Change and Sea Level Rise in Southeast Asia: Current Knowledge and Future Projections, drawing from his experience as Vice Chair of the IPCC Working Group 1 (2008–2015) and on-going involvement in related organisations and projects. He highlights the vulnerability of SEA to climate change due to geographical, socio-economic, and environmental factors. He posits that the region faces heightened risks from climate variability, including increased frequency and intensity of extreme weather events such as floods, storms, and droughts, and witnessed several devastating climate disasters over the past two decades. Rising sea levels pose a critical threat to coastal cities, notably Manila, Jakarta and Bangkok, which are also subsiding. The article presents an overview of the regional dynamics that define climate and its variability, as well as projections of warming and sea level rise based on climate modelling under various scenarios. He highlights the likelihood of the global average temperature exceeding the 1.5 °C threshold and draws attention to the severe consequences predicted by climate modelling. The article ultimately stresses the urgency for stronger national actions to mitigate climate impacts and adapt to changing environmental conditions.

In Chap. 4, Gabriel Yit Vui Yong discusses the Changing Global Order and Climate: A Comparative Analysis of Cartesian and Organismic Perspectives by considering two contrasting paradigms. Cartesian Mechanism is the bedrock of modern scientific thinking and the industrial revolution. It is also the basis of Western capitalism and colonialisation. This is the prevailing dominant paradigm of the modern world. Its principles are instilled in education and academic research, and produced the vast collection of knowledge generated over the past few centuries. The current unipolar global political and economic order and its systems are developed within this paradigm. It however has given rise to the host of environmental problems confronting humanity, including climate change. Organismic systems offer an alternative paradigm that is based on close observation of nature and its behaviour and depicts natural systems and dynamics more accurately. The article examines developments associated with changing global order, such as the rise of BRICS, multipolarity, dedollarisation, and intensifying geopolitical conflicts, as well as climate change from both paradigms, contrasting between the two perspectives. It argues that although cartesian mechanism has been successful in the development of academic knowledge, including understanding and observation of climate change, its application in geopolitics and economics is one-sided because it ignores feedback and organic response. It views people and nature mechanistically, but in reality, all living beings are sentient and possess cognitive abilities. In this regard, the rise of the Global South is a natural response to hegemonic abuses, now that many countries, China in particular, have developed sufficient capacity to break the shackles. Although the transition would be painful, a future multipolar order in inevitable.

In Chap. 5, Norzurianie Kamarulzaman discusses Food Security Using Rice Production in Brunei Darussalam as a Case Study. It draws from a 2010 graduate study and follow-up work a decade later on the Wasan scheme. Food security encompasses the availability, access, utilisation, and stability of food systems. Brunei, despite its small population and oil wealth, struggles with rice self-sufficiency, producing less than 5% of its needs. Various initiatives to bolster local production have faced persistent hurdles, including poor soil conditions, poor management, inadequate technical support, and lack of knowledge on the cultivation of introduced varieties. Cultural factors, notably the Islamic practice of 'sedakah' and subsistence farming, undermine the rice production programme, but contributes to community food security. At the regional level, all countries, apart from Singapore, appear to have sufficient food stored to absorb disruption in the supply chain. An ASEAN plus three system is in place for emergency situations. The article however highlights serious environmental impacts inherent in modern agriculture and proposes a broader understanding of food security from an ecological perspective. It posits that Brunei's tropical environment offers diverse food sources, and advocates for sustainable agricultural practices that integrate agroecology to enhance food security while addressing climate impacts, illustrating the intricate balance between economic development, environmental preservation, and cultural sustainability.

In Chap. 6, Izni Azrein Noor Azalie presented his thoughts in ASEAN's Halal Production Networks: Towards Regional Integration and Effective Industrial Governing, a Brunei case study. He draws from comprehensive research work on Brunei's halal industrial development, focussing on value chain analysis and production network analyses, employing a Foucauldian framework. The case study involved 15 firms, six of which were presented in details to illustrate issues and problems encountered. The research elucidated the different strategies taken by firms with differing outcomes, against shortcomings and inconsistencies of the competent authority and its policies and actions respectively. It highlights a variety of barriers, the importance of regional networks, and demands to achieve success and sustainability, particularly for fledging systems such as in the case of Brunei. The chapter argues that to enhance the halal food industry's sustainability and productivity, a regional and global approach must be adopted. This involves overcoming challenges posed by national borders, protectionism, and varying halal standards among ASEAN countries to create integrated halal production networks. Findings from the Brunei study highlights the diverse strategies that could be employed, including leveraging familial ties and forming joint ventures for outsourcing and market access, and building strategic linkages with local and international partners. It also emphasised the need for a more proactive state role in facilitating the halal industry's growth, strengthening regulations, and creating a cohesive production network that transcends geographic limitations.

Chapter 7, Noor Hasharina Hassan sheds some light on Young Consumers' Understanding of Sustainable Consumption through a survey with first-year university students in Brunei. Her findings suggest that, while knowledgeable of the concept and practice, the students' understanding are superficial and resemble following the collective narratives of the global sustainability bandwagon. The survey explored the students' awareness of sustainable consumption in relation to urban design and lifestyle, climate change, as well as the teachings of Islam, as the majority of the sample were Muslims. It similarly revealed that they had knowledge, but at a superficial level and were unable to see interconnections in the real world. When asked to rate their commitment to sustainable consumption on a 5-level Likert scale, the majority responded with a middle rating, with more females considering themselves as more sustainable, while the opposite was true for males. Young consumers, while exhibiting alignment with the broader collective view on sustainable consumption, also reveal individualistic-hedonistic traits.

In Chap. 8, Ullah provides the relevant theoretical framework to understand the geopolitical dynamics in Politics and Security in Southeast Asia. He presented an analysis of the political and security dynamics issues in the region through the application of the interplay of realism, liberalism and constructivism and shows how realist power struggles, liberal institutional cooperation and constructivist norms have given rise to the present situation, tensions and risks. He relates his analysis to the legacy of Western colonisation, the influences of the Cold War, and the rise of globalisation and regionalism. The chapter further discusses the role of ASEAN as a stabilising force, highlighting how regional cooperation has managed disputes and promoted dialogue despite complex inter-state relations. The chapter also refers to critical theories and the human security paradigm to explore how economic dependency, authoritarianism and internal conflict could affect regional stability. By integrating these frameworks, this chapter contributes to a nuanced understanding of Southeast Asia's unique political landscape and security challenges in a rapidly evolving global order.

In Chap. 9, Mary Dorothy dL. Jose and Anabelle B. Ragsag present a detailed study on disaster management through a gender lens in Unpacking Women's Participation in Disaster Management: Insights from the Philippines and its diaspora. The chapter discussed the development of gendered narratives in disaster management in the Philippines, the influence of feminist organisations and the impact on the development of inclusive disaster management strategies. The authors highlight the incongruence between programmes introduced by foreign agencies from the West and local initiatives, where the former was more interested in addressing gender issues from a policy perspective, while the latter's concern is to respond directly to the needs of women in disaster situations. They further highlighted the role of feminist organisations in incorporating women's voices and experiences at the different stages of disaster management in broader socio-political conditions. The introduction of 'women friendly spaces' is an example of a positive outcome involving the Waray community, which allowed the women to organise and actively participate in the development of their local communities in disaster response. They also highlight the need to improve understand of the role played by the diaspora community in the shaping of political narratives and provision of support. They concluded that disaster risk reduction management should be placed in a broader context that addresses people's livelihoods and takes into account the needs and capacities of all members of the community, including those who are transgender and gender non-conforming. This must be grounded in indigenous and local knowledge.

In Chap. 10, Gabriel Yit Vui Yong and Izni Azrein Noor Azalie discuss the opportunities, challenges and issues associated with China's Belt and Road Initiative (BRI) in Brunei and the Maritime Silk Road, which they compared with the ancient silk road of the tenth–fifteenth century, and examine the contrasting characteristics of the Brunei Sultanate and the present nation-state, and their environmental circumstances. The study found 'lack of capacity' to be a major impediment, and analyse the issue from a governance stand-point, employing a Foucauldian framework, linking nested spheres of governances to desired outcomes. The study uncovered dissonances that impede effective governance, which inadvertently created persistent barriers to development. They present a viable way forward, which addressed the major gaps and disconnects, and highlighted an integrating mechanism as a critical cog needed to get the wheel turning.

In Chap. 11, Diotima Chattoraj examines the intersection of development, mobility and sustainability in Southeast Asia: Mobility, development and long-term plan. She highlights how these interconnected elements shape the region's socio-economic landscape, economic development and contributed to environmental degradation and inequality. Migration from developing to industrialised areas has significant implications on politics, economics, and social integration. Countries like Singapore, Malaysia, and Thailand benefit from labour mobility, while others struggle with migration challenges such as irregular movements and human trafficking. Migration and mobility in Southeast Asia are diverse in form, ranging from rural–urban economic migration, transnational employment, illicit activities, as well as a consequence of political conflicts or environmental impacts. The chapter critiques how short-term economic drivers often overshadow sustainable long-term planning, emphasising the need for a balanced approach to manage migration that considers immediate economic growth alongside social equity and environmental sustainability. It argues that a holistic understanding of their dynamics is essential for developing inclusive policies, ensuring that growth benefits all societal segments.

In Chap. 12, Ullah further developed the understanding of regional migration patterns in Migration Routes in Southeast Asia: How economy shapes the routes. He examined the interplay between migration routes and economic structures, providing an in-depth analysis of the ways in which labour mobility is influenced by economic factors through the lens of Neoclassical Economics, Dual Labour Market theory, New Economics of Labour Migration, and World Systems Theory. The chapter shed light on the motivations, mechanisms and impacts of migration within and across national borders. It identified the Malaysia-Indonesia and Thailand-Myanmar corridors as key migration routes established by the pull forces of wage differentials, labour demand and regional politics on people mobility. The role of remittances in supporting the economies of sending countries such as the Philippines and Cambodia emphasises the profound economic impact of migration. By linking theoretical frameworks with empirical evidence, this chapter highlights the centrality of economic imperatives in shaping the Southeast Asian migration landscape, while also addressing regulatory challenges related to labour rights, integration and regional cooperation.

References

ASEAN (Association of Southeast Asian Nations). (2007). *ASEAN charter*. https://asean.org/wp-content/uploads/images/archive/publications/ASEAN-Charter.pdf

ASEAN. (2024). *Statistical highlights*. https://www.aseanstats.org/publication/ash-2024/

Billé, F., Mehendale, S., & Lankton, J. W. (Eds.). (2022). *The Maritime Silk Road: Global connectivities, regional nodes*. Amsterdam University Press.

Blouet, B., & Blouet, B. W. (2001). *Geopolitics and globalization in the twentieth century* (256 pp.). Reaktion Books. ISBN-10: 1861890850

Cardoso, J. d. C. (2024, March 29). When can timor-leste expect to become a full member of ASEAN? *The Diplomat*. https://thediplomat.com/2024/03/when-can-timor-leste-expect-to-become-a-full-member-of-asean

Chakrabarty, D. (2010). *The ambiguous allure of the West: Traces of the colonial in Thailand* (R. V. Harrison & P. A. Jackson, Eds.; 1st ed.). Hong Kong University Press. http://www.jstor.org/stable/j.ctt1xwbmf

Galimberti, S. (2023, March 27). ASEAN 2050: Towards a new ambition. *The Jakarta Post*. https://asianews.network/asean-2050-toward-a-new-ambition/

Gupta, A. (2005). *The physical geography of Southeast Asia* (468 pp.). Oxford Regional Environments. ISBN: 9780199248025.

Haggart, P. (2001). *Geography: A global synthesis* (833 pp.). Prentice Hall. ISBN: 0582320305, 9780582320307.

IEA (International Energy Agency). (2022). *Greenhouse gas emissions from energy in 2022*. https://www.iea.org/data-and-statistics/data-product/co2-emissions-from-fuel-combustion

Islam, M. (2023, March 27). *The shifting sands of power: Understanding geopolitics in the 21st century*. The Geopolitics. https://thegeopolitics.com/the-shifting-sands-of-power-understanding-geopolitics-in-the-21st-century/

Köppen climate classification.

Kratosa, P. (2021). *South East Asia colonial history, V1–4*. Routledge.

Kittiprapas, S. (2000). Thailand: The Asian financial crisis and social changes. In T. Van Hoa (Ed.), *The social impact of the Asia crisis*. Palgrave Macmillan. https://doi.org/10.1057/9780333978016_4

Laker, B. (2023, March 15). Southeast Asian economies expected to grow. *Forbes* [Video File]. YouTube. https://www.forbes.com/sites/benjaminlaker/2023/03/15/southeast-asian-economies-expected-to-grow/?sh=2458311430ac

Nicholl, R. (Ed.). (1975). *European sources for the history of the Sultanate of Brunei in the sixteenth century*. Muzium Brunei.

Nijman, J., Shin, M., & Muller, P. O. (2020). *Geography: Realms, regions, and concepts* (18th ed.). Wiley. ISBN: 978-1-119-60736-6.

Olson, D. M., Dinerstein, E., Wikramanayake, E. D., Burgess, N. D., Powell, G. V. N., Underwood, E. C., D'Amico, J. A., Itoua, I., Strand, H. E., Morrison, J. C., Loucks, C. J., Allnutt, T. F., Ricketts, T. H., Kura, Y., Lamoreux, J. F., Wettengel, W. W., Hedao, P., & Kassem, K. R. (2001). Terrestrial ecoregions of the world: A new map of life on earth. *Bioscience, 51*(11), 933–938.

Purvis, M., & Grainger, A. (Eds.). (2004). *Exploring sustainable development: Geographical perspectives*. Earthscan. ISBN: 1-85383-472-6.

Runciman, S. (1960). *The White Rajah: A history of Sarawak from 1841 to 1946* (320 pp.). Cambridge University Press.

Sachs, J. D. (2020). *The ages of globalization: Geography, technology, and institutions*. Columbia University Press.

Seitelman, M. (1948). *The cession of Sarawak. Far eastern survey* (Vol. 17, No. 3, 35–37). Institute of Pacific Relations.

Ullah, A. K. M. A., & Chattoraj, D. (2022). *Covid-19 pandemic and the migrant population in Southeast Asia: Vaccine, diplomacy and disparity*. World Scientific.

Ullah, A. K. M. A., Lee, S. W. L., & Hassan, N. H. (2024). *Southeast Asia in pre- and post-COVID-19: Economy, society, mobility and religion*. Routledge.

UN Charter. (n.d.). https://www.un.org/en/about-us/un-charter/full-text

UNESCO. (n.d.). *Cultural selection: The early Maritime Silk Roads and the emergence of stone ornament workshops in Southeast Asian port settlements*. Retrieved 15 March

2024. https://en.unesco.org/silkroad/content/cultural-selection-early-maritime-silk-roads-and-emergence-stone-ornament-workshops
UN General Assembly. (1960). *Declaration on the granting of independence to colonial countries and peoples*. Audio Visual Library of International Law. https://legal.un.org/avl/ha/dicc/dicc.html
WCED (World Commission on Environment and Development). (1987). *Our common future* (416 pp.). Oxford University Press. ISBN: 9780192820808.

Gabriel Yit Vui Yong is a lecturer in the Geography, Environment, and Development Programme at Universiti Brunei Darussalam. With a BSc in Geological Geophysics and an MSc in Environmental Analysis, he has spent three decades examining the complexities of sustainable development in Brunei and the region. Actively involved in fieldwork and national-regional projects, his extensive experience has fostered a view of the world as an intricate, evolving system influenced by conceptual processes of individuals and groups.

A. K. M. Ahsan Ullah is Professor of Geography, Environment and Development at the University of Brunei Darussalam (UBD). Dr Ullah has worked with institutions such as the City University of Hong Kong, IPH at the University of Ottawa, McMaster University, Saint Mary's University, Dalhousie University in Canada, the American University in Cairo (AUC), Osnabruck University, Germany, and the Asian Institute of Technology (AIT), Thailand. His research areas include migration and mobilities, intercultural encounters and development, with a geographic focus on the Asia-Pacific, Africa, and Middle East, and theoretical focus on globalisation and neoliberalism, development and human rights, transnationalism, gender, intersectionality and the everyday life.

Chapter 2
Changing Realities, 2025

Gabriel Yit Vui Yong

Abstract This chapter sets the context for the book, examining how humanity stands at the precipice of a major global shift—one that will have profound implications for Southeast Asia. The primary driver of this transformation is the decline of the West-led unipolar international order and the rise of a multipolar political-economic system, shaped by China and the Global South. Alongside this shift, global warming and the relentless advancement of the global information-communication network and system (ICNS) are collectively reshaping human realities, both individually and collectively. The chapter explores the construction of reality in the human mind, emphasizing the role of media, state agencies, and private enterprises in shaping perceptions. It highlights how these actors have fostered a simulacrum—a world where narratives serving vested interests often overshadow factual realities. Drawing on Vernadsky's theory of the Biosphere-Noosphere, the chapter connects the evolution of mind, planet, and the ICNS, proposing that AI-mediated research within the ICNS could help distinguish fact from disinformation. Methodologically, the analysis relies on trawling the Internet—a repository of timely, first-hand information—while cross-checking sources to verify their credibility. Discussion focuses on contemporary geopolitical tensions, trade and proxy wars, and the policies of the current hegemon, illustrating how they inadvertently resulted in the rise of China and the Global South. The chapter also addresses the looming tipping point, arguing that nuclear war surpasses climate change as existential threat. It concludes by affirming the inevitable emergence of a more equitable and peaceful multipolar world order, though not without the turbulence of transition.

Keywords Realities · Global shift · Simulacrum · Information communication network and system · Multipolar world

G. Y. V. Yong (✉)
Faculty of Arts and Social Science, Universiti Brunei Darussalam, Gadong, Brunei Darussalam
e-mail: gabriel.yong@ubd.edu.bn

G. Y. V. Yong and A.K.M. A. Ullah (eds.), *Brunei and the Future Transformation of Southeast Asia*, https://doi.org/10.1007/978-981-96-7901-0_2

Introduction

Any old person is likely to have said, on more than one occasions, that things have changed in his or her lifetime. Having lived for six decades, I concur. Looking back half a century, my hometown and the landscape have changed drastically. Technological advancement has been astounding, significantly improving public services, such as healthcare, transport and communication. Quality of life has become significantly better in terms of convenience, comfort, and even pain management, which can now be rendered truly painless. The surrounding is safer, well-maintained, and supported by a range of services, provided by professionals, and increasingly, automated systems. Zooming out, the last statement however is only true for a minority in the upper echelon of society, unfortunately. The reality is that inequality is a feature of society, which has become more pronounced over the past decades as wealth becomes more concentrated in the possession of a few.

In the less affluent part of the human world, and particularly in neglected marginal regions, the opposite trajectory is observed. Small rural areas are now urbanised but neglected, promoting illicit activities that threaten the quiet safety of the past with violent contestations of space and property. The built and natural environments are deteriorating and polluted, and quality of life has declined, with little hope for a better future. Modern urban development is inherently destructive to the natural environment, as it entails displacing and replacing it with artificial structures and systems. Without the necessary maintenance, they will fall into disrepair and the function they were meant to serve, hampered to various degree or ceased to exist. Therefore, the changes over the past 50 years have been good for some, but deleterious for others. However, atmospheric warming due to CO_2 emission from the break-neck pace of economic and urban development, which grew to global proportions in the later decades of the last century, affects everyone, poor or wealthy.

However, while these changes are apparent to academics and professionals sharing the same interests, reality is quite different for the majority. The entertainment industry, for example, has developed the capacity to conjure up fantasy worlds that are capable of drawing minds into the constructed reality, allowing them to participate emotionally in the story and building relationships with the characters. The effects are so strong for some that they form groups (fans) to sustain, consolidate and build upon such relationships. This is similarly true with gaming, which is now a global phenomenon as games are played by individuals across the planet in different countries, connected by a global information-communication network and system (ICNS). In the same way, special-interest activist groups form spontaneously to address a variety of concerns, some real, while others, constructed realities that resonate with members of the groups.

Prior to the introduction of multimedia communication technologies, the seemingly mundane written words have also created realities that are different from the physical one in fictions and media, the latter often driven by the propensity to sensationalise incidents to increase sales. Our world and understanding of it is influenced by the interactions of our minds, language, and the external world (Putnam, 1991).

Reality is in part socially constructed, and language is the medium that facilitates its development. Politicians and celebrities routinely hire professionals to 'spin' the truth so as to convey the desired narratives to the public. Constant broadcasting of the same and similar narratives contributes to the development of collective realities. Over the past decade, advancement in audio-visual technology has made it possible to engage in a composite physical-virtual world through augmented, virtual and mixed reality (AR, VR and MR respectively). The consolidation of the global ICNS has ushered in an 'information' or 'knowledge' age, 'knowledge economy' and digitalisation of services, which employs artificial intelligence (AI) and algorithms to execute services. Language and communication have become the most powerful shapers of reality.

Therefore, while there is a physical reality unfolding, they could be perceived or believed to be occurring differently. This chapter will attempt to shed light on the changing realities at this juncture of history, as humanity crosses the quarter mark of the twenty-first century. It will present the case that humanity is standing at the precipice of a seismic shift in global order, using the global ICNS to gather reliable information for analysis. It hopes to navigate the information space, where facts co-exist with misinformation and fabricated lies, to provide some clarity on the transformation that is taking place in the region and the planet as a whole.

Physical Versus Constructed Reality

The physical reality refers to the tangible world, which is observed and experienced through the combination of sensory faculties, supported by current knowledge from cumulative experiences. Perception is most accurate, if the observer and the pool of knowledge referred to, are less influenced by socially constructed knowledge not based on facts, i.e., 'strong' realism (Putnam, 1991). This reality is closely connected to the physical planet and its dynamics. Earth science could aid in improving understanding. It has accumulated extensive knowledge that explains: sudden drastic changes (Catastrophism) and slow, imperceptible processes that build mountain ranges as well as reduce them to valleys and plains over 100's to millions of years (Uniformitarianism); planetary-scale changes, e.g., planetary expansion because the ocean floor is observed to be spreading apart, or comprising of lithospheric 'plates' that are in constant motion, giving rise to volcanoes and earthquakes at margins; and evolution of the physical planet and life (Lovelock & Margulis, 1979; Vernadsky, 1988).

Since the eighteenth century, scientific studies, notably of earth surface processes, living organisms and ecology, and climate and hydrology, have generated a treasure trove of reliable knowledge of the physical environment. Climate and hydrology, in particular, are highly relevant in the understanding of changing climate, which has become a major concern over the past half a century. However, a combination of miscommunication, tendency to sensationalise, and increasing disconnect from physical reality (a consequence of modernisation) has generated much confusion of the physical environmental changes, and specifically, of climate. Indeed, it has led to a common mis-association of the term 'climate change' with disasters. At the extreme end of the spectrum, advocates believe that it will end in a global catastrophe, if humankind does not take the necessary actions to curb CO_2 emissions.

According to many, humanity is presently confronted with a climate crisis, or 'climate emergency' (UNEP, n.d.), to heighten its urgency level. The increase in the concentration of CO_2 in the atmosphere since the twentieth century has altered local, regional and global climates, and raised concerns about a 'global warming' (Lindsey & Dahlman, 2024). In response, the UN had established the Intergovernmental Panel on Climate Change (IPCC) in 1988 to assess the scientific information pertaining to global warming and its predicted consequences. The 1992 Earth Summit in Rio de Janeiro adopted the United Nations Framework Convention on Climate Change (UNFCCC) to facilitate international co-operation on climate issues. In 1995, the first UN Conference of the Parties (COP1) was held in Berlin to discuss viable global responses to climate change. The Kyoto Protocol was put forward and adopted in 1997, but only came into force in 2005. The Protocol was based on 'common but differentiated responsibilities and respective capabilities' and bound only the industrialised countries to climate action, as they were identified as the main contributor and therefore, ought to bear the responsibility to control greenhouse gas emissions (UN Climate Change, n.d.).[1] The Doha Amendment was adopted to extend the Kyoto commitment from 2013 to 2020, but it only came into force in 2020. The Agreement faced numerous issues related to emerging economies and corporate participation, challenging its relevance (Singh, n.d., Earthhood).

The IPCC comprised of 3 working groups. Working Group (WG) 1 focuses on the physical science. To date, it has published six comprehensive assessment reports (AR), the last one was released in 2021. AR6 unequivocally states that human-induced climate change is real and that sea levels will rise by 0.4–0.8 m by the end of the twenty-first century. In Southeast Asia (SEA), sea level rise will have significant impacts due to the coastal and maritime characteristic of the region. The configuration of the coastline will change, the extent of which depends on the rate of sea level rise as well as the magnitude and frequency of associated tidal flooding, storm surges and other associated processes. Coastal cities will be most at risk if the necessary adaptation measures were not taken. Higher energy trapped in the atmosphere is likely to intensify hydro-meteorological circulations and thus the occurrence of extreme weather events. However, contrary to popular belief and the more alarmist of climate activism, climate change would not trigger a global catastrophe. Disasters occur in specific locations, where high-intensity climate hazards coincide with vulnerable communities. Although a global average temperature of 2 °C above pre-industrial levels was set as a threshold at COP21, exceeding this threshold would not cause the global climate system to "collapse", because it is a scientifically nonsensical conjecture. The reality is closer to heightened frequency and magnitude of disruptions and associated incidences of disasters. In the highly emotionally sensitive minds of the 'woke' generation, where superlatives are used routinely to describe mundane

[1] https://unfccc.int/kyoto_protocol.

activities, 'global catastrophe' would not be out of place. However, they do not reflect physical reality.

Environmental change alters the ecosystems and environmental dynamics, which in turn affect people and the built structures and systems. However, scientific studies are increasingly based on conceptualised models, i.e., representations rather than reality (Putnam, 1991), which adds uncertainties because of the constructed parts of reality. These are influenced by a range of factors, including beliefs, priorities and concerns, external pressures, and context, which might not have been adequately accommodated. Increasing adoption of machine learning and remote sensing techniques further adds to the constructed portion of reality, particularly, when ground truthing, a critical part of such methodology, is not adequately carried out. The ICNS, which has become ubiquitous and indispensable, is increasing the opportunity for public participation in scientific studies as 'crowd sources' of information. While this builds up databases rapidly, they could also add confusion and increase social construction of reality. Meanwhile, independent of scientific studies, a plethora of self-styled experts and activists have emerged, enabled by the ICNS. Many have grown in prominence, such that they have no qualms dismissing scientists as 'climate deniers', if the latter do not agree with their emotionally charged alarmist explanations. Being 'influencers', they shape the minds of followers in their social networks, and added to the confusion and anger in many climate change debates.

The vast amount of information that flow continuously through the global ICNS also supports the functioning of governments, institutions, industries and the global economy. Augmented with AI, it has created an alternative reality and 'digital twins', which may deviate slightly or markedly from their physical counterparts. Nevertheless, they are real (or even more real than the actual version) to those who have become reliant on them. Digitalisation has altered reality so significantly that it is regarded as the fourth industrial revolution (Parmentola & Tutore, 2023). In the twenty-first century, reality comprised a virtual and a physical component. It is largely constructed, often deliberately by multiple state and private agencies to further specific agendas, which are consolidated as a consequence of social interaction. Baudrillard (1994) described this world as a simulacrum. In the USA, the enormous volume of commercial and activist contents, together with social media creations by individual boggers and vloggers, add to the complexity of the simulated reality. Society within the simulacrum lives according the different realities they have embraced, many trapped in 'pet interest' mental-emotional prisons. Their interactions within these worlds constantly reinforce the reality; as the motivation guru, Anthony Robbins often say, 'whatever you believe is real' (Robbins, 1997).

The urbanites born and raised in this century are 'digital natives' (Sadiku et al., 2017). They navigate the hybrid world effortlessly via mobile 'smart' devices, sometimes as avatars in fantasies, where the responsibilities of life and reality are not a concern, as most are quite detached from the physical-natural world. Life in a simulacrum resembles the simulated world in the movie, the Matrix[2] (Wachowski, 1999),

[2] Referring to the 1999 movie to highlight the difference between the physical world and simulated reality.

where the digital reality in the 'cloud' is created by the participation of myriads of actors, interacting with the largely constructed world, continuously co-created and sustained by the ICNS. The current trend in ICNS research and development is closing the physical-digital divide, as it seeks to create a 'mixed reality' future. In a 'metaverse', for example, people would be able to interact in a virtual space, while physically separated but also physically interacting in a digital medium (Graylin & Rosenberg, 2024; Walsh, 2020). As the world becomes increasingly digitalised, more older folks are becoming excluded from welfare, public and commercial services as a consequence, because they are unable to navigate the interface of the smart kiosks or applications on smart devices. So, while contemporary society strives to be more equitable and inclusive, digitalisation is excluding the older generations from increasingly wider range of basic services. Ironically, many digital natives claim to be 'eco' and climate warriors, despite having little personal experience of the natural world.

The transformation that has emerged was predicted by Vladimir Vernadsky in his theory on the Biosphere in 1926 (Vernadsky, 1988). From a geochemical perspective, complex systemic processes on the surface of the primordial Earth gave rise to life (creating the biosphere) from the chemical constituents available. As all lifeforms are sentient (a necessary quality for survival) and have the ability to learn, constant interaction between life and environment inevitably led to the emergence of the mind, individual as well as collective. Vernadsky called it the Noosphere. Mind and knowledge co-evolved through communication, continuously refining and correcting knowledge, which eventually lead to emergent coherence (Stacy, 2001) or discord (Yong, 2005), resulting in different schools of thought, alliances and groups. Technological innovation is a natural part of this process, because it is a feature of the learning process, an innate trait of lifeforms. This is mediated through communication and exchanges, fostering system development, or 'autopoiesis' (Maturana & Varela, 1980). However, although communication is key to survival, progress and prosperity, humanity has not always been co-operative, nor has it always strived for harmonious co-existence. Instead, in pre-modern times, clans and kingdoms often fought over territory and resources to ensure their survival and prospects for prosperity. Smaller, weaker groups are conquered and integrated into larger, more powerful ones or form alliances under recognised leaders, increasing diversity and complexity over space and time (Seidl, 2004).

The establishment of the global ICNS greatly enhanced communication and exchange of information, with concomitant outcomes within and across social networks almost in real-time. The consolidation of the Noosphere has significant implications on academic practices and methodologies. Cartesian duality and objectivity are no longer attainable, simply because the observer exists within the frame of analysis in a highly fluid world and therefore, is constantly influenced by the subject. In any case, as large parts of mental realities are constructed or influenced by the simulacrum, which contains mis- and dis-information, the latter often disseminated with malicious intent. In the current world of geopolitical tension, disinformation is ammunition in information warfare, waged against imagined 'enemies' by states or the competition in industry to manufacture consent for unpopular plans and actions (Herman & Chomsky, 2002). Prior to the Information Age, the researcher's world could be described as 'data-poor'. Research methodologies entail practical steps of acquiring data from observation, field surveys and or extracting information from people or information repositories. Instruments were developed for these purposes. Contextural influences are often not considered. However, in the present age, the researcher's world is inundated with information, a large portion of which is fabricated deliberately or consequentially. A novel methodology is therefore needed to sieve out accurate factual information from the mix of information, misinformation and disinformation.

Methodology

The literature in academia represents the accumulation of knowledge on various subject matters. Since the invention of printing, books have become stores of explicit knowledge in physical form and libraries are the repositories. In the digital world, electronic or e-publication are generated and made accessible to the public at or close to real time. They are therefore 'current' and often in multimedia format. Interactive videos streamed 'live' have become popular because they address current issues and allow for public participation. Often, they allow viewers to hear directly from the source, without filter or censorship. The ICNS also offers users a means to determine the veracity of information by cross-referencing multiple sources interactively. Social media reports also include comments from the public, including some with solid insider knowledge, mixed together idiotic remarks from the biased and ignorant at the opposite end of the spectrum. This provides the researcher with a means to assess data quality.

Although all leading media companies have digital channels on the Internet, they have lost the monopoly in shaping world views. There are two reasons. Firstly, the ICNS is a democratising system, allowing anyone with access to participate. Consequently, corporate media and state press agencies are no longer able to control information. Secondly, the Ukraine war and the on-going genocide in Palestine by Israel has exposed the mainstream media's double-standards, lies and complicity in the atrocious acts being carried out. In particular, it has uncovered the reality that political elites in the West are: (i) using Ukraine to fight a proxy war with Russia 'to the last Ukrainian', and (ii) supplying arms to, and diplomatic cover for, Israel as it engages in a genocide. They have therefore lost the trust of increasing number of people. Furthermore, aggressive censorship and gagging of journalists keen to expose facts that contradict the established narratives have also contributed to driving many to seek alternative sources of reliable information from independent social media channels. Finally, it is virtually impossible to censor information or push a false narrative for long on the ICNS because it is a complex system, i.e., it has the capacity for self-regulation and self-organisation (Capra, 1996). Therefore, it will eventually isolate and expose misinformation and disinformation, and experts could eventually identify the hands and faces behind them.

This study will therefore trawl the social media to find reliable information. They tend to be clustered by the organising process of algorithms (self-organisation mechanisms of the system), which eventually will separate information from disinformation. The following information are gleaned from the Internet and social media, rather than solely relying on published articles. They are analysed and discussed below.

Change That Has Not Happened in 100 Years

In a news video from 22 March 2023 showing President Xi Jinping leaving a meeting with his Russian counterpart Vladimir Putin on the Ukraine crisis in Moscow, the former was recorded saying to the latter: 'Right now there are changes like we haven't seen in 100 years, and we are the ones driving these changes together' (Al Jazeera English, 2023; Crooke, 2023; Norton, 2023, March 25). President Putin agreed with this. They were alluding to the imminent restructuring of the global economic system, indeed a change of potentially seismic proportions, which appears to be inevitable and irreversible. This section elaborates on the transition.

The American Century: Unipolar World

The 'American Century' began in the waning period of WW2 (Sachs, 2018). At the Bretton Woods Conference in 1944, the US dollar (USD) was chosen as the world's reserve currency, backed by gold at USD 35 per ounce (Siripurapu & Berman, 2023). However, this arrangement was discarded in August 1971 and the USD became a fiat currency, raising the potential for inflation, currency instability, as well as abuse of monetary power (ibid.). However, as the USA was experiencing stagflation in the 1970s, it pulled off a major deal with Saudi Arabia in 1974 that would grow its economy exponentially into the largest in the world: Saudi Arabia will only sell oil in USD and recycle its vast 'petro-dollar' earnings into US Treasuries in exchange for US military support (Wong, 2016, May 30). The deal allowed the USA to engage in deficit spending because it generated global demand for US Treasuries, as the USD has become inextricably linked to oil, the essential energy fuel of modern society. This established the USD as a stable store of value and placed the USA in a position of 'exorbitant privilege' in the global economy. With the disintegration of the Soviet Union, the USA became the sole hegemon, a superpower backed by the most powerful military in the world.

The current world order is therefore a unipolar one, where the USA dictates the global agenda through institutions it established, such as: World Trade Organization (WTO), the World Bank Group, the International Monetary Fund (IMF), and the Organization for Economic Cooperation and Development (OECD). The USD, together with the global financial exchange system (SWIFT), is only one facet of the current world order. Since the WW2, the USA has adopted a three-pronged strategy to achieve its goals. Apart from economic dominance, it also employed military coercion and intervention, as well as information warfare to achieve its objectives. The former had developed from humble and patriotic beginnings during WW2, when the USA fought to defend its sovereignty against the military expansion of fascist regimes. The companies that supplied Allied Forces with weaponry have evolved into a powerful 'military industrial complex' (MIC) with global influence. Due to its history, the MIC has a close mutualistic relationship with the political elites.

This has led to the emergence of the 'deep state', the network of capitalist elites that own the banks, pharmaceuticals, media, and the MIC. They influence US politicians' propensity for military solutions, many of whom are neo-conservatives, who are hell-bent on crushing communism by force. Jeffrey Sachs has criticised this tendency as thriving on endless wars, because war is 'good business' (Sachs, 2018, 2024, March 21). Independent journalists (Abby Martin, Kim Iverson, etc.), ex-soldiers (e.g., Colonel McGregor; Mike Prysner; and Scott Ritter), whistle blowers (Julian Assange and Edward Snowden), academics (e.g., Jeffrey Sachs; Noam Chomsky) and politicians (e.g., Craig John Murray; George Galloway) have exposed the cosy relationship between political elites and the 'deep state'. Although the USA is a founding member of the United Nations in 1945, which has a mandate to promote and maintain peaceful co-existence, it has instigated numerous wars and regime changes around the globe, disregarding the principle of equal rights and self-determination of nations, and coerced nations to back its agenda, which run counter to the UN Charter (n.d.). Instead, it often refers to an unwritten 'international rules-based order', which Edward Norton, editor-in-chief of Geopolitical Economy Report, often explained quite accurately as the 'USA making the rules and ordering everyone around'.

The impact of information warfare cannot be understated. Its prime objective is to manufacture consent for the policies, agenda and actions of the USA (Herman & Chomsky, 2002). The mainstream media, which comprises the major media corporations with ties to the political elites and is part of the deep state, is the vanguard. They devise narratives, which are deployed in coordination with foreign offices and proxies, to create a desired fake narrative in the information space. This battle is also waged within the ICNS via the digital channels of the mainstream media, 'experts' and vloggers, supported by the leading social media platforms, which are part of, or affiliated with, the 'deep state'. The strategy of a coordinated assault on all fronts in the information war makes false narratives appear true and legitimate.

Within the simulacra established, false narratives would become more compelling than accurate news and statements issues by the accused, because they resonate with the established mainstream narratives constructed for societal consumption, and assimilated by society. This has allowed the state to engaged in illegal wars around the world, while presenting itself as the defender of freedom and human rights (Chomsky, 2003). However, as the INCS evolved and become more powerful, its self-organising capacity becomes more efficient, inevitably connecting people with shared interest to unfiltered raw information, and shedding light on fake narratives by isolating them as such. With regards to the current geopolitical issues, it has drawn together independent journalists, activists on the ground, and the affected parties. YouTube Channels, such as M.O.A.T.S., Empire Files, Democracy Now, Geopolitical Economy Report, Kate Iverson, and many more, are exposing lies used in the information war, and the true intention of the West, which is not dissimilar to the colonial agenda of their imperial past. Invariably, they want to keep developing countries from rising, and punishing those that do not toe the line through economic sanctions, military interventions and regime changes. Communication and research via the ICNS is freeing many from the simulacrum and awakening them to the reality of manufactured narratives.

The Rise of China and the Global South

At the end of WW2, China was a poor, third-world country. It had just emerged from a 'century of humiliation', when its territories were annexed, markets pried open, resources and treasures looted, and its people discriminated against at home as well as dispersed to foreign lands as cheap labour under the coolie scheme, or in search of better future. WW2 was followed by a civil war, and over two decades of western sanctions, which contributed to millions of deaths from starvation, which the West attributed to the failure of communism (Brown, 2012; Chen, 2006). However, China is a civilization with a 5000-year history and therefore the pride and tenacity to rise from the ashes. Following the normalization of relationship with the USA in 1972, a Kissinger ploy to separate China from its communist neighbour (Russia), it was able to trade and rebuild its industry and economy. When Deng Xiaoping introduced capitalism into the communist system, China's economy grew rapidly, aided by western corporation seeking to exploit the low-cost of operation. Spurred on by diligence and intense desire to redeem itself, China's economy inevitablely grew. Over the past decade, it has advanced technologically and in the quality of its products. However, although the USA has benefitted from a technically more advanced China and its large and increasingly affluent market, it regards China's rise and development as a threat. Its response is therefore one of containing China, to ensure that it remains subservient. In contrast, China believes that every nation has the right to develop and modernise (Wave Media, 2024, April 10), and to take its place as an equal in the global community, a sentiment that is shared by many in the developing world. The USA-led unipolar world is however inherently racists, as is evident from historical record.

In 1985, when Japan was emerging as Asia's manufacturing power, the USA launched a trade war when the latter's GDP reached a 'critical' 80% of the USA's, resulting in several 'lost decades' for Japan because it bowed down to US demands (Degterev et al., 2021: 219). Therefore, when China surpassed the USA in Composite Index of National Potential in 1995 and absolute GDP in Purchasing Power Parity (PPP) in 2014, the latter resorted to the same tactics to contain China's economic development (ibid.). The USA is incapable of relinquishing or sharing power. The iconic western movie phrase, 'this town aint big enough for both of us' appears to be allegorical of the American psyche: uncompromising, always seeking to be number one, and coveting all available resources.

This is indeed characteristics of western European settlers in the new world, and western European colonialism in general. This racist mindset regards people of other nations and ethnicities as inferior and unworthy to share the same table. Singapore's founding father, Lee Kuan Yew, stated it bluntly in an interview on Global Viewpoint, 26 September 1995: 'for America to be displaced by an Asian people, long despised and dismissed with contempt as decadent, feeble, corrupt and inept, is emotionally very difficult to accept' (Janssen, 2024, January 21; National Archives of Singapore, 1995). It revealed a sense of cultural supremacy. In the nineteenth–early twentieth century, before 'political correctness', xenophobic ideologies were not veiled. East Asian immigrants, primarily Chinese and Japanese, were referred to as the 'Yellow Peril', regarded as inferior, unassimilable, deceitful, barbaric, and disease-ridden (Li & Nicholson Jr., 2021). In the USA, it led to the enactment of discriminatory laws and policies targeting Asian immigrants (e.g., Chinese Exclusion Act of 1882; internment of Japanese Americans during WW2).

The view of non-white as inferior labour class, unworthy of the same human rights, is the basis of slavery and the coolie scheme. It was instrumental in effecting diaspora of Chinese, Indians, Africans, and others from colonised regions to work in other colonies. In the USA, Chinese coolies were assigned the most hazardous and gruelling tasks in the construction of the transcontinental railway (1863–1869), while being subjected to discrimination and violence without any recourse to justice. Racism and religious bigotry are still rife in the USA and the West in general. Despite promoting and adopting inclusive policies, anti-Asian/Chinese hate crimes flared up in the USA during the first Trump presidency. This is largely due to a combination of ignorance—being misled by 'China bad' disinformation—and declining quality of life at home. Ben Norton highlighted the stark contrast in development between China and the USA and responses by their respective governments in an interview on CGTN's The Point (2024, March 28). However, the trade war launched by Donald Trump against China in 2018 did not resulted in the expected outcome. Instead, China countered with its own sanctions and proposed an alternative to the international community (Degterev et al., 2021).

The stark contrast between China's peaceful and mutually beneficial policy versus political-economic bullying and militarisation by the USA and NATO offered the Global South a way to emancipation from the strangle-hold of US hegemony. China focused on pragmatic solutions and does not impose its will or interfere with the internal affairs of its trading partners—a teaching of Confucius: do not do to others what you do not wish to be done to you. This resonated with most of the Global South, mainly because they share similar experiences of being exploited and discriminated under Western colonialisation in their own land; many continues to be shackled in debt-traps set by West-led international institutions, such as the IMF. The trade bloc established by Brazil, Russia, India, China and South Africa (BRICS) offers the Global South an alternative means to trade and development, while China's Belt and Road Initiative (BRI, McBride et al., 2023, February 2) is building the necessary infrastructure for economic advancement at relatively low cost.

The weaponising of the USD by the Biden administration further highlighted both the risk and cost of transacting in a system that could be used against anyone who opposes the will of the USA. The Global South is rising through the convergence of a number of separate developments towards a new global order. Ironically, the process was initiated by US policies and actions intended to secure its position as the global hegemon (Crooke, 2023). Firstly, instigating and fighting a proxy war against Russia in Ukraine has crippled Europe's economy and pushed Russia and China together to forge an alliance that rivals the West in economic and military might. This has led to the development of an alternative trade and development system. Secondly, the West's unwavering support of Israel in the on-going genocide of Palestinians in Gaza has also undermined its moral superiority. At the UN, it is clear that the West stands against the rest of the world (Geopolitical Economy Report, 2024, April 6).

The Emerging Multi-Polar World

The evolution of humanity is shaped to a large extent by geography, technology, and institutions, and the process is particularly rapid in the current globalised world due to the global ICNS (Sachs, 2020). Specific to the information war against China, Andy Boreham's Report on China YouTube channel contrasts 'China bad' headlines and articles with ground truth from his base in Shanghai. Jerry Grey, a former UK police officer who has cycled extensively throughout rural China, similarly shares what he has learned from personal experience regarding China's development, laws, system of governance (including democracy) on his channel, Jerry's Take on China. He addressed misconceptions and disinformation using facts and logical arguments, referring to data available from western institutions. Cyrus Janssen presents cases of (foreign) people and companies thriving in China and progress made in the BRI and BRICS, to dispel malicious lies about them. Many in the West however still believe China to be under the control of an evil megalomanic CCP[3] out to take over the world, and who is obsessed with robbing the west of its 'freedom', something closer to the villains in their movies than reality. There are increasing number of individuals that are sharing their personal experiences and discoveries to correct blatant lies, such as the continued baseless disinformation of genocide, slave labour, Muslim oppression and deculturalisation of minorities in Xinjiang, China.

A significant portion of the global ICNS is controlled to some degree by the US 'deep state'. One social media platform, however, remains free: TikTok. This is the social media of choice for Generation Z, because of its raw and unfiltered content, and its community-building model, a feature of its algorithm. It is therefore not unexpected when the US government sought to ban it, and attempted to force the parent Chinese company, Byte Dance, to sell it to a US firm on trumped-up charges

[3] The incorrect acronym used by neocons. The correct one is the CPC, or Communist Party of China.

of national security, despite the fact that TikTok is a separate entity, headed by a non-Chinese, and has its server in the USA, and has fully abided with US regulation. In an interview by Myriam Francois with three Gen-Z content creators on the proposed ban on TikTok (Al Jazeera English, 2024, April 2), the guests were unison in identifying the platform as key to reliable information, education and organising protests, such as the ones that calls for liberation and justice for Palestine, peace, etc. Although TikTok is perceived by many older folks as a time-wasting app where people share frivolous stuff, it is also a vibrant 150 million-strong community, where people learn, do business and try to effect positive change. TikTok has been instrumental in exposing the rampant corruption and moral bankruptcy of the US political class, their hand in the declining economy and living condition, and the entities behind the US proxy war in Ukraine and the Gaza genocide. This has resulted in millions of protests in the West calling for justice for Palestine and 'no' to NATO/war. It resembles a revolution to reclaim democracy from the corporations that now control politics and the information space, which serve the elites at the expense of the mass.

Consolidation in the Noosphere is exposing manufactured lies designed to mislead the public to support the agenda of an elite-class minority. It is undermining the propaganda machine of the US 'deep state', and awakening the many who are seeking the truth. The people in the Global South are more aware than their western counterparts, largely because most live outside the western simulacrum and have experienced exploitation and varying degrees of discrimination. This knowledge is more profound among the older generation in SEA, many of whom are understandably determined not to suffer from similar treatment again. In Africa, many are still repressed despite the so-called end of colonialisation. It is therefore not surprising that the Global South responded positively to the emergence of BRICS, China's BRI and the Shanghai Cooperation Organisation (SCO), the Asian Infrastructure Investment Bank (AIIB) and New Development Bank (NDB), despite the continuous disinformation assault by the mainstream media. China and Russia are the two juggernauts in the global economy capable of challenging the USA, and the actions of the West have pushed them together to develop an alternative to the current global system. Countries, including vassals of the USA, are increasingly looking to de-risk against the USD, now in a precarious position because a de-dollarisation process has begun. The illegal use of Russian foreign assets and plans to confiscate them was the Rubicon, which made it clear that the western financial system cannot be trusted.

The future will most certainly be a multipolar one, and the USD is likely to lose its status as the world's reserve currency outside the SWIFT system. Timing is the only uncertainty. It is unclear if 'the change that that world has not seen in 100 years' will happen within the next decade. But the multipolar world would be more peaceful and stable, as nations and economies will operate in complimentary and equitable system. The Global South will emerge as the most active and dynamic region, probably with China as the main trade and manufacturing hub. Commodity-rich countries in Africa, South America and Asia will experience dramatic growth and development that commensurate with the scale of their resource base. However, the world will have to undergo the pangs of childbirth. This might be in the form of war, economic recession, i.e., a brief and painful period of much chaos and uncertainty, because the hegemon will be fighting 'tooth and nail' to retain the status quo. The disinformation machine, military bullying, weaponising of economic and financial instruments, and application of the 'international rules-based order' would be in overdrive. Based on present-day developments, the eventual collapse of the unipolar world and rise of a multipolar one would most likely be a combined effect of (a) demise of the USD as the world reserve currency due to the USA having to face up to its gargantuan debt (presently at 35 trillion, with interest payment of 1 trillion each quarter), and (b) restructuring of the political system and government, in response to a people's revolution. This transition could trigger a global recession, as the sovereign wealth of most countries are inextricably connected the USD. Finally, the possibility of a 'hot war' is very high, as the USA and NATO are already poised to strike in the Middle East, Taiwan and desperately trying to prolong the war in the Ukraine.

Tipping Point

The 'tipping point' is commonly associated with the impending climate catastrophe predicted by alarmist activists. It is also used by politicians, journalists, teachers and many academics to denote the threshold beyond which 'all hell will break loose'. The Paris Agreement has set the threshold at 2 °C, a temperature level that the rise in global average temperature must not exceed, compared to pre-industrial levels. This is of course arbitrary, as there is no way of knowing the climate catastrophe threshold or how it would affect humanity and the planet. The alarmists generally envision a disaster orgy of monstrous proportions, as seen in the videos intended to rouse the public from its apathy. The scientific basis of the tipping point however refers to the situation in which the system is brought so far out of thermodynamic equilibrium that it breaks apart (chaos) and reorganises itself into a new structure. According to the Nobel Prize laureate Ilya Prigogine, a 'dissipative structure' forms spontaneously in order to adapt to a new, higher energy regime (Prigogine & Stengers, 1984). The tipping point is associated with chaos theory (Gleick, 1987), but it is often misunderstood. Chaos does not mean disorder in the colloquial sense, but a higher order of complexity that arises from the interactions of non-linear systems. The apparent unpredictability of chaotic events is due to their sensitivity to minute changes or 'roughness' in system, which can be attributed to the heterogeneity of the context. Beyond the tipping point, however, the old structure continues to co-exist with the emergent features (dissipative structure). By and large, it is a new equilibrium system with a new structure and order that manifests itself as a new reality.

It is unlikely that exceeding the 2 °C limit will trigger a global catastrophe. This statement may provoke much criticism, disbelief or anger, because the narrative is so deeply ingrained in the minds of many in society. However, a sound analysis of the data shows that climate change is to some extent predictable and that governments and communities can prepare for it by adapting appropriately to the anticipated changes. Although this is recognised by many in the scientific community, there is a lack of political will and investment to implement the necessary structural changes, perhaps because of the higher costs compared to the management of CO_2 emissions, which often include loopholes associated with cost. The lack of appropriate action will have an impact on densely populated areas where urban structures are vulnerable to climate hazards. Although the consequences of inadequate preparation could be severe for some people, the tipping point that would hit humanity hardest would be a hot war.

Presently, the geopolitical situation is 'far from equilibrium', i.e., on the verge of chaos (Prigogine & Stengers, 1984). Although a reorganisation would lead to a new order, the transition could be turbulent and violent. The likelihood of war is high, especially because the current hegemon clinging on to power has the largest military and readily resorts to military solutions. This mentality is often portrayed in Hollywood films, where the answer is 'guns! Lots of guns!' and the film ends with an explosive finale, a spectacle of destruction. When the USA became aware of China's economic success, it responded by 'pivoting eastwards' and encircling it with military bases and proxies, prompting the investigative journalist and filmmaker John Pilger to produce a film, The Coming War on China (Pilger, 2016). In the documentary, Pilger warned of the folly of war and pointed out the irony of the capitalist West demonising China just because it is 'good at business'. However, the war mongering nature of the USA is well-known in the geopolitical circle (e.g., Chomsky, 2003; Sachs, 2018) and is treated as fact by independent journalists and media channels, such as M.O.A.T.S., Jacobin, Empire Files, Democracy Now, Geopolitical Economy Report, etc. Former US president, Jimmy Carter, who has been critical of US militarism throughout his life, remarked in an address at Grand Rapids Community College that the USA has been at war in some part of the world since WW2 and that the inability to live in peace is actually a violation of human rights (MLive, 2014, September 23). He also exposed the fact that Israel is an apartheid state and how the lobby group, AIPAC, influenced US politics in an interview with Amy Goodman (2007) in Democracy Now, a topic he has covered in his book on Palestine (Carter, 2006).

Presently, the irrational determination of the USA and its western allies to fight Russia 'until the last drop of Ukrainian blood', crush Russia's economy and thereby the lives and future of the Russian people, maintain its unwavering support for Israel in the on-going slaughter of a civilian population in Gaza, while trying to provoke China into a war using Taiwan as a proxy, all points to its desire for a hot war. Fortunately for humanity, Russia, China and the rest of the 'enemies' have exercised restrain. In the case of Russia, it took military action on February 22, 2022, after eight years of pursuing a diplomatic solution following the brutal overthrow of the legitimate government in Ukraine, to address an existential threat imposed by NATO's expansion to its 'front door' (Carlson, 2024, February 9). The reckless actions of the USA have therefore raised the probability of a hot war with each provocation (Jacobin, 2024, March 5). A tipping point that triggers a nuclear war would truly be catastrophic for humankind and the planet.

However, restructuring and emergence could also occur without violence. This has been the stance of the emerging Global South, which has continued to forge ahead to build the necessary infrastructure under the banner of shared future and common prosperity, against constant disinformation assault, political games, and economic sanctions. In SEA, the USA and its Western allies have attempted to court ASEAN members to join their fight against China. However, the region shares not only historical and cultural ties and geographical proximity with China, but also a common experience of being colonised and humiliated by the West. Malaysian prime minister Anwar Ibrahim echoed the stance of SEA nations (except for the Philippines), that 'if the USA has problems with China, it should not impose it on us' (McGuirk, 2024, March 4). He continued, 'Each country's goal is to develop and trade with others in peace, and would like to be friends with the West and China alike'.

Meanwhile, in the West, the majority are opposed to their government's policies and actions, and have flooded the city centres with mass protests, in particular against the genocide In Gaza. The utter disgust, outrage and helplessness felt by most of the world's population, witnessing the meaningless slaughter of innocent lives in the most brutal and heinous ways, has pushed humanity to the brink, beyond which, there is absolutely no trust in the reigning hegemon, nor respect for it. This could only lead to the end for the current Western global political-economic system. Moreover, the USA has created a 35 trillion-dollar time-bomb of its own devise, which will implode as demands for US bonds wane against an emerging BRICS, which is developing a digital currency based on blockchain technology that is 'free of politics' (Janssen, 2024, April 4). Hopefully, the wanton destruction and senseless slaughter in the Middle East and Ukraine is already the chaotic transition to the new world order.

Conclusion

This chapter examined the changing realities at this juncture of human history. Filtering through the vast information flows in the global ICNS, the following could be treated as reliable trends:

1. Planetary evolution and dynamics are real, but the reality in individual and collective minds are increasingly being constructed and shaped by media, state agencies, private enterprises and influential individuals.
2. Rising temperatures and changes to hydro-meteorological circulation is changing climate across the globe, but their perceived impacts are based more on narratives than physical reality.
3. Disasters are likely to increase, if society remains unprepared with real structural adjustments because they occur when hazards coincide with vulnerable populations. Adaptation is needed more than emissions cuts dictated by policies and programmes based on climate narratives.
4. The Noosphere is consolidating with the advancement of the global ICNS. The process has emancipated increasing number of people from the simulacrum created by modernisation, particularly following WW2 and the collapse of the

USSR, and is reshaping individual and collective realities locally, regionally and globally.

5. The democratising and empowering aspect of the ICNS is uncovering the disinformation disseminated by Western media and political system, enabling many to question established narratives, including about Russia and China, NATO and Ukraine, Israel and Palestine, and has awaken them to the realities on the ground. This, together with US policies and actions, have fostered the rise of China and the Global South.
6. Although the emerging multi-polar world will be more equitable and inclusive and therefore aligns with the ultimate goals of Sustainable Development, the transition will be chaotic because the incumbent will be fighting 'tooth and nail' to retain its hegemonic position, as is evident from recent actions and policies.
7. While reality tends to play out differently to our expectations, current developments indicate that humanity is at the precipice of major change. However, the tipping point, often associated with climate change, is more likely to be a hot, potentially nuclear, war, than CO_2 emission causing global averaged temperature to rise 2 °C above pre-industrial level.
8. Finally, although the timing of the global shift cannot be precisely predicted, it is inevitable and the new global order and the turbulence associated with the transition will have significant implications on the region.

References

Al Jazeera English. (2023, March 22). *Xi tells Putin of 'changes not seen for 100 years'* [Video File]. YouTube. https://youtu.be/bEpTRr7QcWg

Al Jazeera English. (2024, April 2). *How will Gen Z react to the TikTok ban in the US?* | The Stream [Video File]. YouTube. https://youtu.be/K8q4Rrf1A0g

Baudrillard, J. (1994). *Simulacra and simulation.* University of Michigan Press. ISBN 0-472-06521-1.

Brown, C. (2012). China's Great Leap Forward. *Education About Asia, 17*(3), 29–34. https://www.asianstudies.org/publications/eaa/archives/chinas-great-leap-forward/

Capra, F. (1996). *The web of life: A new scientific understanding of living systems.* Anchor Books.

Carlson, T. (2024, February 9). *Exclusive: Tucker Carlson interviews Vladimir Putin* [Video File]. YouTube. https://youtu.be/fOCWBhuDdDo

Carter, J. (2006). *Palestine: Peace not apartheid.* Simon & Schuster.

CGTN: The Point. (2024, march 28). *What's wrong with U.S. society and politics?* [video] YouTube. https://youtu.be/LkfDvMeW6Q8

Chen, X.-Z. J. (2006). China and the US trade embargo, 1950–1972. *American Journal of Chinese Studies, 13*(2), 169–186. https://www.jstor.org/stable/44288827

Chomsky, N. (2003). *Hegemony or survival: America's quest for global dominance.* Metropolitan Books.

Crooke, A. (2023, March 27). L'Entente is a bitter pill for the West. *The Greenville Post.* https://www.greanvillepost.com/2023/03/27/lentente-is-a-bitter-pill-for-the-west/

Degterev, D. A., Ramich, M. S., & Tsvyk, A. V. (2021). U.S. - China: "Power transition" and the outlines of "conflict bipolarity". *Vestnik RUDN. International Relations, 21*(2), 210–231. https://doi.org/10.22363/2313-0660-2021-21-2-210-231

Graylin, A. W., & Rosenberg, L. (2024). *Our next reality: How the AI-powered metaverse will reshape the world.* Nicholas Brealey.
Geopolitical Economy Report. (2024, April 6). *West votes against rest of world in UN Human Rights Council* [Video File]. YouTube. https://youtu.be/VHBIfmxkHh4
Gleick, J. (1987). *Chaos: The making of a new science.* Viking.
Goodman, A. (2007, September 10). *Fmr. Pres. Jimmy Carter on Palestine, Iraq, Iran, selling weapons to Indonesia during the occupation of East Timor & more.* Democracy Now. https://www.democracynow.org/2007/9/10/fmr_president_jimmy_carter_on_palestine
Herman, E. S., & Chomsky, N. (2002). *Manufacturing consent: The political economy of the mass media.* Pantheon. ISBN 0-375-71449-9. https://en.wikipedia.org/wiki/Manufacturing_Consent
Jacobin. (2024, March 5). *How to stop WORLD WAR III with Abby Martin* [Video File]. YouTube. https://youtu.be/FfSS-dzucsU
Janssen, C. (2024, January 21). *Asia rising: Kishore Mahbubani's bold prediction for the 21st century* [Video File]. YouTube. https://youtu.be/AMdHwgrZt_8
Janssen, C. (2024, April 4). *How Russia survived 16,000 sanctions (genius strategy).* https://youtu.be/okYw7Mb3WK0
Li, Y., & Nicholson, H. L., Jr. (2021). When "model minorities" become "yellow peril"—Othering and the racialization of Asian Americans in the COVID-19 pandemic. *Social Compass, 15*(2), Article e12849. https://doi.org/10.1111/soc4.12849
Lindsey, R., & Dahlman, J. (2024, January 18). *Climate change: Global temperature.* NOAA. https://www.climate.gov/news-features/understanding-climate/climate-change-global-temperature
Lovelock, J. E., & Margulis, L. (1979). *Gaia: A new look at life on earth.* Oxford University Press. ISBN 978-0-19-213899-4.
Maturana, H. R., & Varela, F. J. (1980). *Autopoiesis and cognition: The realization of the living.* Reidel Publishing.
McBride, J., Berman, N., & Chatzky, A. (2023, February 2). *China's massive Belt and Road Initiative.* Council on Foreign Relations. https://www.cfr.org/backgrounder/chinas-massive-belt-and-road-initiative
McGuirk, R. (2024, March 4). *Malaysia's prime minister resists US pressure and says Malaysians don't have a problem with China.* Associate Press. https://apnews.com/article/malaysia-china-australia-anwar-ibrahim-b4d75a8423b5265bec1a05d69231965f
MLive. (2014, September 23). *Former President Carter says America has been 'almost constantly at war' since WWII* [Video File]. YouTube. https://youtu.be/LBkEoXgX5HQ
National Archives of Singapore. (1995). *Senior Minister Lee Kuan Yew's interview with Nathan Gardels of global viewpoint on 26 Sep 95 at Istana. Singapore.* Ministry of Information and the Arts. Document no. 017-1995-09-26.
Norton, B. (2023, March 25). *China & Russia pledge 'changes not seen in 100 years': Xi & Putin take aim at US dollar hegemony* [Video File]. YouTube. Eurasia. https://geopoliticaleconomy.com/2023/03/25/china-russia-xi-putin-dollar-economy/
Parmentola, A., & Tutore, I. (2023). *Industry 4.0 technologies for environmental sustainability: Intended and unintended consequences.* Springer.
Pilger, J. (Director, writer). (2016). *The coming war on China.* Dartmouth Films.
Putnam, H. (1991). *Representation and reality* (154 pp.). MIT Press. ISBN: 9780262660747.
Prigogine, I., & Stengers, I. (1984). *Order out of chaos.* Bantam.
Robbins, A. (1997). *Unlimited power: The new science of personal achievement* (448 pp.). Free Press. ISBN-10: 0684845776.
Sachs, J. D. (2018). *A new foreign policy: Beyond American exceptionalism.* Columbia University Press.
Sachs, J. D. (2020). *The ages of globalization: Geography, technology, and institutions.* Columbia University Press.
Sachs, J. D. (2024, March 21). *Jeffrey Sachs Interview—Year of instability.* Jeffrey Sachs of Fans [Video File]. YouTube. https://youtu.be/ZSj8d3Tk0cM

Sadiku, M. N. O., Shadare, A. E., & Musa, S. M. (2017). Digital natives. *International Journal of Advanced Research in Computer Science and Software Engineering, 7*(7), 125. https://doi.org/10.23956/ijarcsse.v7i7.111, https://www.researchgate.net/publication/326071415_Digital_Natives

Seidl, D. (2004). *Luhmann's theory of autopoietic social systems*. Munich Business Research. https://www.researchgate.net/publication/277293382

Singh, K. (n.d.). *New climate deal & no clarity on CDM continuity*. Earthhood. https://www.earthood.com/earthscoop/new-climate-deal-no-clarity-on-cdm-continuity

Siripurapu, A., & Berman, N. (2023, July 19). *The dollar: The world's reserve currency*. Council on Foreign Relations. https://www.cfr.org/backgrounder/dollar-worlds-reserve-currency

Stacy, R. D. (2001). *Complex responsive processes in organizations: Learning and knowledge creation*. Routledge.

UN Charter. https://www.un.org/en/about-us/un-charter/full-text

UN Climate Change. (n.d.) *What is the Kyoto Protocol?* https://unfccc.int/kyoto_protocol

UNEP. (n.d.). *The climate emergency*. Retrieved 15 march 2024. https://www.unep.org/climate-emergency

Vernadsky, V. I. (1988). *The biosphere*. Translated into English by Langmuir, D.B. in 1988. Copernicus.

Walsh, M. (2020, June 8). *The metaverse is the next universe* [Video File]. YouTube. https://youtu.be/0TCGBhY7qfE

Wave Media. (2024, April 10). *How does the Gaza war prompt China's desire for a new world order?* [Video File]. YouTube. https://youtu.be/rKqV-oOqo_8

Wong, A. (2016, May 30). *The untold story behind Saudi Arabia's 41-year U.S. debt secret*. Bloomberg. https://www.bloomberg.com/news/features/2016-05-30/the-untold-story-behind-saudi-arabia-s-41-year-u-s-debt-secret

Yong, G. Y. V. (2005). *Misunderstanding: Natural features in Communication, learning & knowledge sharing*. Conference on Information Literacy - Information, Learning, Innovation: Delivering Global Knowledge. Universiti Brunei Darussalam Library, pp. 57–67.

Gabriel Yit Vui Yong is a lecturer at Universiti Brunei Darussalam with background in Geological Geophysics (BSc, Reading) and Environmental Analysis (MSc, Hull). Over the past three decades, he has taken an inter- and trans-discipliantry path and engaged in a wide range of national and regional projects, which has shaped his view of the world as a complex, sentient and evolving system that is interconnected across multiple dimentions and scale.

Chapter 3
Climate Change and Sea Level Rise in Southeast Asia: Current Knowledge and Future Projections

Fredolin Tangang

Abstract The countries of Southeast Asia are highly exposed and vulnerable to the effects of climate change and climate variability. This article provides an overview of the current understanding of future climate change and sea level rise in the region. It also provides an overview of weather and climate characteristics and past trends. In recent decades, the mean surface air temperature has risen, by about 1.0–1.5 °C between pre-industrial period (1850–1900) and 2020. Extreme heatwaves and intense precipitation events have also become more frequent, which is largely attributed to human activities that have led to increased greenhouse gas concentrations in the atmosphere. Sea levels are rising at a rate of 4.0–8.2 mm per year, which has a significant impact on low-lying areas such as the Mekong Delta. Climate projections show that the tendency for extreme weather events and sea level rise will increase with each degree of global warming. The intensity and frequency of weather and climate extremes will increase significantly with warming above 2.0 °C compared to the conditions expected at 1.5 °C. Sea level rise will also be greater under the same scenario. Countries in the region face increased climate risks, as the current Nationally Determined Contributions (NDCs) under the Paris Agreement are not sufficient to curb warming to below the 1.5 °C target. Instead, they are expected to lead to warming of 2.5–2.9 °C above pre-industrial levels by the end of the twenty-first century.

Keywords Climate change · Climate variability · Sea level rise · Future projection · Emission scenarios · Warming levels

---

F. Tangang (✉)
Geography, Environment and Development (GED) Programme, Faculty of Arts and Social Sciences, Universiti Brunei Darussalam, Gadong, Brunei Darussalam
e-mail: fredolin.tangang@ubd.edu.bn

G. Y. V. Yong and A.K.M. A. Ullah (eds.), *Brunei and the Future Transformation of Southeast Asia*, https://doi.org/10.1007/978-981-96-7901-0_3

Introduction

Southeast Asia is unique in many ways, especially in terms of its geography and socio-economic conditions. It consists primarily of the least developed and developing countries and is the third most populous region in the world, with a current population of about 700 million people, which is expected to reach 750 million by 2050 (Hirschman & Bonaparte, 2012). With a large proportion of the population living in coastal regions, river basins and urban areas, and an economy largely dependent on agriculture, Southeast Asia is considered one of the most vulnerable regions to climate change and climate variability. Although some countries are more vulnerable than others, all nations in the region have already been affected by climate-related disasters (Carnegie, 2024). Over the past two decades, many countries have experienced extreme weather events, including larger magnitude floods, storms, heat waves and droughts, which have caused significant socio-economic damage. These extreme weather and climate events are likely to become more frequent and intense as climate change continues. According to a report by the Asian Development Bank (ADB), this could cost Indonesia, the Philippines, Thailand and Vietnam the equivalent of 6.7% of their combined gross domestic product (GDP) annually by 2100—more than double the global average (ADB, 2009).

In 2011, Thailand was hit by a massive flood disaster that caused economic losses of USD 46.5 billion and over 800 deaths (Supharatid et al., 2016). Due to their geographical location in the typhoon belt of the north-west Pacific, Vietnam and the Philippines are more affected by typhoons than other countries in the region. In 1997, Vietnam was hit by Typhoon Linda, which caused widespread flooding, affected 500,000 people and claimed 3000 lives (Anh et al., 2019). Super Typhoon Haiyan devastated the Philippines in 2013, killing more than 6000 people and causing economic losses of USD 1.8 billion (Hernandez Jr. et al., 2015). Myanmar, which is more susceptible to cyclones that originate in the Indian Ocean and the Bay of Bengal, is not spared from these disasters (Alejandria et al., 2024). In 2008, Cyclone Nargis hit Myanmar and killed 138,000 people (Fritz et al., 2010). The most recent Typhoon Yagi in early September 2024 was a similarly deadly and extremely destructive tropical storm, affecting the Philippines, Vietnam, southern China, Macau, Hong Kong, Laos, northern Thailand and Myanmar. In Vietnam, the typhoon caused massive flooding and landslides, resulting in 291 deaths and 38 missing persons, injuring almost 2000 people and damaging 251,000 hectares of cultivated land (UNCT, 2024).

Although Malaysia, Singapore, Indonesia and Brunei do not lie in the typhoon belt, they too have been affected by extreme weather events. In December 2014, the north-eastern region of Peninsular Malaysia was hit by the worst flooding on record during the northeast monsoon, affecting 500,000 people and causing economic losses of around USD 500 million (Hai et al., 2017). Selangor and Kuala Lumpur, located in the western part of Peninsular Malaysia, were severely hit by the worst recorded flood in this region, affecting 50,000 people and causing 50 deaths, with total economic losses of around USD 1.5 billion (Rahman, 2022). Sabah and Sarawak in Borneo were also regularly hit by floods. In Brunei, flooding is a common occurrence (Islam et al., 2023).

Indonesia, the largest and most populous country in the region, faces a similar dilemma (Nugraheni et al., 2022). However, besides floods, the region is also increasingly affected by droughts and heatwaves. In 2015, Indonesia experienced a prolonged drought that led to extensive forest fires and a prolonged period of haze, which impacted neighbouring countries Malaysia, Singapore and Brunei (Lohberger et al., 2018). While these climate-related disasters occur annually, records show that the frequency of occurrence has steadily increased in recent decades, with concomitant socio-economic impacts (Beirne et al., 2021; Dagli & Ferrarini, 2019).

The region is not only confronted with extreme weather and climate events. More and more countries are also affected by rising sea levels, coastal erosion and the intrusion of salts into agricultural land. Vietnam has already suffered economic losses due to the effects of sea level rise and salt intrusion in the Mekong Delta (Hai et al., 2024; Thach et al., 2023; Van Tho, 2022). Indonesia, the Philippines and Malaysia are particularly vulnerable due to their many islands and extensive coastlines (NAHRIM, 2017; Siringan & Maria, 2024; Triana & Wahyudi, 2020). As a small coastal state with a relatively short coastline of only about 161 km, sea level rise is a potential threat to Brunei (Michalak et al., 2021; Ratnayake et al., 2023). Additionally, rising sea temperatures in Southeast Asian waters also pose a major risk to the region's marine ecosystems (Kay et al., 2023), which could have serious impacts on livelihoods and food security.

As limiting global warming to less than the 1.5 °C target remains uncertain, countries in the region must also focus on adaptation measures to increase climate resilience. Therefore, understanding the magnitude of climate risks at the local level is crucial for climate action, especially for local adaptation measures. However, the compounded risk is not only determined by the type and severity of climate hazards, but also by exposure and vulnerability (IPCC, 2022). Exposure and vulnerability, which are not climate factors, are not the focus of this review. Instead, this chapter examines the current understanding of future climate change in Southeast Asia. This, however, requires knowledge of the weather and climate characteristics of the region, including observed trends over recent decades. This chapter also provides an overview of observed and projected sea level rise in the region.

IPCC Report on Climate Change

In its Sixth Assessment Report (AR6), the United Nation's (UN) Intergovernmental Panel on Climate Change (IPCC) concluded that recent climate change is widespread, rapid, intensifying and unprecedented for thousands of years (IPCC, 2021). The global mean surface temperature (GMST) has risen steadily, with each of the last four decades being successively warmer than any previous decade. On average over the years 2011–2020, the temperature increase was 1.1 °C higher than in the pre-industrial period (1850–1900), representing a warming that is unprecedented for at

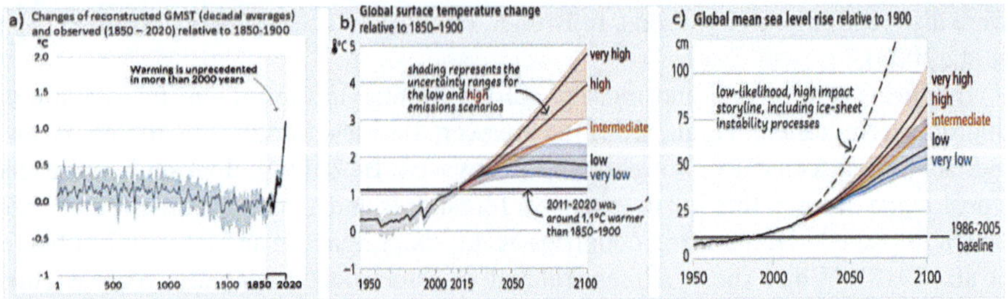

Fig. 3.1 **a** The GMST changes relative to pre-industrial level 1860–1900 of observed temperature (1850–2020) and reconstructed (prior 1850), **b** the projected GMST relative to 1860–1900, and **c** the projected global mean sea level rise relative to 1900, according to different GHG emission scenarios. Figures are extracted from the IPCC WG1 (IPCC, 2021) and IPCC Synthesis AR6 (IPCC, 2023) reports

least 2000 years (Fig. 3.1a). The IPCC attributed this warming trend to the increasing concentration of greenhouse gases (GHGs) in the atmosphere due to human activities since the beginning of the industrial period in the early 1900s. The concentration of carbon dioxide (CO_2) has reached 410 ppm, well outside the natural range of 180–300 ppm (Petit et al., 1999), and is higher than it has been for at least a million years (IPCC, 2021).

The Earth's climate system has been in a state of 'radiative imbalance' since 1970, in which it continues to retain excess heat, driving global warming and climate change (Forster et al., 2021). More than 91% of this additional heat has been absorbed by the oceans, leading to an increase in ocean temperatures and heat content, as observed over the last three decades. These systematic changes have been detected not only in surface air and ocean temperatures, but in all components of the climate system, including glacier mass loss, extreme temperature and precipitation events, sea level rise (Chen et al., 2021) and in natural ecosystems (Gulev et al., 2021). These have impacted human systems around the world (IPCC, 2022). The magnitude of change is expected to increase in the future as GHG concentrations continues to rise. Under the worst-case scenario, the global mean surface temperature (GMST) is projected to rise by 4–5 °C above pre-industrial levels, with sea levels rising by up to 1 m by 2100 (Fig. 3.1b, c) and weather extremes becoming more intense and frequent (IPCC, 2021). Climate risk is expected to increase significantly if global warming exceeds 1.5 °C (IPCC, 2023). To contain warming to below 1.5 °C, rapid and deep emission cuts are needed to reduce current annual emissions by around 50%, i.e., from around 60 $GtCO_2$-eq to around 20–30 $GtCO_2$-eq by 2030; Net Zero CO_2 must be achieved by 2050 (IPCC, 2023). However, the IPCC has also pointed out that the current pace and scope of climate protection measures are not sufficient to combat climate change (IPCC, 2023).

Weather and Climate Characteristics in Southeast Asia and Changing Trends: An Overview

Situated between the Indian and Pacific Oceans and flanked by two large continents—Asia to the north-west and Australia to the south—the weather climate in Southeast Asia is mainly determined by the Asian-Australian monsoon system (Wang & Chang, 2008). Within the maritime continent, the subcomponent of this system is sometimes referred to as the Maritime Continental Monsoon (Robertson et al., 2011) and the Australian-Maritime Continental Monsoon (IPCC, 2021). The region is generally characterised by hot and humid conditions, but its weather and climate exhibit considerable temporal and spatial variability due to the influence of two oceans, surrounding seas, scattered land masses and the complex topography of the region.

Weather and Climate Pattern

The boreal summer component of the Asian-Australian monsoon, commonly known as the Asian summer monsoon, influences the Southeast Asian mainland and brings increased rainfall during the boreal summer. In some parts of the maritime continent, weather conditions during the Asian summer monsoon are hot and dry. During the boreal winter, which accompanies the Australia–maritime continental monsoon, the region tends to experience more rainfall. However, certain parts of the maritime continent, such as the central Philippines, southern Thailand and south-eastern Vietnam, receive significant rainfall during both monsoons. The annual rainfall cycles in a particular location can also be influenced by the local topography. On the east coast of Peninsular Malaysia, for example, the Asian winter monsoon dominates the annual rainfall cycle, while on the west coast the cycle has double peaks during the inter-Monsoon periods (Tangang et al., 2012). In Indonesia, several different patterns of the annual precipitation cycle can be observed (Aldrian & Dwi Susanto, 2003).

Depending on the location and season, Southeast Asia can experience tropical storms, extreme rainfall events, floods, heat waves and droughts. The Philippines, where around 20 typhoons occur each year—most frequently between June and November—is particularly at risk. Some tropical storms move westwards and intensify as they pass over the South China Sea, causing catastrophic flooding in Vietnam, Laos, Cambodia and occasionally Thailand. Myanmar, on the other hand, is affected by tropical storms that develop over the Indian Ocean and the Bay of Bengal, although these occur less frequently than those that hit the Philippines.

During the boreal winter monsoon, Southeast Asia can also be affected by cold surges, i.e. episodes of strong winds from the advection of cold air masses caused by the strengthening of the Siberian High, usually 4–6 episodes per winter season. While most countries in mainland Southeast Asia may experience extreme cold events, regions such as Java, southern Thailand, the eastern Philippines and eastern parts of Peninsular Malaysia are more likely to experience extreme precipitation and flooding due to the enhanced moisture convergence caused by these cold surges (Olaguera et al., 2022; Siswanto et al., 2017; Tangang et al., 2008). The influence of cold surges can also be modulated by the Borneo vortices, cyclonic circulations that form during the winter monsoon (Chang et al., 2005 ; Chen et al., 2013 ; Juneng & Tangang 2010). A Borneo vortex near the east coast of Peninsular Malaysia can enhance moisture convergence in this area (Juneng et al., 2007). In addition, Borneo vortices can intensify into tropical depressions, which can have even more severe impacts (Liang et al., 2023).

The weather and climate in Southeast Asia are further complicated by the influence of various modes of climate variability that modulate atmospheric conditions in the tropical region. From a risk perspective, in addition to anthropogenic climate change, the role of natural climate variability in the development of climate hazards must also be considered (IPCC, 2022). On intra-seasonal time scales of 20–60 days, weather and climate in the region can be influenced by the Madden–Julian Oscillation (MJO), a large-scale atmospheric disturbance that spreads eastwards from the Indian Ocean into the Pacific Ocean and passes over Southeast Asia (Crook et al., 2024; Da Silva & Matthews, 2021; Zhang, 2005). During the boreal summer, the MJO shifts northeastwards over Southeast Asia and is referred to as the Boreal Summer Intra-Seasonal Oscillation (BSISO) (Xavier et al., 2024). Depending on the phase, the MJO and BSISO can either enhance or suppress convection over Southeast Asia (Da Silva & Matthews, 2021) and alter the diurnal cycle of precipitation (Peatman et al., 2014). During periods of suppressed convection, the MJO and BSISO can produce anomalously dry conditions and extreme heat.

Dry periods associated with the MJO have been linked to wildfires and air quality degradation (Jud et al., 2020; Koplitz et al., 2018). In combination with other factors, the MJO has also been linked to extreme precipitation events and widespread flooding (Tangang et al., 2008). On interannual time scales, weather and climate in Southeast Asia are influenced by the El Niño–Southern Oscillation (ENSO) and the Indian Ocean Dipole (IOD) (Juneng & Tangang, 2005; Salimun et al., 2014; Supari et al., 2018). During an El Niño event, especially an extreme one, Southeast Asia often experiences long-lasting droughts and periods of extreme heat (Supari et al., 2018; Tangang et al., 2017; Zakaria et al., 2024). Conversely, La Niña events create favourable conditions for heavy rainfall, which can lead to flooding (Tangang et al., 2017). The simultaneous occurrence of the IOD can also exacerbate the influence of ENSO over Southeast Asia (Amirudin et al., 2020).

Trends in Temperatures, Rainfall and Extremes

Numerous studies have identified significant long-term changes in weather and climate indicators across the region. Both the Asian-Australian monsoon has shown broad changes (e.g., IPCC, 2021; Loo et al., 2015; Zuo et al., 2013). The IPCC AR6 indicated that the South and Southeast Asian monsoon has weakened in the second half of the twentieth century, primarily due to anthropogenic aerosol forcing (IPCC,

2021). Despite the considerable inter-annual variability associated with ENSO, many studies report significant trends in observed temperature, rainfall, and extreme weather events. According to the IPCC WG1 AR6, the surface air temperature in Southeast Asia has increased by 1.0–1.5 °C in 2020 compared to the pre-industrial period (1850–1900) (Chen et al., 2021). While there are spatial variations, indices of temperature extremes indicate increasing trends across the region (Cheong et al., 2018; Cinco et al., 2014; Limsakul & Goes, 2008; Manton et al., 2001; Mohd Salleh et al., 2015; Muhammad et al., 2024; Nguyen et al., 2017; Supari et al., 2017; Tangang et al., 2007). The rate of increase in daily minimum temperatures has been greater than that of maximum daily temperatures, resulting in declining trends in the diurnal temperature range (DTR) across the region (Cheong et al., 2018). The IPCC AR6 indicates with high confidence that extreme heat events in Southeast Asia have been increasing, while extreme cold events have been decreasing (IPCC, 2021). These trends in extreme heat and cold events are attributed to human influence (IPCC, 2021).

Observed precipitation trends across Southeast Asia, including extremes, show spatial variation, with some areas experiencing significant increases while others show no discernible trends (Cheong et al., 2018; Cinco et al., 2014; Endo et al., 2009; Limsakul & Singhruck, 2016; Manton et al., 2001; Supari et al., 2017). The lack of significant trends in precipitation-related indices in some areas could be due to a low signal-to-noise ratio resulting from the significant interannual variability associated with ENSO (Amirudin et al., 2020; Juneng & Tangang, 2005; Salimun et al., 2014; Singh & Qin, 2020; Supari et al., 2018; Zakaria et al., 2024). However, a recent study by Edirisinghe et al. (2023) reported increasing trends in summer precipitation over mainland Southeast Asia based on a gridded precipitation dataset from 1981 to 2021. The IPCC AR6 report also found a significant increase in extreme precipitation events in Southeast Asia, although confidence in these trends and their attribution to human influence was relatively low (IPCC, 2021). In addition, the IPCC AR6 also pointed out that there is no significant trend in terms of tropical cyclones hitting the region, but that more extreme tropical cyclones have hit the region (Ranasinghe et al., 2021).

Future Climate Change in Southeast Asia

A number of studies have projected broad changes in the Asian-Australian monsoon in future periods of a warmer climate (e.g., Chevuturi et al., 2018; Wang et al., 2014, 2021a, 2021b). These changes are likely to influence future climate patterns in the region, particularly in terms of mean temperature, precipitation, and extreme weather events. According to the IPCC AR6, Southeast Asia is projected to continue experiencing warming trends, although the magnitude of this warming is likely to be slightly less than the global average (IPCC, 2021).

Wang et al. (2014) found that under the RCP4.5 scenario, which represents a stabilisation of GHG and 2–3 °C warming by the end of the twenty-first century, precipitation will increase significantly in the region affected by the Asian-Australian monsoon. The uneven warming, where the northern hemisphere be increasing more than the southern hemisphere, is likely to lead to hemispheric sea level pressure differences, which may result in an increase in Asian summer monsoon precipitation but a decrease in Australian summer monsoon precipitation (Wang et al., 2014). This result is consistent with the projected changes in seasonal precipitation based on the coordinated regional climate downscaling method within the Coordinated Regional Climate Downscaling Experiment (CORDEX) for Southeast Asia (CORDEX-SEA) (Tangang et al., 2019, 2020). The CORDEX-SEA downscaling simulations used several global climate models (GCMs) from the Coupled Model Intercomparison Project Phase 5 (CMIP5), which were downscaled to a horizontal resolution of 25 km using several regional climate models (RCMs) (Tangang et al., 2020).

The CORDEX-SEA projections show that under the RCP4.5 and RCP8.5 emission scenarios, which include a strong reliance on fossil fuels, mean boreal winter precipitation (December-January–February) on the Southeast Asian mainland will increase by 10–20% at the beginning (2011–2040), middle (2041–2070) and end (2071–2100) of the twenty-first century compared to the historical period (1976–2005) (Fig. 3.2). For some areas of the maritime continent, however, a minimal increase or slight drying out is projected. For the boreal summer precipitation (June–July–August), the projections show a different pattern. While an increase in summer precipitation is expected for Myanmar and northern Thailand, especially under RCP8.5, the rest of mainland Southeast Asia—including central and southern Thailand, Cambodia, Laos and Vietnam—is projected to dry out (Fig. 3.2). The maritime continent, particularly Indonesia, is also likely to become drier.

In contrast, the projections of the latest CMIP6 GCMs indicate a drying trend for the boreal winter over the Southeast Asian mainland under the SSP2-4.5, SSP3-7.0 and SSP5-8.5 scenarios (Ona et al., 2024), which contradicts the CORDEX-SEA simulations. Nevertheless, the projected changes in boreal summer precipitation are consistent in both simulation series. In particular, both simulations predict a drying trend for the maritime continent, especially for Indonesia. This trend is consistent with the significant increase in projected consecutive dry days (CDD) by 10–50% on the maritime continent, especially in Indonesia, by the end of the twenty-first century, as shown by the CORDEX-SEA simulations (Supari et al., 2020). The most recent analysis of the CMIP6-GCM simulations by Ghosh and Shepherd (2023) consistently showed a drying trend over the southwestern maritime continent, which can be linked to the intensity of basin-wide sea surface temperature warming in the equatorial Pacific. Supari et al. (2020) also pointed to a projected intensification and more frequent extreme precipitation under the RCP8.5 scenario by the end of the twenty-first century, a conclusion supported by other studies (e.g. Ge et al., 2021; Hariadi et al., 2024; Tangang et al., 2018; Try & Qin, 2024). However, despite the projected intensification of tropical cyclones in the western Pacific, the frequency of events affecting the Philippines, Vietnam and countries on the mainland in the South China Sea is expected to decrease (Ranasinghe et al., 2021).

According to the IPCC AR6, the warming trend in Southeast Asia will continue, although the extent of warming is likely to be slightly lower than the global average

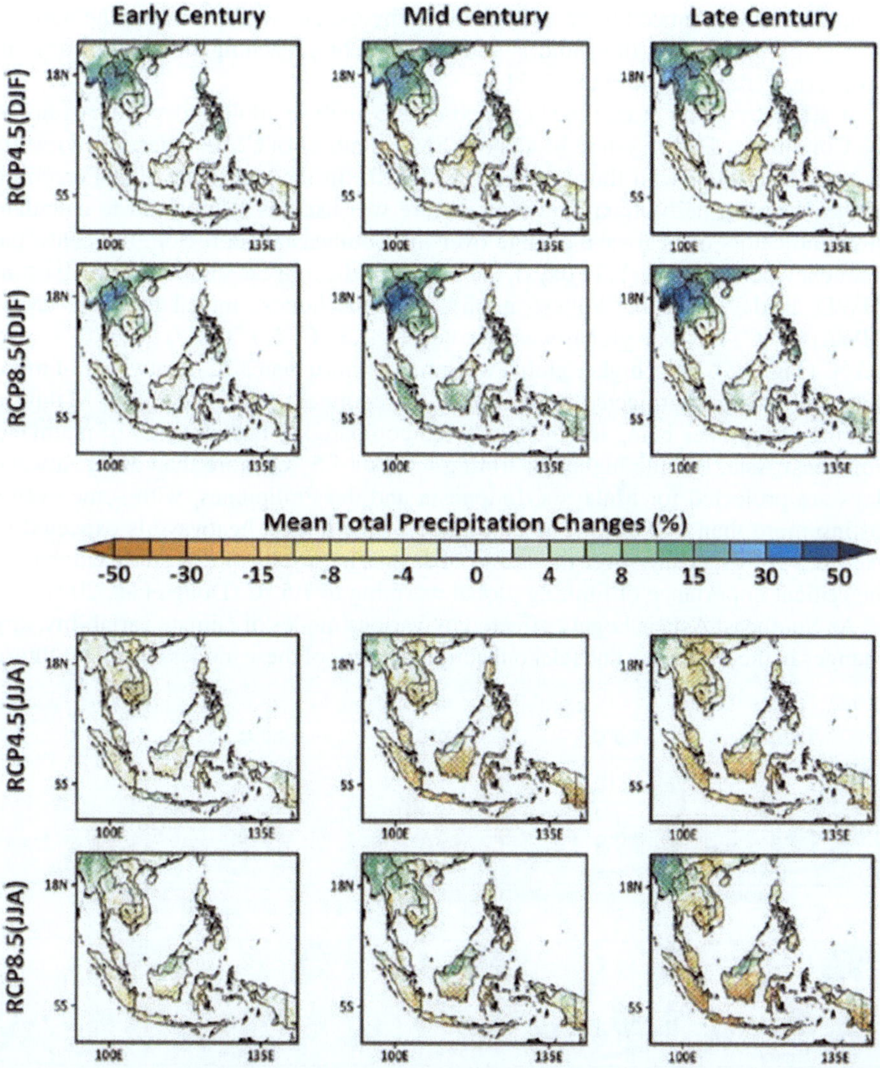

Fig. 3.2 Projected changes of CORDEX Southeast Asia based on multi-model simulations under two GHG scenarios, RCP4.5 and RCP8.5, for early (2011–2040), middle (2041–2070) and end (2071–2100) of the twenty-first century relative to the historical period of 1976–2005 (adapted from Tangang et al., 2020)

(Ranasinghe et al., 2021). Extreme heat events are expected to increase in intensity and frequency by the end of the twenty-first century, particularly under the high-emission scenario RCP8.5/SSP5-8.5, under which the annual exceedance of the 41 °C threshold for the heat index is expected to reach 250 days (Ranasinghe et al., 2021). Even in the lower-emission scenario SSP1-2.6, where the increase in global average temperature is estimated to be around 2.0 °C by the end of the century, the number of days exceeding this threshold is still considerable, reaching around 100 days in some areas (Ranasinghe et al., 2021).

A study by Dong et al. (2021), based on the analysis of 40 independent runs of the Community Earth System Model (CESM) version 1 of CMIP5, generated results that are consistent with those of the IPCC AR6. In the study, the 90th Percentile of the observed daily maximum temperature was used as a threshold to calculate three indicators of heatwave change over the Southeast Asia region: the heatwave frequency per year (HWF, in days), the length of the longest annual heatwave event (HWD, in days) and the hottest amplitude of the hottest annual heatwave event (HWA, in °C) for four global warming levels: 1.5 °C, 2.0 °C, 3.0 °C, 4.0 °C and 4.5 °C (Fig. 3.3). For a higher global warming of more than 3 °C, heatwaves of more than 200 days are projected on the maritime continent, with Indonesia, Malaysia and the Philippines being the most affected compared to the countries of mainland Southeast Asia. For the highest warming level of 4.5 °C, more than 300 heatwave days are projected for Malaysia, Indonesia and the Philippines, with some events lasting more than 100 days. The amplitude of the hottest heatwave is expected to exceed 5 °C, especially over the island of Java. The results of this study emphasise the critical importance of limiting global warming to 1.5 °C (Dong et al., 2021).

As Southeast Asia is largely affected by various modes of climate variability, any changes in the frequency and teleconnection patterns of these modes can affect future

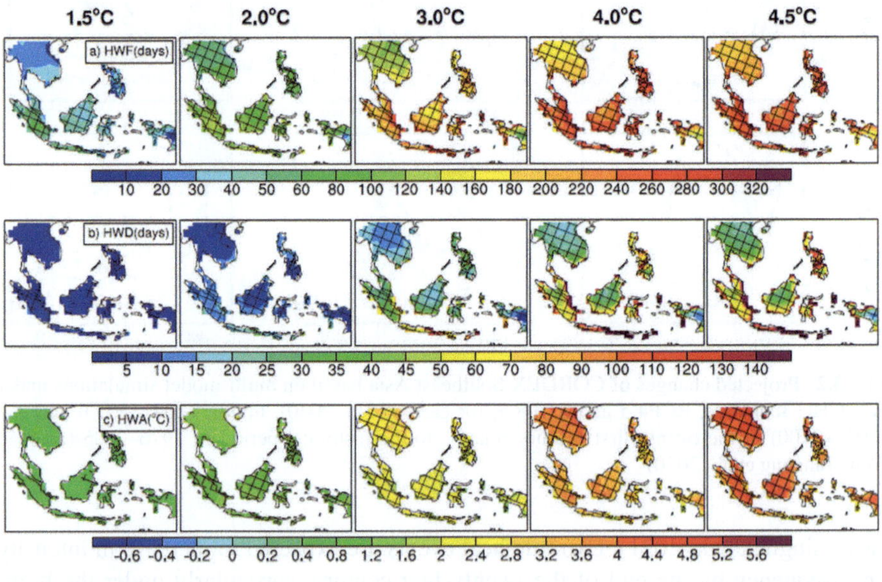

Fig. 3.3 Projected changes in **a** heatwave yearly frequency (unit: days), **b** duration of the yearly heatwave event (unit: days), and **c** hottest amplitude of the hottest yearly heatwave event (unit: °C) under 1.5, 2.0, 3.0, 4.0, and 4.5 °C global warming levels relative to the current climate. Hatched areas indicate changes exceeding the 95% confidence level (adapted from Dong et al., 2021)

weather and climate extremes in the region. A recent study by Liang et al., (2022) detected that nearly 60% increase in extreme rainfall over Asia and Australia by the end of the twenty-first century under the SSP5-8.5 scenario, and 84% out of this can be associated with the MJO, affecting countries including Malaysia and Indonesia. The characteristics of Borneo vortex could also change in a warmer world (Liang et al., 2023). Although there is no strong model consensus indicated in the IPCC AR6, some studies have suggested possible changes of IOD and ENSO characteristics and their teleconnection over Southeast Asia in a warmer climate, which could affect how these phenomena affect the region (Cai et al., 2014, 2021; Chen et al., 2023; Nguyen-Le, 2024).

Sea Level in Southeast Asia Seas: Observed and Projected Changes

Strassburg et al. (2015) estimated sea level rise (SLR) trends for the period from 1993 to 2010, based on satellite altimeter data produced and distributed by the Archiving, Validation, and Interpretation of Satellite Oceanographic (AVISO) system. In the Andaman Sea and the Gulf of Thailand, the estimated SLR rate ranged between 4.0 and 4.5 mm yr^{-1}, while in the Strait of Malacca, the South China Sea, and the Java Sea, the estimated rates were between 4.5 and 5.0 mm yr^{-1}. The estimated SLR rates in the Sulu Sea, Celebes Sea, Makassar Strait, and the seas of eastern Indonesia were slightly higher, ranging from 6.0 to 8.0 mm yr^{-1}. SLR estimates based on tidal gauge data from various locations, covering the period from 1993 to 2019, provided by the NASA Sea Level and Assessment Tool (https://sealevel.nasa.gov/data_tools/16), reveal consistent trends. For Penang Island, located in the Strait of Malacca, the estimated SLR rate was 5.02 ± 2.12 mm yr^{-1}. However, in Bangkok, the estimated SLR was 8.19 ± 3.61 mm yr^{-1}, nearly twice the rate estimated by Strassburg et al. (2015) for the Gulf of Thailand. The discrepancy between the two estimates could be due to large contribution of land subsidence. In the Sulu Sea area, estimates for several locations along the northeast coast of Borneo showed lower rates, around 4.0–5.0 mm yr^{-1}. These were substantially lower than the rates estimated by Strassburg et al. (2015), which they explained could be affected the Pacific Decadal Oscillation (PDO). Nevertheless, these regional SLR estimates are generally higher than the global mean sea level (GMSL) increase of 3.7 mm yr^{-1} estimated by the IPCC (AR6) for the period from 2006 to 2018.

The projected sea level rise by 2100 in Southeast Asian seas for different global warming levels are tabulated in Table 3.1. It is based on the IPCC AR6 sea level projections (extracted from the NASA Sea Level Projection tool; https://sealevel.nasa.gov/ipcc-ar6-sea-level-projection-tool). Using the SSP5-8.5 emissions scenario, and accounting for contributions from stereodynamic sea level, glaciers, Greenland, Antarctica, land water storage, and vertical land movement (Fox-Kemper et al., 2021), the projections ranged from 0.78 to 0.84 m relative to the 1995–2014 baseline. The waters surrounding the northern Philippines is forecasted to rise by 0.95–1.12 m compared to the baseline. This is due to a greater contribution from vertical land movement. A global warming of 4 °C by 2100 would likely lead to a sea level rise of 0.60–0.85 m. These estimates are somewhat lower compared to the GMSL under SSP5-8.5, which is between 0.98 and 1.88 m, with a best estimate of 4.4 °C rise in global mean surface temperature (GMST) compared to pre-industrial levels (IPCC, 2021). For Bangkok and Manila, however, the projected sea level rise by 2100 at 4 °C warming is significantly higher: 1.82 m (1.67–2.05 m) and 1.27 m (1.10–1.53 m) respectively (Table 3.1). The high values for these cities could be due to land subsidence caused by groundwater extraction and earthquakes (Cao et al., 2021; Jaroenongard et al., 2021; Tay et al., 2022).

Jakarta is another megacity where significant land subsidence is occurring (Cao et al., 2021; Tay et al., 2022; Triana & Wahyudi, 2020). However, the low estimates for Jakarta could be due to the fact that they were derived from a point in the Java Sea close to the city. Therefore, they are unlikely to be affected by land subsidence. The total sea level rise for Jakarta, including land subsidence, could be much higher than shown in Table 3.1. At 1.5 °C warming, sea level rise is projected to be between 0.38 and 0.62 m for most locations, with the exception of Bangkok and Manila, which have equal values across all warming levels. These estimates increase to 0.46–0.69 m and 0.55–0.79 m, respectively, compared to the baseline year under 2.0 °C and 3.0 °C warming scenarios, respectively.

Conclusion

In recent decades, the frequency and intensity of weather and climate extremes in Southeast Asia have increased, a trend that is attributed to human influences on the climate system. Climate projections indicate that extreme events will intensify. The impacts of the associated hazards on exposed and vulnerable populations and on natural systems are likely to worsen (IPCC, 2022). Sea level rise, which is already affecting coastal regions and cities, is a growing concern, particularly in Bangkok, Manila and Jakarta, where groundwater extraction is exacerbating the problem. The decline of coastlines in many Southeast Asian countries will have severe impacts coastal populations, resources, infrastructure and economies, should global average temperature increase by more than 2 °C (Asuncion & Lee, 2017; Dong et al., 2024).

One crucial question remains: Can the world limit global warming to below 1.5 °C? Current policies and NDC pledges under the Paris Agreement are inadequate to achieve the GHG reductions needed. The implementation gap, based on policies in place as of the end of 2020, is projected to lead to global warming of approximately 3.2°C by 2100 (IPCC, 2023). To avoid exceeding the 1.5 °C threshold, stronger commitment than what was achieved at the last United Nations Framework Convention on Climate Change (UNFCCC) Conference of the Parties (COP 28) is needed (UNFCCC, 2023). Countries in Southeast Asia will need to review and improve their NDCs to fulfil these requirements. To reduce the impacts of climate change

Table 3.1 Projected sea level rise for several locations in Southeast Asia by 2100 according to different global warming levels relative to 1995–2014 baseline

Locations	Projected sea level rise in meters relative to 1995–2014 baseline				VLM (Meters)
	Global warming levels by 2100				
	1.5 °C	2.0 °C	3.0 °C	4.0 °C	
Yangon, Myanmar	0.62 (0.42–0.84)	0.69 (0.50–0.92)	0.79 (0.63–1.02)	0.85 (0.67–1.11)	0.17 (0.08–0.26)
Penang. Island, Malaysia	0.39 (0.19–0.63)	0.47 (0.27–0.71)	0.58 (0.39–0.83)	0.65 (0.44–0.92)	−0.08 (−0.19–0.02)
Port Klang, Malaysia	0.38 (0.18–0.61)	0.46 (0.24–0.71)	0.56 (0.37–0.82)	0.63 (0.41–0.92)	−0.10 (−0.22–0.01)
Singapore (West Tuas)	0.42 (0.21–0.66)	0.50 (0.28–0.76)	0.60 (0.39–0.87)	0.68 (0.46–0.96)	−0.07 (−0.20–0.07)
*Padang, West Sumatra	0.46 (0.13–0.80)	0.53 (0.21–0.89)	0.64 (0.33–1.01)	0.72 (0.38–1.10)	−0.03 (−0.25–0.20)
*Pontianak, West Kalimantan	0.47 (0.15–0.81)	0.55 (0.23–0.91)	0.65 (0.33–1.02)	0.73 (0.40–1.10)	−0.03 (−0.25–0.20)
*Jakarta, Indonesia	0.46 (0.14–0.80)	0.54 (0.20–0.90)	0.64 (0.33–1.00)	0.72 (0.39–1.11)	−0.03(−0.25–0.20)
*Bali, Indonesia	0.49 (0.29–0.72)	0.56 (0.35–0.81)	0.65 (0.48–0.90)	0.75 (0.54–1.03)	−0.01 (−0.09–0.07)
Kuantan (Tanjung Gelang), Malaysia	0.43 (0.24–0.65)	0.51 (0.32–0.75)	0.61 (0.42–0.86)	0.68 (0.49–0.95)	−0.06 (−0.16–0.04)
Kota Bharu (Geting), Malaysia	0.38 (0.18–0.60)	0.46 (0.25–0.71)	0.55 (0.36–0.81)	0.63 (0.43–0.90)	−0.11 (−0.22–0.01)
Ko Lak, Thailand	0.38 (0.26–0.56)	0.49 (0.33–0.70)	0.58 (0.43–0.81)	0.66 (0.51–0.90)	−0.06 (−0.10–−0.02)
Bangkok (Phrachula Chomklao) Thailand	**1.55 (1.43–1.72)**	**1.64 (1.49–1.85)**	**1.75 (1.60–1.97)**	**1.82 (1.67–2.05)**	**1.11 (1.07–1.16)**
Ho Chi Minh (Vungtau), Vietnam	0.49 (0.30–0.72)	0.57 (0.37–0.81)	0.67 (0.48–0.92)	0.74 (0.55–1.01)	0.01 (−0.10–0.12)
Danang, Vietnam	0.52 (0.34–0.74)	0.59 (0.40–0.82)	0.70 (0.51–0.94)	0.78 (0.59–1.04)	0.05 (−0.06–0.15)
Hondau, Vietnam	0.45 (0.30–0.63)	0.52 (0.36–0.72)	0.63 (0.47–0.86)	0.71 (0.55–0.95)	−0.01 (−0.07–0.06)
Bintulu, Sarawak, Malaysia	0.45 (0.29–0.65)	0.53 (0.36–0.76)	0.63 (0.48–0.86)	0.70 (0.53–0.96)	−0.03 (−0.09–0.03)
*Bandar Seri Begawan, Brunei	0.47 (0.31–0.67)	0.55 (0.38–0.78)	0.65 (0.50–0.89)	0.72 (0.55–0.98)	−0.02 (−0.07–0.03)
Kota Kinabalu, Sabah, Malaysia	0.46 (0.30–0.66)	0.54 (0.37–0.76)	0.64 (0.49–0.87)	0.71 (0.54–0.97)	−0.03 (−0.08–0.01)
Manila, Philippines	**1.02 (0.86–1.22)**	**1.09 (0.92–1.31)**	**1.19 (1.04–1.41)**	**1.27 (1.10–1.53)**	**0.52 (0.50–0.55)**

(continued)

Table 3.1 (continued)

Locations	Projected sea level rise in meters relative to 1995–2014 baseline				VLM (Meters)
	Global warming levels by 2100				
	1.5 °C	2.0 °C	3.0 °C	4.0 °C	
*Ilocos Norte, Luzon, Philippines	0.74 (0.55–0.95)	0.80 (0.61–1.04)	0.91 (0.74–1.16)	0.99 (0.80–1.26)	0.22 (0.15–0.29)
Davao, Davao Gulf, Philippines	0.50 (0.32–0.71)	0.59 (0.41–0.82)	0.68 (0.54–0.91)	0.76 (0.57–1.02)	0.01 (−0.03–0.05)

Note The estimates are in meters and are presented along with their likely ranges. Locations marked with an asterisk indicate that the estimates were obtained from the nearest sea location. VLM refers to the vertical land movement component, with its contribution to the total sea level change also provided in meters and increase climate resilience, Southeast Asian countries will also need to pursue sustainable development and implement adaptation measures to limit exposure and vulnerability to climate risks.

References

Aldrian, E., & Dwi Susanto, R. (2003). Identification of three dominant rainfall regions within Indonesia and their relationship to sea surface temperature. *International Journal of Climatology, 23*, 1435–1452. https://doi.org/10.1002/joc.950

Alejandria, M. C., Grace, R., Gloria Cajilig, P., & Smith, W. (2024). Guest editorial: Complex disasters, complex solutions: Advancing inclusive governance and decolonization in Southeast Asia. *Southeast Asia: A Multidisciplinary Journal, 24*(3), 167–170. https://doi.org/10.1108/SEAMJ-11-2024-089

Amirudin, A. A., Salimun, E., Tangang, F., Juneng, L., & Zuhairi, M. (2020). Differential influences of teleconnections from the Indian and Pacific Oceans on rainfall variability in Southeast Asia. *Atmosphere, 11*(9), 886. https://doi.org/10.3390/atmos11090886

Anh, L. T., Takagi, H., & Thao, N. D. (2019). Storm surge and high waves due to 1997 Typhoon Linda: Uninvestigated worst storm event in southern Vietnam. *Journal of Japan Society of Civil Engineers, Series B3 (Ocean Engineering), 75*(2), I73–I78.

Asian Development Bank (ADB). (2009). *The economics of climate change in Southeast Asia: A regional review.* https://www.adb.org/sites/default/files/publication/29657/economics-climate-change-se-asia.pdf

Asuncion, R. C., & Lee, M. (2017). *Impacts of sea level rise on economic growth in developing Asia* (ADB Economics Working Paper Series No. 507). https://hdl.handle.net/10419/169338

Beirne, J., Renzhi, N., & Volz, U. (2021). *Bracing for the typhoon: Climate change and sovereign risk in Southeast Asia* (ADBI Working Paper 1223). Asian Development Bank Institute. https://www.adb.org/sites/default/files/publication/684891/adbi-wp1223.pdf

Cai, W., Borlace, S., Lengaigne, M., Van Rensch, P., Collins, M., Vecchi, G., Timmermann, A., Santoso, A., Mcphaden, M. J., Wu, L., England, M. H., Wang, G., Guilyardi, E., & Jin, F.-F. (2014). Increasing frequency of extreme El Niño events due to greenhouse warming. *Nature Clinical Practice Endocrinology & Metabolism, 4*, 111–116.

Cai, W., Santoso, A., Collins, M., et al. (2021). Changing El Niño-Southern Oscillation in a warming climate. *Nature Reviews Earth & Environment, 2*, 628–644. https://doi.org/10.1038/s43017-021-00199-z

Cao, A., Esteban, M., Valenzuela, V. P. B., Onuki, M., Takagi, H., Thao, N. D., & Tsuchiya, N. (2021). Future of Asian Deltaic Megacities under sea level rise and land subsidence: Current adaptation pathways for Tokyo, Jakarta, Manila, and Ho Chi Minh City. *Current Opinion in Environmental Sustainability, 50*, 87–97. https://doi.org/10.1016/j.cosust.2021.02.010

Carnegie, P. J. (2024). Disaster, risk and conducting human security research in Southeast Asia. *Southeast Asia: A Multidisciplinary Journal, 24*(3), 185–197. https://doi.org/10.1108/SEAMJ-02-2024-0015

Chang, C. P., Harr, P. A., & Chen, H. J. (2005). Synoptic disturbances over the equatorial South China Sea and western maritime continent during boreal winter. *Monthly Weather Review, 113*, 489–503.

Chen, C., Sahany, S., Moise, A. F., Chua, X. R., Hassim, M. E., Lim, G., & Prasanna, V. (2023). ENSO–rainfall teleconnection over the maritime continent enhances and shifts eastward under warming. *Journal of Climate, 36*(14), 4635–4663. https://doi.org/10.1175/JCLI-D-23-0036.1

Chen, D., Rojas, M., Samset, B. H., Cobb, K., Diongue Niang, A., Edwards, P., Emori, S., Faria, S. H., Hawkins, E., Hope, P., Huybrechts, P., Meinshausen, M., Mustafa, S. K., Plattner, G.-K., & Tréguier, A.-M. (2021). Framing, context, and methods. In V. Masson-Delmotte, P. Zhai, A.

Pirani, S. L. Connors, C. Péan, S. Berger, N. Caud, Y. Chen, L. Goldfarb, M. I. Gomis, M. Huang, K. Leitzell, E. Lonnoy, J. B. R. Matthews, T. K. Maycock, T. Waterfield, O. Yelekçi, R. Yu, & B. Zhou (Eds.), *Climate change 2021: The physical science basis. Contribution of Working Group I to the Sixth Assessment Report of the Intergovernmental Panel on Climate Change* (pp. 147–286). Cambridge University Press. https://doi.org/10.1017/9781009157896.003

Chen, J.-D., Yen, M.-C., & Matsumoto, J. (2013). The winter rainfall of Malaysia. *Journal of Climate, 26*, 936–958.

Cheong, W. K., Timbal, B., Golding, N., Sirabaha, S., Kwan, K. F., Cinco, T. A., Archevarahuprok, B., Vo, V. H., Gunawan, D., & Han, S. (2018). Observed and modelled temperature and precipitation extremes over Southeast Asia from 1972 to 2010. *International Journal of Climatology, 38*(7), 3013–3027. https://doi.org/10.1002/joc.5479

Chevuturi, A., Klingaman, N. P., Turner, A. G., & Hannah, S. (2018). Projected changes in the Asian-Australian Monsoon region in 1.5°C and 2.0°C global-warming scenarios. *Earth's Future, 6*(3), 339–358, https://doi.org/10.1002/2017EF000734

Cinco, T. A., de Guzman, R. G., Hilario, F. D., & Wilson, D. M. (2014). Long-term trends and extremes in observed daily precipitation and near surface air temperature in the Philippines for the period 1951–2010. *Atmospheric Research, 145–146*, 12–26. https://doi.org/10.1016/j.atmosres.2014.03.025

Crook, J., Morris, F., Fitzpatrick, R. G. J., Peatman, S. C., Schwendike, J., Stein, T. H., Birch, C. E., Hardy, S., & Yang, G.-Y. (2024). Impact of the Madden–Julian oscillation and equatorial waves on tracked mesoscale convective systems over southeast Asia. *Quarterly Journal of the Royal Meteorological Society, 150*(760), 1207–1808. https://doi.org/10.1002/qj.4667

Da Silva, N. A., & Matthews, A. (2021). Impact of the Madden–Julian Oscillation on extreme precipitation over the western Maritime Continent and Southeast Asia. *Quarterly Journal of the Royal Meteorological Society, 147*(739), 3049–3492. https://doi.org/10.1002/qj.4136

Dagli, S., & Ferrarini, B. (2019). *The growth impact of disasters in developing Asia* (ADB Economics Working Paper Series No. 585). ADB. https://doi.org/10.22617/WPS190224-2

Dong, W. S., Ismailluddin, A., Yun, L. S., et al. (2024). The impact of climate change on coastal erosion in Southeast Asia and the compelling need to establish robust adaptation strategies. *Heliyon, 10*(4), e25609. https://doi.org/10.1016/j.heliyon.2024.e25609

Dong, Z., Wang, L., Sun, Y., Hu, T., Limsakul, A., Singhruck, P., & Pimonsree, S. (2021). Heatwaves in Southeast Asia and their changes in a warmer world. *Earth's Future, 9*, e2021EF001992. https://doi.org/10.1029/2021EF001992

Edirisinghe, M., Alahacoon, N., Ranagalage, M., & Murayama, Y. (2023). Long-term rainfall variability and trends for climate risk management in the summer monsoon region of Southeast Asia. *Advances in Meteorology, 2023*, 2693008. https://doi.org/10.1155/2023/2693008

Endo, N., Matsumoto, J., & Lwin, T. (2009). Trends in precipitation extremes over Southeast Asia. *SOLA, 5*, 168–171.

Forster, P., Storelvmo, T., Armour, K., Collins, W., Dufresne, J.-L. Frame, D., Lunt, D. J., Mauritsen, T., Palmer, M. D., Watanabe, M. Wild, M., & Zhang, H. (2021). The earth's energy budget, climate feedbacks, and climate sensitivity. In V. Masson-Delmotte, P. Zhai, A. Pirani, S. L. Connors, C. Péan, S. Berger, N. Caud, Y. Chen, L. Goldfarb, M. I. Gomis, M. Huang, K. Leitzell, E. Lonnoy, J. B. R. Matthews, T. K. Maycock, T. Waterfield, O. Yelekçi, R. Yu, & B. Zhou (Eds.), *Climate change 2021: The physical science basis. Contribution of Working Group I to the Sixth Assessment Report of the Intergovernmental Panel on Climate Change* (pp. 923–1054). Cambridge University Press. https://doi.org/10.1017/9781009157896.009

Fox-Kemper, B., Hewitt, H. T., Xiao, C., Aðalgeirsdóttir, G., Drijfhout, S. S., Edwards, T. L., Golledge, N. R., Hemer, M., Kopp, R. E., Krinner, G., Mix, A., Notz, D., Nowicki, S., Nurhati, I. S., Ruiz, L., Sallée, J.-B., Slangen, A. B. A., & Yu, Y. (2021). Ocean, cryosphere and sea level change. In V. Masson-Delmotte, P. Zhai, A. Pirani, S. L. Connors, C. Péan, S. Berger, N. Caud, Y. Chen, L. Goldfarb, M. I. Gomis, M. Huang, K. Leitzell, E. Lonnoy, J. B. R. Matthews, T. K. Maycock, T. Waterfield, O. Yelekçi, R. Yu, & B. Zhou (Eds.), *Climate change 2021: The physical science basis. Contribution of Working Group I to the Sixth Assessment Report of the*

Intergovernmental Panel on Climate Change (pp. 1211–1362). Cambridge University Press. https://doi.org/10.1017/9781009157896.011

Fritz, H. M., Blount, C., Thwin, S., Thu, M. K., & Chan, N. (2010). Cyclone Nargis storm surge flooding in Myanmar's Ayeyarwady River Delta. In Y. Charabi (Ed.), *Indian Ocean tropical cyclones and climate change*. Springer.

Ge, F., Zhu, S., Luo, H., Zhi, X., & Wang, H. (2021). Future changes in precipitation extremes over Southeast Asia: Insights from CMIP6 multi-model ensemble. *Environmental Research Letters, 16*, 024013. https://doi.org/10.1088/1748-9326/abd7ad

Ghosh, R., & Shepherd, T. G. (2023). Storylines of Maritime Continent dry period precipitation changes under global warming. *Environmental Research Letters, 18*, 034017. https://doi.org/10.1088/1748-9326/acb788

Gulev, S. K., Thorne, P. W., Ahn, J., Dentener, F. J., Domingues, C. M., Gerland, S., Gong, D., Kaufman, D. S., Nnamchi, H. C., Quaas, J., Rivera, J. A., Sathyendranath, S., Smith, S. L., Trewin, B., von Schuckmann, K., & Vose, R. S. (2021). Changing State of the Climate System. In V. Masson-Delmotte, P. Zhai, A. Pirani, S. L. Connors, C. Péan, S. Berger, N. Caud, Y. Chen, L. Goldfarb, M. I. Gomis, M. Huang, K. Leitzell, E. Lonnoy, J. B. R. Matthews, T. K. Maycock, T. Waterfield, O. Yelekçi, R. Yu, & B. Zhou (Eds.), *Climate change 2021: The physical science basis. Contribution of Working Group I to the Sixth Assessment Report of the Intergovernmental Panel on Climate Change* (pp. 287–422). Cambridge University Press. https://doi.org/10.1017/9781009157896.004

Hai, D. H., Nam, V. Q., Samadhiya, A., Kumar, A., Gupta, S., & Jagtap, S. (2024). Unravelling the economic impact of climate change in Vietnam's Mekong River Delta and Southeast region. *Discov Sustain, 5*, 125. https://doi.org/10.1007/s43621-024-00323-1

Hai, O. S., Samah, A. A., Chenoli, S. N., Subramaniam, K., & Ahmad Mazuki, M. Y. (2017). Extreme rainstorms that caused devastating flooding across the East Coast of Peninsular Malaysia during November and December 2014. *Weather and Forecasting, 32*, 849–872. https://doi.org/10.1175/WAF-D-16-0160.1

Hariadi, M. H., van der Schrier, G., Steeneveld, G.-J., Sutanto, S. J., Sutanudjaja, E., Ratri, D. N., Sopaheluwakan, A., & Klein Tank, A. (2024). A high-resolution perspective of extreme rainfall and river flow under extreme climate change in Southeast Asia. *Hydrology and Earth System Sciences, 28*, 1935–1956. https://doi.org/10.5194/hess-28-1935-2024

Hernandez, J. Y., Jr., Aquino, R. E. R., Pacheco, B. M., & Cruz, E. C. (2015). Damage caused by Typhoon Haiyan in the Philippines, review of structural regulations and practice, and research developments in wind engineering. *Wind Engineers JAWE, 40*(3), 270–274.

Hirschman, C., & Bonaparte, S. (2012). Demography of Southeast Asia (L. Williams & P. Guest, Eds.). Southeast Asia Program. Cornell University.

IPCC. (2021). Summary for policymakers. In V. Masson-Delmotte, P. Zhai, A. Pirani, S. L. Connors, C. Péan, S. Berger, N. Caud, Y. Chen, L. Goldfarb, M. I. Gomis, M. Huang, K. Leitzell, E. Lonnoy, J. B. R. Matthews, T. K. Maycock, T. Waterfield, O. Yelekçi, R. Yu, & B. Zhou (Eds.), *Climate change 2021: The physical science basis. Contribution of Working Group I to the Sixth Assessment Report of the Intergovernmental Panel on Climate Change* (pp. 3–32). Cambridge University Press. https://doi.org/10.1017/9781009157896.001

IPCC. (2022). Summary for policymakers [H.-O. Pörtner, D. C. Roberts, E. S. Poloczanska, K. Mintenbeck, M. Tignor, A. Alegría, M. Craig, S. Langsdorf, S. Löschke, V. Möller, & A. Okem, Eds.]. In H.-O. Pörtner, D. C. Roberts, M. Tignor, E. S. Poloczanska, K. Mintenbeck, A. Alegría, M. Craig, S. Langsdorf, S. Löschke, V. Möller, A. Okem, & B. Rama (Eds.), *Climate change 2022: Impacts, adaptation and vulnerability. Contribution of Working Group II to the Sixth Assessment Report of the Intergovernmental Panel on Climate Change* (pp. 3–33). Cambridge University Press. https://doi.org/10.1017/9781009325844.001

IPCC. (2023). Summary for policymakers. In Core Writing Team, H. Lee, & J. Romero (Eds.), *Climate change 2023: Synthesis report. Contribution of Working Groups I, II and III to the Sixth Assessment Report of the Intergovernmental Panel on Climate Change* (pp. 1–34). IPCC. https://doi.org/10.59327/IPCC/AR6-9789291691647.001

Islam, S. N., Reinstädtler, S., & Ibrahim, K. H. (2023). Floods in Tutong District, Brunei Darussalam and the role of Tutong River: Approaches for mitigation measures. In S. Eslamian & F. Eslamian (Eds.), *Disaster risk reduction for resilience*. Springer. https://doi.org/10.1007/978-3-031-43177-7_8

Jaroenongard, C., Babel, M. S., Shrestha, S., Weesakul, S., Nitivattananon, V., & Khadka, D. (2021). Projecting relative sea level rise under climate change at the Phrachula Chomklao Fort Tide Gauge in the upper Gulf of Thailand. *Water, 13*, 1702. https://doi.org/10.3390/w13121702

Jud, M. A., Juneng, L., Tangang, F. T., Latif, M. T., Chung, J. X., & Ahamad, F. (2020). Madden Julian oscillation modulation for surface ozone in Peninsular Malaysia. *Atmospheric Environment, 233*, 117577. https://doi.org/10.1016/j.atmosenv.2020.117577

Juneng, L., & Tangang, F. T. (2005). Evolution of ENSO-related rainfall anomalies in Southeast Asia region and its relationship with atmosphere–ocean variations in Indo-Pacific sector. *Climate Dynamics, 25*, 337–350.

Juneng, L. & Tangang, F. (2010). Long-term trends of winter monsoon synoptic circulations over the maritime continent: 1962-2007. *Atmospheric Science Letter*, 11, 199–203. https://doi.org/10.1002/asl.272

Juneng, L., Tangang, F. T., & Reason, C. J. C. (2007). Numerical case study of an extreme rainfall event during 9–11 December 2004 over the east coast of Peninsular Malaysia. *Meteorology and Atmospheric Physics, 98*, 81–98.

Kay, S., Avillanosa, A. L., Cheung, V. V., Dao, H. N., Gonzales, B. J., Palla, H. P., Praptiwi, R. A., Queirós, A. M., Sailley, S. F., Sumeldan, J. D. C., Syazwan, W. M., Then, A.Y.-H., & Wee, H. B. (2023). Projected effects of climate change on marine ecosystems in Southeast Asian seas. *Frontiers in Marine Science, 10*, 1082170. https://doi.org/10.3389/fmars.2023.1082170

Koplitz, S. N., Mickley, L. J., Jacob, D. J., Marlier, M. E., DeFries, R. S., Gaveau, D. L. A., et al. (2018). Role of the Madden-Julian Oscillation in the transport of smoke from Sumatra to the Malay Peninsula during severe non-El Niño haze events. *Journal of Geophysical Research: Atmospheres, 123*, 6282–6294. https://doi.org/10.1029/2018JD028533

Liang, J., Catto, J. L., Hawcroft, M. K., et al. (2023). Borneo Vortices in a warmer climate. *npj Climate and Atmospheric Science, 6*, 2 https://doi.org/10.1038/s41612-023-00326-1

Liang, S., Wang, D., Ziegler, A. D., et al. (2022). Madden–Julian Oscillation-induced extreme rainfalls constrained by global warming mitigation. *npj Climate and Atmospheric Science, 5*, 67. https://doi.org/10.1038/s41612-022-00291-1

Limsakul, A., & Goes, J. I. (2008). Empirical evidence for international and longer period variability in Thailand surface air temperature. *Journal of Atmospheric Research, 87*, 89–102.

Limsakul, A., & Singhruck, P. (2016). Long-term trends and variability of total and extreme precipitation in Thailand. *Atmospheric Research, 169*, 301–317. https://doi.org/10.1016/j.atmosres.2015.10.015

Lohberger, S., Stängel, M., Atwood, E. C., & Siegert, F. (2018). Spatial evaluation of Indonesia's 2015 fire affected area and estimated carbon emissions using Sentinel-1. *Global Change Biology, 24*, 644–654.

Loo, Y. Y., Billa, L., & Singh, A. (2015). Effect of climate change on seasonal monsoon in Asia and its impact on the variability of monsoon rainfall in Southeast Asia. *Geoscience Frontiers, 6*(6), 817–823. https://doi.org/10.1016/j.gsf.2014.02.009

Manton, M. J., Della-Marta, P. M., Haylock, M. R., et al. (2001). Trends in extreme daily rainfall and temperature in Southeast Asia and the South Pacific: 1961–1998. *International Journal of Climatology, 21*(3), 269–284. https://doi.org/10.1002/joc.610

Michalak, P., Patsili, A., Carmen, O., & Keßler, C. (2021). Impact of sea level rise on the land cover structure in Southeast Asia. *AGILE GIScience Series, 2*, 36. https://doi.org/10.5194/agile-giss-2-36-2021

Mohd Salleh, N. H., Hasan, H., & Kassim, S. (2015). Trends in temperature extremes across Malaysia. *Advances in Environmental Biology, 9*(27), 174–181.

Muhammad, M. K. I., Hamed, M. M., Harun, S., et al. (2024). Heatwaves in Peninsular Malaysia: A spatiotemporal analysis. *Science and Reports, 14*, 4255. https://doi.org/10.1038/s41598-024-53960-x

NAHRIM. (2017). *Sea level rise projection for Malaysia.* https://www.nahrim.gov.my/wp-content/uploads/CSIRO_Malaysia_SeaLevelRiseReport_FinalReport_2017.pdf

Nguyen, V. T., Mai, V. K., Vu, V. T., Nguyen, D. M., Nguyen, N. B. P., Le, D. D., Ha, T. M., & Luu, N. L. (2017). Changes in climate extremes in Vietnam. *Vietnam Journal of Science, Technology and Engineering, 59*(1), 79–87. https://doi.org/10.31276/VJSTE.59(1).79

Nguyen-Le, D. (2024). Projected ENSO teleconnection on the Southeast Asian climate under global warming. *Environmental Research Letters, 19*(1), 014001. https://doi.org/10.1088/1748-9326/ad0d3e

Nugraheni, I. L., Suyatna, A., Setiawan, A., Abdurrahman. (2022). Flood disaster mitigation modeling through participation community based on the land conversion and disaster resilience. *Heliyon, 8*(8), e09889.

Olaguera, L. M. P., Cruz, F. A. T., Dado, J. M. B., Villarin, J. R. T. (2022). Complexities of extreme rainfall in the Philippines. In A. Unnikrishnan, F. Tangang, & R. J. Durrheim (Eds.), *Extreme natural events.* Springer. https://doi.org/10.1007/978-981-19-2511-5_5

Ona, B. J., Raghavan, S. V., Nguyen, N. S., Ngai, S. T., & Nguyen, T. H. (2024). Changes in future rainfall over Southeast Asia using the CMIP6 multi-model ensemble. *Journal of Atmospheric Science Research, 7*(2), 62–82. https://doi.org/10.30564/jasr.v7i2.6335

Peatman, S. C., Matthews, A. J., & Stevens, D. P. (2014). Propagation of the Madden–Julian Oscillation through the Maritime Continent and scale interaction with the diurnal cycle of precipitation. *Quarterly Journal Royal Meteorological Society, 140*(680), 814–825.

Petit, J. R., et al. (1999). Climate and atmospheric history of the past 420,000 years from the Vostok ice core, Antarctica. *Nature, 399*(6735), 429–436.

Rahman, S. (2022). *Malaysia's floods of December 2021: Can future disasters be avoided?* Perspective, ISEAS, 26, https://www.iseas.edu.sg/wp-content/uploads/2022/03/ISEAS_Perspective_2022_26.pdf

Ranasinghe, R., Ruane, A. C., Vautard, R., Arnell, N., Coppola, E. E., Cruz, F. A., Dessai, S., Islam, A. S., Rahimi, M., Ruiz Carrascal, D., Sillmann, J., Sylla, M. B., Tebaldi, C., Wang, W., & Zaaboul, R. (2021). Climate change information for regional impact and for risk assessment. In V. Masson-Delmotte, P. Zhai, A. Pirani, S. L. Connors, C. Péan, S. Berger, N. Caud, Y. Chen, L. Goldfarb, M. I. Gomis, M. Huang, K. Leitzell, E. Lonnoy, J. B. R. Matthews, T. K. Maycock, T. Waterfield, O. Yelekçi, R. Yu, & B. Zhou (Eds.), *Climate change 2021: The physical science basis. Contribution of Working Group I to the Sixth Assessment Report of the Intergovernmental Panel on Climate Change* (pp. 1767–1926). Cambridge University Press. https://doi.org/10.1017/9781009157896.014

Ratnayake, U., Amirah Arun, S. N., Abdul Rahman, E. K., & Shams, S. (2023). Impact of sea-level rise on a coastal catchment of Brunei Darussalam. In R. K. Bhattacharjya, B. Talukdar, & K. L. Katsifarakis (Eds.), *Sustainable water resources management*. Advances in Sustainability Science and Technology. Springer. https://doi.org/10.1007/978-981-16-7535-5_4

Robertson, A. W., et al. (2011). The maritime continent monsoon. In C.-P. Chang, Y. Ding, N.-C. Lau, R. H. Johnson, B. Wang, & T. Yasunari (Eds.*), The global monsoon system: Research and forecast* (2nd ed., pp. 85–98). World Scientific. https://doi.org/10.1142/9789814343411_0006.

Salimun, E., Tangang, F. T., Juneng, L., Behera, S. K., & Yu, W. (2014). Differential impacts of conventional El Nino versus El Nino Modoki on Malaysian rainfall anomaly during winter monsoon. *International Journal of Climatology, 34*, 2763–2774.

Singh, V., & Qin, X. (2020). Study of rainfall variabilities in Southeast Asia using long-term gridded rainfall and its substantiation through global climate indices. *Journal of Hydrology, 585*, 124320. https://doi.org/10.1016/j.jhydrol.2019.124320

Siringan, F. P., & Sta. Maria, M. Y. Y. (2024). Sea level rise and coastal erosion in the Philippines: Impacts and adaptation strategies for coastal communities. In K. B. Berse, J. M. Pulhin, & A.

G. M. La Viña (Eds.), *Climate emergency in the Philippines. Disaster risk reduction.* Springer. https://doi.org/10.1007/978-981-99-7804-5_3

Siswanto, S. G. v. d., Oldenborgh, G. J. v., Hurk, B. v. d., Aldrian, E., Swarinoto, Y., Sulistya, W., & Sakya, A. E. (2017). A very unusual precipitation event associated with the 2015 floods in Jakarta: An analysis of the meteorological factors, *Weather and Climate Extremes, 16*, 23–28. https://doi.org/10.1016/j.wace.2017.03.003

Strassburg, M. W., Hamlington, B. D., Leben, R. R., Manurung, P., Gaol, J. L., Nababan, B., Vignudelli, S., & Kim, K. Y. (2015). Sea level trends in Southeast Asian seas. *Climate of the Past, 11*, 743–750. https://doi.org/10.5194/cp-11-743-2015

Supari, S., Tangang T. F., Juneng, L., Cruz, F., Chung, J. X., Ngai, S. T., Salimun, E., Mohd, M. S. F., Santisirisomboon, J., Singhruck, P., Phanvan, T., Ngo-Duc, T., Narisma, G., Aldrian, E., Gunawan, D., & Sopaheluwakan, A. (2020). Multi-model projections of precipitation extremes in Southeast Asia based on CORDEX Southeast Asia simulations. *Environmental Research, 184*, 109350.

Supari, S., Tangang, F., Salimun, E., Aldrian, E., Sopaheluwakan, A., & Juneng, L. (2018). ENSO modulation of seasonal rainfall and extremes in Indonesia. *Climate Dynamics, 51*, 2559–2580.

Supari, S., Tangang, T. F., Juneng, L., & Aldrian, E. (2017). Observed changes in extreme temperature and precipitation over Indonesia. *International Journal of Climatology, 37*, 1979–1997. https://doi.org/10.1002/joc.4829

Supharatid, S., Aribarg, T., & Supratid, S. (2016). Assessing potential flood vulnerability to climate change by CMIP3 and CMIP5 models: Case study of the 2011 Thailand great flood. *J Water Climate Change, 7*(1), 52–67. https://doi.org/10.2166/wcc.2015.116

Tangang, F., Chung, J. X., Juneng, L., Supari, S., Salimun, E., Ngai, S. T., Jamaluddin, A. F., Mohd, M. S. F., Cruz, F., Narisma, G., Santisirisomboon, J., Ngo-Duc, T., Van Tan, P., Singhruck, P., Gunawan, D., Aldrian, E., Sopaheluwakan, A., Grigory, N., Remedio, A. R. C., … Kumar, P. (2020). Projected future changes in rainfall in Southeast Asia based on CORDEX–SEA multi-model simulations. *Climate Dynamics, 55*, 1247–1267.

Tangang, F., Farzanmanesh, R., Mirzaei, A., et al. (2017). Characteristics of precipitation extremes in Malaysia associated with El Niño and La Niña events. *International Journal of Climatology, 37*(51), 696–716. https://doi.org/10.1002/joc.5032

Tangang, F., Juneng, L., & Ahmad, S. (2007). Trend and interannual variability of temperature in Malaysia: 1961–2002. *Theoretical and Applied Climatology, 89*, 127–141. https://doi.org/10.1007/s00704-006-0263-3

Tangang, F., Santisirisomboon, J., Juneng, L., Salimun, E., Chung, J., Supari, C. F., Ngai, S. T., Ngo-Duc, T., Singhruck, P., Narisma, G., Santisirisomboon, J., Wongsaree, W., Promjirapawat, K., Sukamongkol, Y., Srisawadwong, R., Setsirichock, D., Phan-Van, T., Aldrian, E., Gunawan, D., … Yang, H. (2019). Projected future changes in mean precipitation over Thailand based on multi-model regional climate simulations of CORDEX Southeast Asia. *International Journal of Climatology, 39*, 5413–5436. https://doi.org/10.1002/joc.6163

Tangang, F., Supari, S., Chung, J., Cruz, F., Salimun, E., Ngai, S., Juneng, L., Santisirisomboon, J., Santisirisomboon, J., Ngo-Duc, T., Phan-Van, T., Narisma, G., Singhruck, P., Gunawan, D., Aldrian, A., Sopaheluwakan, A., Nikulin, G., Yang, H., Remedio, A. R. C., … Hein-Griggs, D. (2018). Future changes in annual precipitation extremes over Southeast Asia under global warming of 2 °C. *APN Science Bulletin, 8*(1). https://doi.org/10.30852/sb.2018.436

Tangang, F. T., Juneng, L., Salimun, E., Sei, K. M., Le, L. J., & Muhamad, H. (2012). Climate change and variability over Malaysia: Gaps in science and research information. *Sains Malaysiana, 41*(11), 1355–1366.

Tangang, F. T., Salimun, E., Juneng, L., Vinayachandran, P. N., Yap, K. S., Reason, C. J. C., Cehera, S. K., & Yasunari, T. (2008). On the roles of Northeast cold surge, the Borneo Vortex, the Madden-Julian Oscillation and the Indian Ocean Dipole during the worst 2006/2007 flood in Peninsular Malaysia. *Geophysical Research Letters, 35*, L14S07.

Tay, C., Lindsey, E. O., Chin, S. T., et al. (2022). Sea-level rise from land subsidence in major coastal cities. *Nat Sustain, 5*, 1049–1057. https://doi.org/10.1038/s41893-022-00947-z

Thach, K. S. R., Lee, J. Y., Ha, M. T., Cao, M. T., Nayga, M. R., Yang, J. E. (2023). Effect of saline intrusion on rice production in the Mekong River Delta. *Heliyon, 9*(10), https://doi.org/10.1016/j.heliyon.2023.e20367

Triana, K., & Wahyudi, A. J. (2020). Sea level rise in Indonesia: The drivers and the combined impacts from land subsidence. *ASEAN Journal on Science & Technology for Development, 37*(3), 115–121, https://doi.org/10.29037/ajstd.627

Try, S., & Qin, X. (2024). Evaluation of future changes in climate extremes over Southeast Asia using downscaled CMIP6 GCM projections. *Water, 16*, 2207. https://doi.org/10.3390/w16152207

UNCT. (2024). *Viet Nam: Typhoon Yagi and floods situation thematic update No.1: Gender and protection.* https://vietnam.un.org/sites/default/files/2024-09/Typhoon%20Yagi_Thematic%20Update%20Protection%20and%20Gender%20-%20Number%201%20%2823%20Sept.%202024%29.pdf

UNFCCC. (2023). https://unfccc.int/sites/default/files/resource/Summary_GCA_COP28.pdf

Van Tho, N. (2022). Salinity intrusion in the Vietnamese Mekong Delta, a threat: Possible causes, effects on people's life and production, and temporary solutions and adaptable strategies. In H. Y. Jeon (Eds.), *Sustainable development of water and environment*. Environmental Science and Engineering. Springer. https://doi.org/10.1007/978-3-031-07500-1_1

Wang, B., Biasutti, M., Byrne, M. P., et al. (2021a). Monsoons climate change assessment. *Bulletin of the American Meteorological Society, 102*(1), E1–E19. https://doi.org/10.1175/BAMS-D-19-0335.1

Wang, B., Yim, S. Y., Lee, J. Y., et al. (2014). Future change of Asian-Australian monsoon under RCP 4.5 anthropogenic warming scenario. *Climate Dynamics, 42*, 83–100. https://doi.org/10.1007/s00382-013-1769-x

Wang, H., Ji, Z., Zhu, X., Lin, L., & Dong, W. (2021b). Future changes in the Asian-Australian monsoon system with 1.5°C and 2°C rise in temperature. *Journal of Geophysical Research: Atmospheres, 126*, e2020JD032872. https://doi.org/10.1029/2020JD032872

Wang, Z., & Chang, C. P. (2008). Mechanism of the asymmetric monsoon transition as simulated in an AGCM. *Journal of Climate, 21*(8), 1829–1836. https://doi.org/10.1175/2007JCLI1920.1

Xavier, P., Diong, J. Y., Bin Abdullah, M. F. A., et al. (2024). The influence of boreal summer intraseasonal oscillations on precipitation extremes and their characteristics in Southeast Asia. *npj Climate and Atmospheric Science, 7*, 112. https://doi.org/10.1038/s41612-024-006 58-6

Zakaria, N. A. B., Tangang, F., Salimun, E., Amirudin, A. A., Chung, J. X., Juneng, L., Tan, M. L., Zulkafli, Z., Marzuki, M., Santisirisomboon, J., Akhir, M. F., Abdullah, M. F. A., Jamaluddin, A. F., & Faisal, S. (2024). Spatio-temporal variation and El Nino modulation of meteorological droughts in Malaysia. *International Journal of Climatology, 44*(15), 5560–5579

Zhang, C. (2005). Madden–Julian Oscillation. *Reviews of Geophysics, 43*. RG2003.

Zuo, Z., Yang, S., Zhang, R., Jiang, P., Zhang, L., & Wang, F. (2013). Long-term variations of broad-scale Asian summer monsoon circulation and possible causes. *Journal of Climate, 26*(22), 8947–8961. https://doi.org/10.1175/JCLI-D-12-00691.1

Fredolin Tangang is an Associate Professor of Climate Change and Adaptation at the Faculty of Arts and Social Sciences, Universiti Brunei Darussalam. Previously, he served as a Professor of Climatology at the National University of Malaysia from 2007 until his retirement in March 2023. His research focuses on climatology, climate variability, climate change, and climate modeling. He earned his Ph.D. in 1997 from the University of British Columbia in Vancouver, Canada, where he developed a model to forecast the El Niño–Southern Oscillation (ENSO) phenomenon.

Chapter 4
Changing Global Order and Climate: A Comparative Analysis of Cartesian and Organismic Perspectives

Gabriel Yit Vui Yong

Abstract Humanity is currently confronted with a changing climate and major shifts in the global order. Although significant negative impacts are to be expected, the future is fraught with uncertainty. This chapter, examines these developments from a higher perspective, at the paradigm level: by comparing between Cartesian Mechanism (CM) and Organismic Systems (OS). It begins with an introduction of the main tenets of CM and its role in shaping current realities, and the relationship between worldview and paradigms. It then presents key works that are pivotal in the OS development. They include Capra's Web of Life, Vernadsky's Biosphere, Lovelock and Margulis' Gaia, Maturana and Varela's Autopoiesis, and Soleri's Arcology. The chapter discusses the changing global order and climate change by comparing the perspectives from the two paradigms, and how they contributed to the changes, organic responses and outcomes. It highlights that policies, plans and actions, based on CM, do not necessarily yield expected results, because people and environment are part of a planetary complex system, which responds to perturbations through a self-organisation process to attain new equilibrium. They are more accurately depicted by OS. The chapter concludes that: (i) a multipolar world is an inevitable outcome, which will be more sustainable because it aligns with the planetary and OS; (ii) a hot war is probable because of the CM worldview adopted by the West; (iii) there will be deleterious fallout due to CM actions but recovery due to OS response; (iv) climate policies based on CM will be ineffective, and that OS response would be more sustainable and climate-resilient; and (v) the knowledge society that has emerged is more aware of sustainability but increasingly ecologically illiterate. The chapter shows that comparative analysis by considering both CM and OS improves understanding of current developments, issues and concerns.

Keywords Cartesian mechanism · Organismic system · Paradigm · Comparative analysis · Worldview

G. Y. V. Yong (✉)
Faculty of Arts and Social Science, Universiti Brunei Darussalam, Gadong, Brunei Darussalam
e-mail: gabriel.yong@ubd.edu.bn

G. Y. V. Yong and A.K.M. A. Ullah (eds.), *Brunei and the Future Transformation of Southeast Asia*, https://doi.org/10.1007/978-981-96-7901-0_4

Introduction

At the quarter mark of the twenty-first century, the world is on the brink of a climate catastrophe, escalating geopolitical conflict and possibly, a third world war. The coming decades are therefore fraught with uncertainty as the current unipolar political-economic world order gives way to a multipolar one in which the global South will have a fairer share, more say and genuine sovereignty. The uncertainty stems from how the incumbent leadership will respond to this transition. In the wake of recent events, the United States of America (USA), together with its allies, the collective West, appears to be fighting tooth and nail to maintain its privileged position. This includes spreading disinformation, silencing dissenting voices, economic bullying and instigating or waging proxy wars. Meanwhile, deficit spending in the USA continues to rise unabated while the domestic economy shrinks due to structural weaknesses resulting from fiscal delinquency, globalisation, the weaponisation of the US dollar (USD), and the funding of numerous wars and over 750 military bases across the world. At the time of writing, it is evident that the United States' (US) economy is on the verge of a major correction. This situation arises from a combination of factors, the most important of which are: massive debt and the interest payments that come with it, de-dollarisation and a foreign policy that is inadvertently leading to the decline of the US as the global hegemon.

Although this is clear to many outside the Western sphere, the political elites within the West inexplicably cling on to the agenda with greater vigour, as if ramping up threats and bullying could restore the situation. They actually seem to believe that reality has not changed since colonial times, that the world is better under their leadership, and that their economies are 'too big to fail', regardless of the metrics that suggest otherwise. In the meantime, their actions have an impact on the entire world because the existing financial, economic and political systems are closely interlinked and based on the US dollar (USD) as the reserve currency. As the majority of international trade is conducted in the USD via the SWIFT system, all nations are exposed to the risk of economic shocks, if the USD loses its pre-eminence and or is devalued. This scenario is becoming increasingly likely with the consolidation of BRICS (alliance comprising of Brazil, Russia, India, China and South Africa), and the fact that the USA will have to account for its enormous debt at some point. The spectre of nuclear war due to escalating military action in the Middle East and Ukraine is also possible and has existential implications.

In any case, the repercussions arising from the imminent change in the world order and restructuring of international systems could be damaging and serious. However, they could also be relatively amicable if leaders adopt more rational and conciliatory positions. The uprising of the masses in many Western cities and universities protesting against the policies and actions of politicians (particularly, in relation to Israel and Palestine) offers a glimmer of hope for a peaceful global transition, even though they have subsided in the final months of 2024. Furthermore, China and Russia, the two main perceived enemies of the West, have stated on several occasions that they want to pursue and work towards a peaceful and mutually beneficial future in which the sovereignty and cultures of all nations are respected. However, the 'labour pangs of (re)birth' may be inevitable.

Meanwhile, the backdrop to this geopolitical drama is also changing. The Intergovernmental Panel on Climate Change (IPCC, 2021) has clearly stated in its latest assessment report that man-made climate change is happening. Concerns were initially raised at the 1992 Earth Summit. This launched the United Nations (UN) Conference of the Parties (COP), which still meets annually to address issues and agree on collective actions. Tackling climate change was part of the international agenda that emerged from the Earth Summit and is at the heart of the UN's thirteenth Sustainable Development Goal (SDG 13) in the current stage of the programme. The global community, including political and business leaders, activists and various stakeholders, commits to specific actions at each summit. These could be summarised as programmes to (i) mitigate climate change by curbing greenhouse gas emissions, especially carbon dioxide (CO_2), and (ii) help vulnerable communities adapt to climate change. However, the alarmist faction among climate change advocates considers these efforts insufficient and too slow, as they view climate change as an existential threat and have successfully raised its urgency to the status of an "emergency" at the Climate Action Summit in 2019.[1] Therefore, while international relations are in high tensions, on the brink of economic decoupling and a global hot war, the parties concerned need to enhance cooperation on SDG 13.

This chapter aims to shed light on the complexity of the issues associated with these global changes. The analysis compares and contrasts the Cartesian Mechanism (CM) and Organismic System (OS) perspectives, as the former is the dominant paradigm of the contemporary modern world, while the latter is increasingly gaining ground as an alternative that more accurately reflects physical reality. CM refers to the philosophy of Renee Descartes and the related thoughts and ideas of the Enlightenment period in Western European history, which served as the basis for the subsequent scientific and industrial revolutions, colonisation of the world, and "modernisation" in general. The modern economic system that emerged from the CM worldview, which is founded on building the wealth of (Western) nations, has evolved into today's global system that governs international relations and economics. OS has evolved from observations of nature and their development in multiple disciplines. It offers an understanding that takes into account the geographical context, the history and cultures of places, as well as the relationships between people and environment. Its explanations and conclusions are therefore nuanced, as they incorporate the inherent complexity of places and regions.

---

[1] https://www.un.org/sg/en/content/sg/speeches/2019-09-23/remarks-2019-climate-action-summit.

Methodology

Cartesian mechanism (CM), the dominant paradigm of the modern world, originated in Western Europe. Its worldwide adoption is a consequence of colonisation and globalisation that began in the fifteenth century and peaked in the nineteenth century, when most countries in Asia, the Americas, Africa and the Caribbean were under the control of Western nations, including the USA, itself a former colony. With the civilisation of the world as a primary goal, the populations in the colonies (including the 'protectorates') were educated, inculcated and governed over time to adopt the same paradigm. A prominent feature of CM is the conscious separation of mind and body (the physical world), known as Cartesian dualism. It assumes that physical reality is completely amenable to human analysis and manipulation. The enlightenment introduced a purely mechanistic world, in which everything could be objectively understood through scientific enquiry. CM also asserts that man has a God-given responsibility to subdue nature for his pleasure and purpose. Observing nature and its interconnections from its origin and evolution, reveals a different reality. The reason for addressing global issues at the paradigm level lies in the fact that our perceptions, worldviews, fears and therefore, values, priorities and choices, are strongly influenced or controlled by our assumptions about reality, i.e., our paradigm. In general, people are not aware of the paradigm that determines their worldviews and guides their decisions. However, they control our thoughts and actions and are reinforced by the cumulative perceptions, experiences and knowledge that we acquire through our participation in society and institutions through our lives. They also form the basis of scientific epistemology and accepted facts and 'truths'. Interestingly, several paradigms coexist in our minds. We use them separately, for example in religion, in education, in scientific research and in various life situations, without realising it. However, few are conscious of how different paradigms lead to the different perspectives, values and decisions taken. Furthermore, an account or explanation presented from a particular paradigm is highly likely to be misunderstood by someone whose views are formed by another paradigm, even academics, due to their different assumptions of reality.

Analytical Approach

A system's view recognizes that all phenomena, issues, or problems arise from multiple interacting elements. In CM, systems are mechanically connected in simple traceable connections, with components seen as material entities that respond only to physical laws. Their behaviour could therefore be understood and controlled by design. Consequently, the scientific method emphasises high accuracy and precision in measurements, based on the belief that the researcher could study and manipulate the system objectively and is unaffected by context. Analysis relies on identifying specific parameters and relationships, and generating formulas that are applicable to other similar situations. In contrast, natural systems, examined through in-depth observations, are complex and responsive to environmental dynamics. They can self-regulate and or self-organise in response to external changes. Due to their complexity, tracing individual connections is impossible due to the vast number and dense, networked structure, and the fact that each element is part of larger systems. Analysis of complex systems focuses on the relationships between parts, which are in reality sub-systems, that also comprises smaller systems or are part of other systems. Relationships may be competitive, exploitative, cooperative, mutualistic, etc. Importantly, it must be noted that all systems on Earth are living systems, because they contain living elements (i.e., they are ecosystems) from microscopic to human and planetary dimensions. In CM perspective, the human world is separate from the natural environment, and living elements are objects without the capacity to think, desire or choose. In contrast, they are sentient, cognitive, and capable of learning, choosing, independent action and responding in reality. Historical context also shapes perception, making it essential to consider in analysis involving relationships, e.g., in OS.

Contextual Influence

This brief account of my life is meant to provide the context for the analysis and conclusions. I am a second-generation Chinese immigrant born in Brunei in 1964, when the country began its development as a modern state. I attended a Catholic mission school at the primary level and a government school at the secondary level, which developed a predominantly Western, Christian outlook, partly due to my introverted nature and sheltered upbringing. The presence of foreign nations during the early development phase of the oil and gas industry further exposed the community to Western influences, which were reinforced by the British system of administration introduced since 1906 under the British Residency. When I graduated from university in 1986, shortly after Brunei's independence, I struggled to find a job as racial tensions was pronounced in the few years leading to and following independence. I eventually became a teacher in a private Chinese school that had recently switched over to teaching in the English medium. In 1994, I joined the national university as part of a local human resources development programme under a fortuitous circumstance, which offered me the opportunity for further education at the master's level. Like many in my hometown, I had hope to work in the petroleum industry, which was why I chose to study geological geophysics.

As a geoscientist trained in the British system, my mindset was firmly grounded in CM. However, working in the geography department led me to question the validity of these methods, especially as my specialisation involves modelling and field studies. Models are simplified representations of reality, and therefore, models developed must be continuously revised and refined through validation of outputs against observations. If the models fail to reach an accepted level of accuracy after exhaustive fine-tuning, the underlying paradigm needs to be reconsidered. As geographical research focusses primarily on human–environment interrelationship and their patterns, I soon learned that they, including the human world (economy, management system, international relations, globalisation, etc.) could not be accurately depicted mechanistically, as human factors such as priorities, politics and greed are not captured in the mechanistic paradigm. People do not always act predictably. This realisation led me to question the prevailing epistemology in relation to the environment and human relationships, which ultimately led to a necessary paradigm shift.

Paradigm Shift

As my background is in Earth-environmental science and Western scientific methodology, I will highlight five influential works that has led me to adopt OS as the more accurate theoretical framework in environmental and geographical studies. They are:

i. The Web of Life, introduced by Fritjof Capra;
ii. The Biosphere, according to Vladimir Ivanovic Vernadsky;
iii. The Gaia Theory of James Lovelock and Lynn Margulis;
iv. Autopoiesis, developed by Humberto Maturana and Francesco Varela;
v. Arcology, by Paolo Soleri.

A. The Web of Life

Fritjof Capra is an Austrian-American physicist and founding director of the Centre for Ecoliteracy in Berkeley, California, USA. The Centre promotes awareness of humanity's connection to nature. I got to know OS through his book, The Web of Life (1996). While CM has advanced engineering and industrial development, it fails to accurately portray and understand environmental systems that contain myriads of life forms, including humans and bacteria. OS recognises that natural systems are complex, with dense interconnections that respond holistically to inputs. These insights have emerged from interdisciplinary and transdisciplinary dialogues between scientists from various disciplines, artists and independent keen observers, which recognised that natural systems, such as mature forest biomes, become better organised and more resilient as they mature over time, unlike machines, which deteriorates. CM, in contrast, considers living organisms (and people) as mindless objects, which could be categorised and understood with predictable behaviour and amenable to manipulation. With the rise of environmental concerns in the 1970s, CM views the Earth as injured and fragile, and that responsible humans must intervene to prevent ecosystem collapse and save the planet, humanity's home. In contrast, older civilisations, e.g., 'Eastern philosophy', emphasise immersion in nature and cyclical and dynamic character of reality.

Capra attempted to bridge the gap between Western science and Eastern philosophy in The Tao of Physics (1975) by drawing parallels between relativistic physics and Eastern thought on nature. In The Turning Point (1982), he criticised the reductionist nature of CM and applied systems thinking to explain global crises, arguing that conventional academic approaches, which were founded on CM, was unsuitable for understanding the emerging cultures associated with ecology and social justice. Furthermore, the complex and fractal nature of the natural world defies mechanistic modelling, which treat components essentially as homogenous/uniform units. An organismic view sees human existence as a network of nested systems with the natural capacity for self-regulation and self-organisation as responses to maintain stability. Natural systems always tend towards dynamic equilibrium; they reorganise structurally to restore equilibrium when continuous perturbations significantly unsettle their internal states. Despite advances in systems science, the Western CM paradigm remains the dominant paradigm in modern society, academe and politics, treating the humans-in-charge as saviours and decision-makers, giving rise to anthropocentric perspectives of sustainability. This is reflected in the categorisation of thriving migrant species as "invasive", which has a corresponding view in the human world on human migrants and refugees.

In contrast, the Deep Ecology philosophy of Arne Naess (1989) views the living world from a truly ecocentric perspective. Naess asserts that all life possesses intrinsic value and advocates an ethic that respects all life (and people) and environmental relationships. This framework is congruent with the goals of the United Nations SDGs, most of which focus on social justice and equitable access to resources. This differs significantly from the CM perspectives, which sees human as saviour (or destroyer) of the planet, and capable of controlling even the global climate.

B. The Biosphere

Vladimir Vernadsky (1965) coined the term "biosphere" in his 1926 work, which was based on extensive geochemical research, a field that he pioneered. The biosphere is a realm created by the complex interactions between organisms and their environment on the earth's surface. It is characterised by geochemical cycles of oxygen, carbon, nitrogen and other elements that are driven by the feedback loops between life (defined as the totality of organic life, including microbes and humankind) and the planet. Vernadsky's concept challenges the classical view that life is passive and must adapt to environmental conditions controlled by physiochemical processes. On the contrary, this 'thin film' on the planet has altered the composition of the atmosphere, and drives planetary geochemical cycles, and therefore, the development, evolution and stability of the Earth's surface, the environment of humankind. Vernadsky's work however did not reach the Western academic world until much later, after its translation into English. Despite its prevalent use environmental science, the idea that life is passive continues to be assumed, and many are not aware of the implications. This is a CM legacy.

Vernadsky's biosphere is a complex system, with life as an active and evolving component that continuously shapes the physiochemical planet and in turn, causes it to continually adapt, as the entire system constantly receives energy and material inputs. An interesting and essential feature of the Vernadsky's biosphere is the existence of the noosphere, which represents consciousness, knowledge creation and communication among human beings, a trait that exists in all lifeforms in different degree of development and capacity. The noosphere, which emerged from the biosphere, is coupled with it and the non-living geosphere. Consequently, humankind generates knowledge of their world from interaction with its environment. The decisions and actions taken are determined by perception, state of knowledge, assumptions, beliefs and values, viz. paradigms. Since the late twentieth century, advances in information and communication technology have provided the noosphere with a global infrastructure, which facilitated its development at the planetary scale, ushering humanity into the 'knowledge age', where the economy and management systems are increasingly dependent on data, information and knowledge. Packaged knowledge (narratives), created through human agency, are increasingly being co-created and disseminated using artificial intelligent systems, and consumed in social networks across the planet, continuously. Innovations such as blockchain, which function without a single controlling entity, enhances ecological integrity of knowledge system. The consolidation of the noosphere is shaping the physical-virtual reality of younger generations, while older generations, many of whom still dominate existing institutions, are at risk of being excluded. From a CM perspective, artificial intelligence is seen as a threat that must be controlled.

C. Gaia: The Living Planet

In the 1960s and 70s, James Lovelock, a British atmospheric chemist, corroborated Vernadsky's view of life as an active element in the Earth system with a theory of a self-organising, self-regulating Earth. The Gaia theory regards the entire planet as a living system. Lovelock's theory developed from his studies of atmospheric homeostasis in 1965, in which he traced the stability of the planet's climate to the presence of life (lovelock, 1972). He explained his controversially proposal of the Earth as a 'superorganism' by comparing it to the giant redwood forest. Although the living tissue exists only in the thin outer layer, the entire tree is regarded as the living organism. The Gaia hypothesis was given a boost when Lovelock collaborated with microbiologist Lynn Margulis, whose work on cells (Margulis, 1967) and microorganisms offered important insights into planetary evolution and complex system dynamics. However, it also added further controversy because her findings and theory of endosymbiogenesis were at odds with Darwin's theory of evolution, which was widely accepted by the scientific community. Nevertheless, their joint publications, in particular Gaia: A New Look at Life on Earth (Lovelock & Margulis, 1979), provided a solid scientific framework of a planetary system that is capable of self-regulation and evolution, actively co-created by life.

Despite its scientific rigor, the Gaia theory met with considerable resistance in the academic world, partly due to the use of the name, Gaia, the mythological Greek goddess, Mother Earth, a suggestion of Lovelock's neighbour, the novelist William Golding. It misled some to incorrectly associate the theory with meta-science or teleology, reflecting a bias against ideas that challenge conventional CM. The popularity of the theory among artists and spiritual communities, including the 2001 animated film, Final Fantasy: The Spirit Within (Sakaguchi, 2001), originally titled Gaia, further complicated its reception and led many to dismiss it as pseudoscience. With the emergence of the Internet, however, the work of Lovelock and Margulis became more widely recognised and established the Gaia theory as an important foundational work in environmental and systems research. The theory regards the planet as a complex living system co-created by the evolution of life and the environment over 4 billion years. It describes the evolution the Earth, supported by extensive scientific data, from an extremely hot, highly volatile, anoxic and acidic environment with simple lifeforms over 4 billion years ago, to a cool, watery, oxygen-rich and self-regulating planet today that is teeming with life. Lovelock concluded that life on Earth created the conditions that are 'comfortable' for it to thrive, a condition that reflect dynamic balance between organisms and environment.

D. Autopoiesis: Self-Making

In 1980, Humberto Maturana, who studied cognition in biology, published Autopoiesis and Cognition: The Realization of the Living with his PhD student, Francisco Varela. This work describes self-organisation as an inherent trait of living organisms and identifies cognition, an essential ability for survival, invariably leads to the development of a system, enabling organisms to navigate their environment effectively. Through active communication and interaction, organisms form communities (systems) organically, a process they described as autopoiesis or 'self-making.' The internal organisation of these systems are manifested as emergent structures that regulate interactions and exchanges, consolidating the system's stability in the process. Maturana and Varela's concept of autopoiesis has influenced social sciences and provided a framework for understanding complex natural, human, and knowledge systems, emphasising cognition and complexity in contrast to conventional mechanistic views of simplistic, linear, cause-and-effect interaction and external control (Seidl, 2004).

Two key concepts in autopoiesis are embodiment and enactivism. Embodiment acknowledges the physical structures that facilitate bodily interaction and cognition, while enactivism describes the mutual influence and co-adaptation between interacting systems, such as, organisms and their environment, akin to co-evolution in the Gaia theory. In 1984, Maturana and Varela applied their theory to artificial intelligence, and in The Tree of Knowledge: The Biological Roots of Human Understanding (1987), they linked autopoiesis to epistemology, challenging the conventional notion of objective reality. They argued that knowledge arises from our interactions as structured beings, highlighting the significance of context and experience. By incorporating mental, emotional, and knowledge dimensions into systems analysis, autopoiesis offers a more nuanced understanding of contemporary issues, including geopolitical conflicts and the rise of artificial intelligence.

E. Arcology

Arcology, a work by Paolo Soleri, criticises the current development paradigm, which he sees as 'sclerotic' and 'doomed' He describes urban development in the USA, characterised by suburban sprawl and dictated by transport planning, as "grotesque" and contributing to a host of environmental problems, including deforestation, climate change, pollution and habitat loss, as well as social problems such as the high cost of living and traffic congestion. In the 1960s, Soleri introduced arcology, a blend of architecture and ecology, which culminated in his seminal work, The City in the Image of Man (Soleri, 1969). He argued that nature's imperative is to "miniaturise or die". Soleri was alluding to the increasing reduction of the physical space–time that separates interaction, which promotes efficiency as the biome or city grows. This is antithetical to urban sprawl, which increases distance-time ('gigantism'), consumption—and wastage—of resources and time, and the potential for congestion.

Miniaturisation is achieved through complexification over time. This can be seen in nature, for example, in the development of the rainforest: from bare ground to its climax stage through a series of successions. Natural architecture is compact, internally connected and open to environmental flows, tapping resources, such as energy and water directly. Soleri describes this relationship between built and natural environment as 'coherent', stressing the importance of urban planning to consider the entire urban area (forest) instead of individual projects (orchids) (McCullough, 2012). Soleri refers to the natural crowding process in the built environment, as the "urban effect", which he posits is an indicator of sustainability. In natural systems, energy and materials are recycled, eliminating the need for waste disposal. Arcology promotes compact, modular, self-contained, urban development that are three-dimensional, rising vertically upwards and outwards, using multi-functional "hyper-structures" (Stein, 2012, November 10; Yong, 2022). This minimises its footprint on the ground, and therefore interaction with earth surface processes, as well as exposure to climate risks.

Despite its potential, arcology is regarded by its critics as utopian. Soleri's urban experiment, Arcosanti, has yet to reach its goal of 5,000 residents since its inception in 1970. Nevertheless, it supports a thriving, albeit small, community of 50–150 people. Renowned architects such as Vincent Callebaut, Stephan Boeri and Ken Yeang have incorporated elements of arcology into their work. Unfortunately, these are merely orchids in sprawling concrete jungles, as urban planners, developers and investors are still unsure how arcologies could be realised with current technologies and materials. However, due to its ecological design principles, arcology is certainly needed to build a sustainable future for the ever-growing human population (Johnson, 2010). To transition from current unsustainable practises to truly sustainable design, there is a need to combine arcology and biomimicry (Benysus, 1997) (see, e.g., Yong, 2022). This approach would involve development at the local and regional level by mimicking successful organisms and biomes.

Confronting Global Changes

Western CM and OS perspectives of global changes and issues are drastically different, sometimes even diametrically opposite. The differences could be broken down as shown in Table 4.1. Their contrasts are elaborated and discussed with regards to (a) the changing world order and (b) climate change and sustainable development.

Table 4.1 Cartesian Mechanism vs. Organismic System paradigms

Aspects	Western, Cartesian Mechanism (CM)	Organismic System (OS)
1.System characteristics	Simple, linear/no feedback	Complex, dense network, autopoietic
2. Perspective	Anthropocentric, utilitarian	Ecocentric, ecological
3. Drivers	Man, Western	Interacting elements/parties
4. Approach	Control, manage, with force, if necessary; unilaterism	Mutual interest and benefit; multilaterism
5. Goals	Maintain control; high standard of living/wealth for self	Sustainable development, Respect each members' rights
6.International relations	Grow sphere of influence/ control; secure alliance of subordinates	Ecological; grow network of cooperating partners
7.Self-preservation	Strengthen offensive capability, and secure strategic positions	Build defence capability; promote diplomacy
8. Expected system response	Compliance with planned actions; targets achieved	Self-regulation/ self-organisation to maintain or achieve long-term stability

a. **Changing World Order**

The current world order is the product of CM worldview, which could be traced back to the emergence of the modern economic paradigm based on Adam Smith's 1776 work, The Wealth of Nations. It is the basis of colonisation as Western nations sought to seize and control resources found beyond their borders. The creation and manipulation of the international order and systems is characteristic of modern mechanistic thinking, where the capitalist elites view the world as theirs to subdue and exploit for their pleasure and purpose, without regard to the people and environment in lands they colonised. The Bretton Woods institutions founded in 1944 by the same Western powers after the Second World War (WW2) was just another stage in this development. Since then, the USA has taken over as the leader of the unipolar West-led world order. In 1971, it consolidated its dominance by unilaterally abandoning the gold standard and linked the USD as a fiat currency to its economy. The USA is the main architect of the current global order governing international politics, economy, development, communication and information. However, the existing patterns, especially of trade, international relations and alliances, and human mobility, need to be examined from an OS perspective because it offers a more accurate understanding. This is because complex living systems (including people and nations) develop in response to external perturbations, and they do not necessary align with the visions and intentions of the colonial architects' plans (i.e., the perturbations). These organic responses are driven by self-regulatory and self-organisation tendencies to attain equilibrium, as consequences of myriads of independent decisions and actions/reactions. OS also takes into account the historical context, as existing structures influence organic responses. This is ignored in CM, which is only concerned with planning and manipulating the situation to achieve the intended goals.

Capitalism is undoubtedly a major factor in the shaping the current world order and worldview, which promote colonisation as a means of building wealth and power. It inevitably led to competition among Western powers, triggering conflicts (and wars) for supremacy and to establish order among nations. In this process, alliances were formed to increase competitive advantage. This is an organismic response to threats, identified by Margulis (1998) in her theory on endosymbiogenesis. The same process can be observed in the evolution of corporations into transnational giants, which operates in similar way as colonisation, by the same actors, but without use of weapons, except in isolated cases. Although the Western powers employed a CM worldview to achieve its capitalist goals, the responses are organic and organismic. The immediate reaction is to fight back. However, subsequent generations of the colonised would be integrated into the management and governance systems established to maintain stability. This response is autopoietic, and therefore successful in moulding the locals to fit with the systems design by the colonial masters. The colonisers were however able to, firstly, penetrate the indigenous system because they engaged the native party under the guise of friendship and trade, a feature of human system developed to facilitate communication and collaborative existence. However, the colonising party subsequently manipulated the relationship to its advantage, resulting in the latter unwittingly forfeiting its sovereignty and assets or surrendering them, often under military coercion. Brunei suffered such a fate in the nineteenth–twentieth centuries.

The West is currently confronted with an emerging political-economic system in the global South, notably BRICS and the rise of China and its flagship Belt and Road Initiative (BRI). This is a natural development and response to globalisation in the economic boom period following WW2. The reaction of the hegemon however is as expected. From a CM point of view, it will regard the actions of China, the obvious leader and instigator, and that of the BRICS alliance, as an affront that must be curbed and punished, should they not fall in line. Under the CM, the US/West is not part of the world, but above it. Consequently, they have continued to intensify punitive measures to contain and incapacitate rogue nations (as well as individuals and private firms) without realising and acknowledging why or how they are counterproductive. From an OS perspective, the natural response of organisms facing a serious threat would be to band together to improve their chances of survival. The rise of BRICS is a reaction to the existential threat posed by the West, and their consolidation is part of their ongoing response in the search for an alternative path for economic survival and peaceful coexistence, as they are confronted with increasingly aggressive measures and actions (economic sanctions, confiscation of state assets, military actions) by the same nations that had colonised most of the countries in the global South.

As the transition becomes increasingly irreversible, hindsight suggests that it stems from two contradictory unilateral actions by the US to achieve short-term goals: (i) to slow China's economic rise and keep it subservient; and (ii) to bring Russia to its knees. At the time, both nations were prepared to exist peacefully within/under the West-led unipolar world order. The US strategy to contain communist bloc since 1971 was to keep China and Russia apart because of their size and potential to disrupt Western hegemony. To this end, the US facilitated China's entry into the World Trade Organisation (WTO). This organically led to Western capitalists moving their factories to China to take advantage of its cheap labour and huge market potential. In CM worldview, the situation presents an opportunity to be exploited to grow wealth; the consequence was not important. However, China's participation in the global economy inevitably led to its growth and rise due to its inherent capabilities, character, history (a century of humiliation) and the wealth and knowledge accrued in the process; the dominant Western view of China is still that it cannot innovate and can only copy, and that it is only capable of making cheap goods of inferior quality. The outcome is a natural development from an OS perspective, which is not considered in CM. The USA had probably fooled itself by its handling of a similar situation with Japan, when the latter's economy had grown to a level that was considered 'too strong'. With the Plaza Accord of 1985, the US reined in Japan's economic development. However, from an OS perspective, Japan, as a vassal, was already part of the system (Western alliance) in which the US is the leader. Its compliance is simply self-regulation. This would not apply to China or Russia, which are separate entities/systems.

The war in Ukraine also arises as a reaction to an existential threat. The Soviet Union had dissolved in response to Western manipulations that promised peaceful relations. However, in the following decades, the North Atlantic Treaty Organisation (NATO), led by the USA, moved to encircle Russia by recruiting neighbouring states. The aim was to subdue the communist giant and open it up for exploitation (colonisation). After numerous Russian protests in international forums (especially at the UN), NATO agreed not to move an inch further east. However, it did not keep its word and continued to move closer to the Russian border, until its arsenal of weapons are positioned at Russia's doorstep. Ukraine was the red line for Russia, which the West simply ignored. The USA had overthrown the legitimate government in 2014 and installed a puppet. In February 2022, Russia launched a special military operation in Ukraine to force it to the negotiating table. However, the West countered and escalated the war. In the meantime, it also launched an information attack to generate support for its role by comparing Russia to Nazi Germany. It also imposed a series of economic sanctions in an attempt to decimate the Russian economy, and confiscated USD 300 billion worth of Russian state assets. However, these measures were counterproductive as the majority of the world sided with Russia, as OS response to the threat (as they had developed mutually beneficial and cooperative alliances with Russia).

The divergence between CM and OS is that CM views non-Western people as mindless assets, while OS recognizes all living systems, including humans, as cognizant and capable of feeling and learning. This disconnect may explain why the West often overlooks that fact that people around the world can see injustice, manipulation of facts, double standards, and hypocrisy. As Western mainstream media continues to disseminate the same manufactured narrative, it only underscores its untrustworthiness and accelerates the formation of alternative alliances to determine their own future and development, free from Western shackles. This shift contributes to the fragmentation of the unipolar world order, leading to new alliances as part of a broader global reorganisation. The process is irreversible because it is a natural response, which is intensifying and consolidating due to continued unilateral Western actions, designed according to CM worldview and assumptions, deeply rooted in the psyche of the collective West.

b. **Climate Change and Sustainable Development**

According to climate alarmists, climate change (CC) is an existential threat to humanity. Society has been fed this narrative for decades and therefore considered it as the truth. This is especially true for the younger generation, who spends most of their time in the virtual electronic world and, contrary to their belief (that they are environmentally aware), are largely quite ecologically illiterate. In fact, many believe that environmental issues are a derivative of CC and not the other way round. The CC narrative claims that the Earth's climate is changing (based on the rise in global average temperature) and that this has led to environmental problems such as desertification, forest fires, floods, extreme weather events and even tsunamis. However, climate/weather describes the state of the air in terms of the amount of energy, moisture and their movements. These are the product of (i) the amount of heat released from land and water, (ii) the speed of air and moisture in convection flows driven by (iii) thermal gradients caused by the uneven heating of the land/ocean. Life on the surface actively interacts with the air, land and water to produce hydro-meteorological patterns. Over time, different types of soil and vegetation covers, biomes and ecosystems have formed. They drive the geochemical cycles in the wider regional environment (Vernadsky, 1965). Environmental problems, including CC, are the outcome of large-scale development. They affect the specific areas where the activities were conducted, not globally.

The current development paradigm, which is based on CM, does not take into account complex ecological relationships, but only considered the occurrence of resources, space and development and economic potentials. As a result, soil and vegetation, along with the living communities they contain, are being removed and replaced by man-made materials as modern societies systematically replaced the natural with built environment. As urban development spreads over the land surface, the impacts on local and regional hydro-meteorological circulations and patterns are amplified, concomitant with scale and intensity of change. Interactions with the soil surface and organic life, which has been drastically altered in its structure, distribution and composition, are also changing. This leads to a change in the climate pattern, i.e., CC. Earth and environmental scientists, whether with CM or OS worldview, are generally saddened by the wanton destruction of nature in the pursuit of profit and the current trend of development. Although some have been accused as climate deniers, most have an emotional attachment to nature, quite unlike the angry activists who demand justice from corporations or entire industries.

In climate science, CC describes significant changes in climate patterns based on long-term observations (>30 years). Geoscientists, who use a much longer timeframe, know that the climate fluctuates between greater extremes than is currently the case. The current geological epoch, the Holocene, is a warm phase that began 11,700 years ago after the last glacial retreat of the Pleistocene. Sea levels initially rose rapidly, creating today's estuaries, bays, continental shelves and coastlines. In Brunei, bathymetric data indicates sea level had risen by 100 m. Many geoscientists therefore consider the current observations to be part of climate variability. Climate change advocates are unhappy with this term because they believe it absolve industry as the culprit. The question, however, is whether to approach CC from a simplistic CM logic or an OS understanding of the planet. In CM or OS science, the sun is the main driver of weather/climate and the ocean is the main moderator. Greenhouse gases (GHGs), which includes carbon dioxide (CO_2), is only one of many climate parameters. However, West-led climate strategies suggest that GHGs is the only factor, and CO_2 is the obvious link between modern hydrocarbon-fuelled industries and the climate system. This is an oversimplification, and the view that humanity could manage global climate simply through human ingenuity and technology is rooted in CM. In contrast, OS takes into account the complex interrelationships in the environment.

A warming climate is the natural consequence (OS response) of the modern development paradigm, which had resulted in large-scale, irreversible destruction of natural land cover, emissions of GHG and pollutants, voracious and wasteful consumption of energy and material resources. These have altered local–regional hydrometeorological patterns, leading to an apparent 'global warming' and a host of other environmental problems (Yong, 2022). There are also extra-terrestrial forces acting on Earth's climate. Climate change is manifested differently in different parts of the world. It could lead to drier conditions, resulting drought and bush/forest fires, or to wetter patterns and a higher frequency and magnitude of floods, storms and extreme weather events. In coastal regions, powerful waves could erode and inundate the coastline, while storm surges flood low-lying areas close to the shore. Although these are destructive events, disasters only occur when they coincide with a vulnerable human population (Lavell et al., 2012). Contrary to CM's assumption, geohazards cannot be contained. However, vulnerability can be drastically reduced or even eliminated through human action. This is demonstrated by the fact that populations can be kept safe by structures built following design standards that could withstand geohazards of specific magnitudes. This is also evident in many traditional settlement designs because they took environmental dynamics into account (see e.g., Hassan & Yong, 2023; Yong, 2021; Yong & Hassan, 2024). Currently, insufficient effort and resources have been allocated to this (Kamiya et al., 2020; UN Climate Change, n.d.). Meanwhile, strategies and programmes that focus primarily on carbon reduction address the problem but not the reality. They consume (waste) funds that could have been used for much-needed adaptation projects.

The current modern development paradigm is unsustainable, even 'doomed' (Soleri, 1969). It received global attention at the 1992 UN Conference on Environment and Development in Rio de Janeiro, commonly known as the Earth Summit. It was the culmination of earlier initiatives, notably (a) the 1972 UN Conference on the Human Environment in Stockholm and (b) the 1985 Brundtland Commission, which examined the sustainability of current development. The outcome, Our Common Future (WCED, 1987), identified three interdependent goals as critical to achieving sustainable development: (i) sustained economic growth to support the growing population; (ii) protection of the natural environment and its ecological integrity, because it is the life support system and the resource base of humankind; and (iii) equity and social justice, because pronounced inequality undermines sustainability. However, development within the framework of the modern market economy leads to the opposite results. While it appears to be very effective in sustaining economic growth, its disregard for feedback loops combined with the unbridled pursuit of maximum profit has led to unsustainable use and exploitation of resources and the associated pollution and responses of the natural environment. In the late 1970s to early 1980s, economists began to consider feedback in economic systems. The Harwick-Solow model, for example, emphasised the need to reinvest economic profits back into society and the environment to ensure economic sustainability. In relation to pollution, Ulrich Beck (1992, originally in German 1986) described the antagonistic exchanges between society, industry and authorities that eventually gave rise to risk awareness and environment considerations in development. In the early 1970s, the movement gained enough momentum to push the government to pass a series of environmental laws.

Of the three Sustainable Development Goals, however, not enough attention has been paid to equality or social justice, mainly because it contrasts diametrically with capitalism, which produces inequality in society and exploits it for the profit of a minority of elites. CM sees consequences as immediate reactions that can be anticipated and addressed. However, the natural response of OS is less obvious, but systemic processes tend to equalise the imbalance (self-regulation). As a consequence of Western corporations expanding their production base across land and ocean to take advantage of opportunities in other countries, they increased income inequality globally and within Western societies, concentrating wealth in à small group of oligarchs (Turner, 2024). However, the economies of some of these countries eventually grew to compete with those of the West because Western corporations had moved their production facilities there. The trajectory of natural systemic process is always towards dynamic equilibrium. This is reflected in the human-economic arena as developing economies grow in stature and capacity. At the firm level, there is a tendency towards more equitable distribution in corporations. Following the Earth Summit, as humanity sought a more sustainable future, it continued its reliance on innovation and artificial systems. In the twenty-first century, a "knowledge society" has emerged, which is paradoxically concerned about the planet and its sustainability, while at the same time becoming increasingly disconnected from the physical natural world. The noosphere is characterised by continuous flows of 'packaged knowledge' created for convenient consumption. Many are confident to claim that they can see clear evidence of CC but cannot point to any.[2]

[2] Based on the many discussions in class over 2–3 decades.

SD cannot be achieved with CM, which favours a small elite group and whose actions have significant negative impacts on humanity and nature, because it is exploitative, destructive and promotes inequality. In contrast, the natural world is an OS. Its inherent ability to regulate and organise itself in order to maintain system stability will cause built structures to collapse, which can have disasterous consequences. Development is therefore more sustainable if it is controlled by OS (Hassan & Yong, 2023; Yong, 2022). Biomimicry, i.e., the in-depth study of the characteristics and life processes of organisms that thrive in the specific locations of interest (i.e. "genius of place", Benysus, 1997) or the established characteristics of natural areas (Benysus' "genius of biomes"), could be used to find sustainable solutions that are best suited to the location in question. Urban development could be designed in harmony with environmental context so as to minimise their physical and ecological footprints (e.g., Soleri et al., 2011; Knippers et al., 2016 and Pawlyn, 2016). In China, sponge cities that mimic wetlands have been found to be effective in mitigating flash floods (Li & Zhang, 2022). As sea levels rise in the coming decades, estuaries that have been filled in by natural and anthropogenic siltation would be revitalised. In contrast to the doomsday scenario propagated by climate alarmists, a proactive OS perspective on climate change sees an opportunity to create a more sustainable future. In the case of Brunei and Borneo, and in much of SEA, CC would revitalise the coastal estuarine and marine ecosystem and open up the prospect of blue urbanism and economy (Yong & Hassan, 2024).

Conclusion

Our perspectives of global changes are determined by our worldviews. They guide our thought process and shape our conclusions. As a non-Western person, who had experienced the consequence of colonialisation, and was brought up and educated in the Western CM paradigm, my acadėmic foray into complex systems provided a means of comparing of CM and OS philosophies and uncovering significant points of divergence. They are the basis of many conflicts and uncertainties, because CM is still prominent in discourse, policy-making, education, industries and development, while the reality is more accurately depicted by OS. The racial tensions in SEA, particular during the period a round Independence, is an example of an OS response to CM action by the colonial administrators.The following are some of the main findings and their implications.

a. Based on current developments, a multipolar world is a certainty in the near future. This will contribute positively to sustainable development as it promotes a more equitable and peaceful world that respects all peoples and environment, reflecting a system in equilibrium. In contrast, the unipolar world created using CM is a hegemonic structure that favours the Western colonial powers. The system is not in equilibrium and is sustained by manipulation and coercion.

b. In the meantime, a hot war is probable, because the West, governed by CM paradigm, appears to believe that escalating coercion through military force could curb the emerging alternative system. However, the OS response is unstoppable, because the West, through their own actions, have undermined their own hegemonic position and facilitated the rise of China, Russia, BRICS and the Global South.
c. The impact of the fallout due to CM worldview and actions (by the West) and the OS response to them on the global and regional economies, such as Southeast Asia, would include short-term shocks and disruptions to supply chains, as well as wild fluctuation in the value of national currencies, all of which will have longer-term repercussions. Recovery is however inevitable because self-regulation is inherent in OS, and therefore shocks will diminish as system stability is restored. The emergence of an alternative financial-trade system would hasten recovery. Nations, firms and individuals that remain in the existing system will experience greater shocks in direct proportion to the amount of USD held.
d. In CM view, global climate could be controlled, by managing CO_2 concentration, because it is the link between the current hydrocarbon-powered economy and climate. Meanwhile, CM is unaware that CC and the host of environmental problems confronting humanity is the outcome of development based on its paradigm, which disregards feedback and organic actions. The implication of ignoring these processes, which is characteristic of CM thought, is that current CC strategies and programmes are unlikely to be effective because they are overly simplistic and over-estimated their ability to control the global climate. Adaptation projects, tailored for specific contexts, are needed more urgently, especially in areas while large vulnerable populations are found.
e. The rise of the knowledge age reflects consolidation of the Noosphere. This will awaken greater proportion of humankind to ecological interconnections from knowledge on OS. It is hoped that this would lead to a transition of development from its current CM basis to OS paradigm, so that truly sustainable development could be achieved. However, while this development has created greater environmental awareness among the younger generations, it has also increasingly isolated them from physical natural reality. There is therefore a need to reconnect them, and humanity as a whole, with the natural world to build eco-literacy, which are important for a sustainable climate-resilient future.

CM in science and technology has significantly improved our understanding of the world, including CC. However, its influence on politics and the mentality of modern society is the main cause of the developments that threaten humanity because it does not adequately consider the systemic effects of the actions taken. This applies to all aspects of human existence, from geopolitics to climate and sustainability. The impacts are experienced at all levels, from personal-individual to communal-societal, national, regional and global. OS provides a more accurate model of the planet, environment and humanity in its diversity and level of organisation, and therefore promotes more equitable, inclusive and sustainable policies and relations.

References

Beck, U. (1992). *Risk society: Towards a new modernity*. Sage.

Benysus, J. (1997). *Biomimicry: Innovation inspired by nature*. Harper Collins.

Biomimicry 3.8. (n.d.). *Genius of Biome*. Retrieved on April 3, 2018, from https://biomimicry.net/the-buzz/resources/genius-of-biome/

Capra, F. (1975). *The Tao of physics: An exploration of the parallels between modern physics and eastern mysticism*. Shambhala Publications.

Capra, F. (1982). *The turning point: Science, society and the rising culture*. Bantam Books.

Capra, F. (1996). *The web of life: A new scientific understanding of living systems*. Anchor.

Hassan, N. H., & Yong, G. Y. V. (2023). Climate change and sustainable development of Kampong Ayer: Its history, people, and environment. In M. Hamdan, M. Anshari, N. Ahmad, & E. Ali (Eds.), *Public policy's role in achieving sustainable development goals* (pp. 96–114). ICI Global.

IPCC. (2021). *Climate change 2021: The physical science basis. Contribution of Working Group I to the Sixth Assessment Report of the Intergovernmental Panel on Climate Change* (V. Masson-Delmotte, P. Zhai, A. Pirani, S. L. Connors, C. Péan, S. Berger, N. Caud, Y. Chen, L. Goldfarb, M. I. Gomis, M. Huang, K. Leitzell, E. Lonnoy, J. B. R. Matthews, T. K. Maycock, T. Waterfield, O. Yelekçi, R. Yu, & B. Zhou, Eds.). Cambridge University Press, 2391. https://doi.org/10.1017/9781009157896

Johnson, B. D. (2010). *Miniaturize or die!: Paolo Soleri: A study into his influence and relevance on today's sustainable urban future*. University of Sheffield, School of Architecture.

Kamiya, M., Prakash, M., & Berggren, H. (2020). *Financing sustainable urbanization: Counting the costs and closing the gap*. UN Habitat. https://unhabitat.org/sites/default/files/2020/02/financing_sustainable_urbanization_-_counting_the_costs_and_closing_the_gap_february_2020.pdf

Knippers, J., Nickel, K. G., & Speck, T. (2016). *Biomimetic research for architecture and building construction: Biological design and integrative structures*. Springer.

Lavell, A., Oppenheimer, M., Diop, C., Hess, J., Lempert, R., Li, J., Muir-Wood, R., & Myeong, S. (2012). Climate change: New dimensions in disaster risk, exposure, vulnerability, and resilience. In C. B. Field, V. Barros, T. F. Stocker, D. Qin, D. J. Dokken, K. L. Ebi, M. D. Mastrandrea, K. J. Mach, G.-K. Plattner, S. K. Allen, M. Tignor, & P. M. Midgley (Eds.), *Managing the risks of extreme events and disasters to advance climate change adaptation*. A Special Report of Working Groups I and II of the Intergovernmental Panel on Climate Change (IPCC) (pp. 25–64). Cambridge University Press.

Li, F., & Zhang, J. (2022). A review of the progress in Chinese Sponge City programme: Challenges and opportunities for urban stormwater management. *Water Supply, 22*(2), 1638–1651. https://doi.org/10.2166/ws.2021.327

Lovelock, J. E. (1972). Gaia as seen through the atmosphere. *Atmospheric Environment, 6*(8), 579–580. https://doi.org/10.1016/0004-6981(72)90076-5

Lovelock, J. E., & Margulis, L. (1979). *Gaia: A new look at life on earth*. Oxford University Press. ISBN: 978-0-19-213899-4

Margulis, L. (1967). On the origin of mitosing cells. *Journal of Theoretical Biology, 14*(3), 225–274. https://doi.org/10.1016/0022-5193(67)90079-3

Margulis, L. (1998). *The symbiotic planet: A new look at evolution*. Basic Books.

Maturana, H. R., & Varela, F. J. (1980). *Autopoiesis and cognition: The realization of the living*. D. Reidel Publishing Company. https://doi.org/10.1007/978-94-009-8947-4

Maturana, H. R., & Varela, F. J. (1984). Autopoiesis, structural coupling, and machine organization. In F. J. Varela & P. Bourgine (Eds.), *Towards a practice of autonomy: Essays on self-organization and decentralization* (pp. 29–48). North-Holland.

Maturana, H. R., & Varela, F. J. (1987). *The tree of knowledge. The biological roots of human understanding*. New Science Library.

McCullough, L. (Ed.). (2012). *Conversations with Paolo Soleri*. Princeton Architectural Press.

Naess, A. (1989). *Ecology, community, and lifestyle: Outline of an ecosophy* (D. Rothenberg, Trans.). Cambridge University Press.

Pawlyn, M. (2016). *Biomimicry in architecture.* RIBA Publishing.
Sakaguchi, H. (2001). *Final fantasy: The spirit within* [Film]. Columbia Pictures.
Seidl, D. (2004). *Luhmann's theory of autopoietic social systems.* Ludwig- Maximilians-Universität München.
Soleri, P. (1969). *Arcology, city in the image of man.* MIT Press.
Soleri, P., Kim, Y., Anderson, C., Nordfors, A., Riley, S., & Tamura, T. (2011). *Lean linear city: Arterial arcology.* Cosanti Press.
Stein, J. (2012, November 10). *Arcology: Sustainable hyperstructures: Jeff Stein at TEDxMission TheCity2.0.* YouTube [video]. https://youtu.be/XCqdvLaBx50
Turner, S. (2024, August 16). *US or China, which is truly democratic? American oligarchy vs Chinese 'consultative democracy'.* Geopolitical Economy report. https://geopoliticaleconomy.com/2024/08/16/us-china-oligarchyconsultative-democracy/
United Nations Climate Change. (n.d.). *COP26 outcomes: Finance for climate adaptation.* Retrieved on December 27, 2021, from https://unfccc.int/process-and-meetings/the-paris-agreement/the-glasgow-climate-pact/cop26-outcomes-finance-for-climate-adaptation#eq-2
Vernadsky, V. I. (1965). *The biosphere: The living envelope of the Earth* (L. G. Sears, Trans.). Harvard University Press. (Original work published 1926)
WCED (World Commission on Environment and Development). (1987). *Our common future.* Oxford University Press.
Yong, G. Y. V. (2021). Identifying cultural traits of the historic Kampong Ayer of Brunei Darussalam using biomimetic analysis. *International Journal of Environment, Architecture, and Societies, 1*(2), 83–94.
Yong, G. Y. V. (2022). Climate-resilient urban design from a biomimicry-arcology perspective. *International Journal of Environment, Architecture, and Societies, 2*(1), 1–15. https://doi.org/10.26418/ijeas.2022.2.1.1-15
Yong, G. Y. V., & Hassan, N. H. (2024). Blue urbanism: An insight of navigating climate change in humid tropical cities in Brunei Darussalam. *International Journal of Environment, Architecture, and Societies, 4*(01), 23–36. https://doi.org/10.26418/ijeas.2024.4.01.23-36

Gabriel Yit Vui Yong is a lecturer at Unversiti Brunei Darussalam. He has a BSc in Geological Geophysics with Mathematics (1986) and a MSc in Environmental Analysis & Dynamics (Hull, 1999). Over three decades, his perspective of natural and human-environment dynamics shifted from physical systems founded on Cartesian Mechanism to that of a complex, sentient, planetary system that is capable of self-organisation and co-evolving with life. He takes a grounded approach in research and has been involved in numerous national-regional projects.

Chapter 5
Food Security: A Closer Look Using Brunei Darussalam as Case Study

Norzurianie Kamarulzaman

Abstract The food crisis of 2007 focussed attention on food security. In the region, ASEAN took a number of initiatives to improve the food security of its members and secured co-operation under the ASEAN Plus Three for the Emergency Rice Reserve (APTERR). Brunei is not an agricultural country. The discovery of oil and gas deposits has enabled the country to flourish as a rentier state, although it introduced economic diversification early on in its national development programme. The small population of less than half a million people combined with the oil wealth suggests that food security should not be a problem. However, dependence on imports, particular of rice, exposes the country to supply chain disruptions. This chapter examines food security in the region using Brunei as a case study, with a focus on rice production. It draws on in-depth knowledge acquired through a 2010–2011 graduate study and subsequent follow-up research. Secondary data was sourced from the ASEAN Food Security Information System. The study has shown that despite its importance, the issue of food security is not given due attention by the relevant agencies at the operational level and society in general because it is inconspicuous. So, life goes on without giving it a second thought. As a result, rice production in Brunei continues to be plagued by poor agricultural management primarily due to inadequate technical and technological support and environmental problems. The chapter discussed the concept of food security and its application in relation to sustainable development.

Keywords Food security · Rice · Brunei · AFSIS · Sustainability

Introduction

According to the former Secretary-General of the United Nations (1997–2006), Kofi Annan, the right to food is an inherent human right that is non-negotiable. Food security is the availability of sufficient food to meet people's nutritional needs and to lead an active and healthy life. It includes access to food, the use of food and the

N. Kamarulzaman (✉)
Faculty of Arts and Social Science, Universiti Brunei Darussalam, Gadong, Brunei Darussalam
e-mail: norzurianie.kamarul@ubd.edu.bn

G. Y. V. Yong and A.K.M. A. Ullah (eds.), *Brunei and the Future Transformation of Southeast Asia*, https://doi.org/10.1007/978-981-96-7901-0_5

sustainability of this system (Food and Agriculture Organisation (FAO), 1996). In Southeast Asia (SEA), food security became a priority issue for the Association of Southeast Asian Nations (ASEAN) in the 1970s and 1980s as the region faced significant challenges in agricultural development and food supply, exacerbated by rapid population growth. ASEAN member states recognised the need to work together to improve regional food security and initiated various frameworks and programmes (Tham & Chin, 2024). The ASEAN Ministers of Agriculture and Forestry (AMAF) was established in 1978 to address this issue. Initiatives include the ASEAN Integrated Food Security (AIFS) Framework and the Strategic Plan of Action on Food Security (SPA-FS) to promote the sustainability of food production, the facilitation of cross-border trade and co-operation on agricultural technologies.

Since the turn of the millennium, there have been several major shocks to food security. These include: (i) food price crises (2007–2008, 2010–2011) caused by soaring staple food prices due to increased demand, unfavourable weather conditions, rising oil prices and diversion of crops for biofuels; (ii) the 2008–2009 global financial crisis; (iii) the COVID-19 pandemic that disrupted supply chains; and (iv) geopolitical conflicts that have introduced uncertainties to trade and peace. In 2011, the ASEAN Plus Three Emergency Rice Reserve (APTERR) was established to provide food assistance in emergency situations caused by disasters and to strengthen food security, stabilise rice price and improve farmers' income and welfare. ASEAN also initiated: (i) the ASEAN Food, Agriculture and Forestry Cooperation (AFT), which includes rice as a major crop in its food security and agricultural development strategies; (ii) the ASEAN Food Security Information System (AFSIS), which covers major food crops (rice, maize, sugar and cassava); and (iii) the ASEAN Good Agricultural Practises (ASEAN GAP) to improve regional farming practises.

In 2007, Brunei launched a programme to achieve rice self-sufficiency in response to the food price crisis. The country set itself the ambitious target of increasing local production from the current 3% to 20% by 2010 and 60% by 2015 (Department of Agriculture, 2009a; Hajah Aidah, 2011). However, according to AFSIS statistics, local production reached only 3.3% and 4.8% self-sufficiency in 2010 and 2015, respectively, i.e. modest increases but far from the targets. Rice production averaged 1,591 metric tonnes (MT) between 2008 and 2021, while average demand (based on food consumption data provided by AFSIS) was 33,224 MT (see Table 5.1). Supply was calculated here by combining production with average imports, which averaged 34,289 MT during this period. The data shows that while production has increased, local consumption and rice imports have fluctuated somewhat, with the latter peaking in 2012 and 2014 (see Fig. 5.1). The country is therefore still dependent on rice imports for this important staple food. Since the discovery of oil and gas in 1929 and 1963, respectively, Brunei has been a rentier state. Despite the DAA's efforts, agriculture together with forestry and fisheries contributed 1.2% to gross domestic product (GDP) (Department of Economic Planning & Statistics (DEPS), 2023), while agriculture alone contributed 0.6% (DAA, 2024). So, food security is a problem. However, Brunei's apparent complacency could be due to its oil wealth combined with a very small population (450,500, DEPS, 2023).

Table 5.1 Brunei rice self-sufficiency, 2008–2021

Year	Production (MT)	Import (MT)	Demand (MT)	Supply–demand (MT)	% Self-sufficiency (%)
2008	880	30,410	30,960	330	2.8
2009	890	31,710	31,790	810	2.8
2010	1,070	28,550	32,360	−2,740	3.3
2011	1,390	31,060	33,630	−1,180	4.1
2012	1,760	46,940	34,720	13,980	5.1
2013	1,660	33,650	33,440	1,870	5.0
2014	1,380	46,390	34,130	13,640	4.0
2015	1,640	32,940	34,020	560	4.8
2016	1,680	33,920	35,040	560	4.8
2017	1,630	33,300	35,000	−70	4.7
2018	1,570	33,620	33,770	1,420	4.6
2019	1,570	31,890	32,970	490	4.8
2020	2,510	34,080	30,520	6,070	8.2
2021	2,650	31,580	32,790	1,440	8.1
Mean	1,591	34,289	33,224	2,656	4.8
Std. Dev	490	5,268	1,367	4,911	1.6

Source ASEAN Food Security Information System

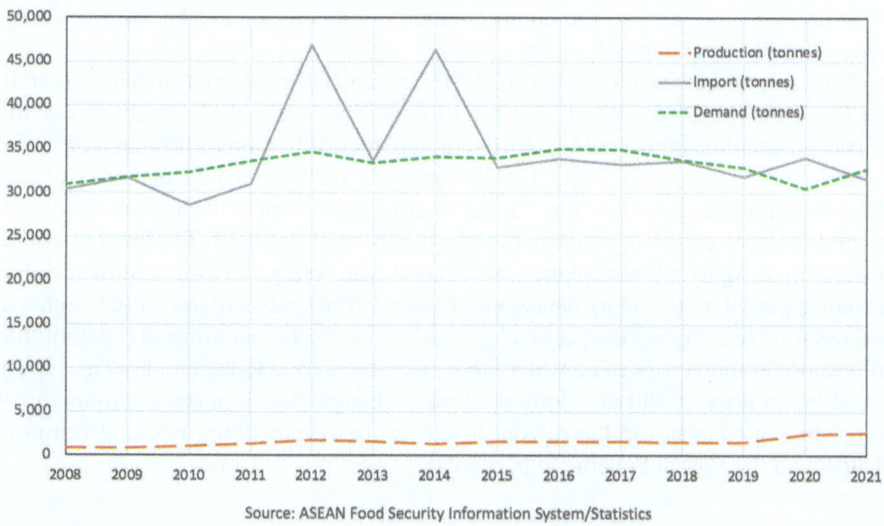

Fig. 5.1 Rice production, demand and import (in MT), Brunei, 2008–2021

This chapter examines food security using Brunei as an example. It begins with details of Brunei's environmental conditions in relation to rice cultivation, followed by research findings conducted as part of the country's flagship rice cultivation project in 2010/2011 and subsequent follow-up activities. Food security in the region is then discussed in terms of sustainability aspects to find alternative means that are more suitable for the region and Brunei in particular.

Methodology

The study analyses food security using the example of rice production in Brunei. The reasons for the selection of Brunei are complex. Firstly, as a small, prosperous country, food security is an unlikely problem. Nevertheless, it is exposed to the risks of supply chain disruption. In the region, Singapore would have similar problems because it is similarly small and wealthy. Secondly, Brunei is not traditionally a food-producing country. Instead, the country's economy in the 15th to nineteenth centuries was based on trade and in modern times on oil rent, which has fuelled fairly unrestricted consumption by the upper echelon of society. In today's age of globalised economy, hyper-consumption and sustainability, Brunei shares many of the region's concerns. The majority of society is also unaware of the threat to food security. However, rice subsidisation further shields the public from food security concerns. After all, the country is located in a humid, tropical region, like most of Southeast Asia, and so the environmental conditions should be favourable for agriculture. But the opposite is the true.

The study will include analyses of AFSIS data and information from field research, the DAA and the DEPS. It will also refer to findings from a study conducted on the Wasan programme between May 2010 and January 2011 (Kamarulzaman, 2017) and follow-up field surveys in 2020. The Wasan study included: (i) an analysis of DAA data on rice yield, farm size and the number of farmers in 2010 and (ii) fieldwork involving interviews with 52 farmers and a number of DAA staff. The latter provided interesting insights into the nuances associated with paddy cultivation in Brunei that are not apparent in secondary datasets and reports. The study will use these insights to discuss food security at a deeper level, going beyond the production and consumption of traditional food crops to consider what constitutes food today and how its supply could be secured. It will also look at climate change, which poses a serious threat to producing countries in the region. It will begin by examining rice cultivation in Brunei and the Wasan Scheme in particular.

Environmental Context and Paddy Cultivation

Brunei is a small coastal state with a land area of only 5,765 km^2. The country consists of three relatively small river basins (catchments, approximately 1,000 km^2)

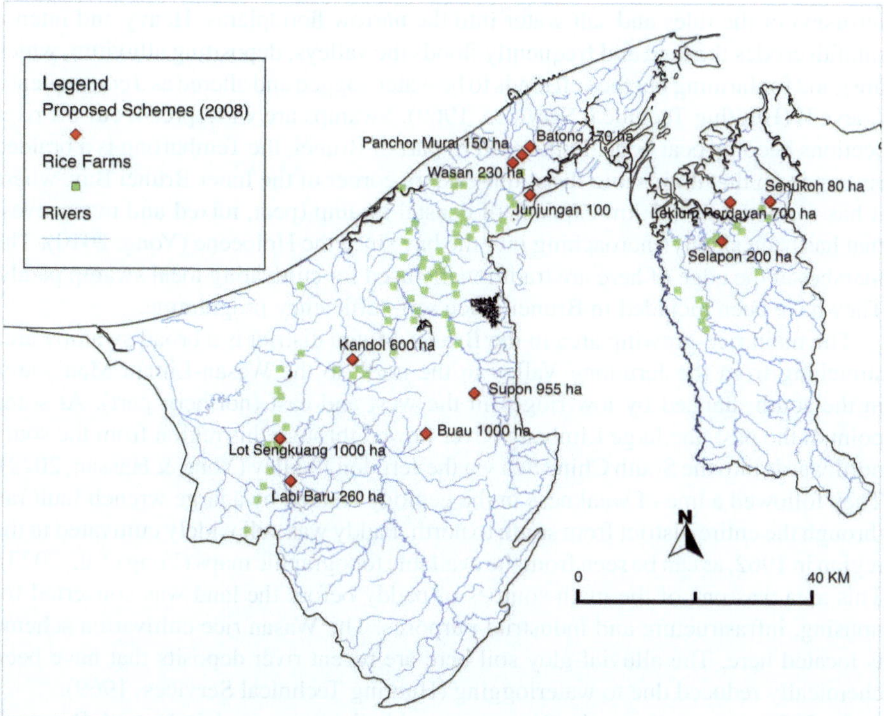

Fig. 5.2 Paddy cultivation schemes in Brunei, 2010. *Source* DAA

with modest freshwater runoff as a result (Sandal, 1996; Yong, 2010). The largest river, the Belait, drains into a large peatland equal in size to its catchment area, before emptying into the South China Sea at the western end of the coastline, 24 km east of the mouth of the Baram. The Belait River is part of the large Baram Delta, which is actively expanding into the sea (ibid.). The delta is also the source of high-grade oil and gas, and much of the industry's assets, including its headquarters, are concentrated along this stretch of coast. At the end of the river basin, before the river enters the vast peatland, lies Kandol, the site of the country's newest and largest rice cultivation project, which became operational in 2019 (see Fig. 5.2). Presently, the area developed for cultivation is 174 ha. The project is managed by a government-linked company (GLC) and is expected to contribute 11–15% of Brunei's rice demand.[1] The site is however prone to flooding, which has disrupt rice production.

The Tutong Basin in the central region is the most agriculturally active area, but contributes little to agricultural GDP (Integrated Environmental Consultants, 2011). However, these are small subsistence farms where local communities grow a wide range of local tropical vegetables and fruits, as well as local rice varieties. However, Brunei's generally subdue topography and relatively low river discharges favour the

[1] Personal communication with DAA.

intrusion of the tides and salt water into the narrow floodplains. Heavy and intense rainfall erodes the land and frequently floods the valleys, depositing alluvium, which are good for farming but the soil tends to be water-logged and altered as a consequence (gleyed) (Hunting Technical Services, 1969). Swamps are widespread; cut-off river sections become peat bogs. In the eastern part of Brunei, the Temburong is a braided river system that drains into the south-eastern corner of the Inner Brunei Bay, where it has formed a 20–25 km expanse of coastal swamp (peat, mixed and mangroves) that has been slowly encroaching into the bay since the Holocene (Yong, 2010). The marshes at the edge of here are traditionally used for cultivating local swamp paddy. They have been included in Brunei's rice self-sufficiency programme.

The main rice-growing area in the Brunei-Muara district is a broad swampy area stretching from the Jerudong Valley in the north to the Wasan-Limau Manis area in the south, flanked by low ridges in the west and east (northern part). At some point in the past, the large Limbang River flowed through this region from the south northwards into the South China Sea via the Jerudong Valley (Yong & Hassan, 2022). They followed a line of weakness in the geology created by a large wrench fault ran through the entire district from south to north. Paddy was still widely cultivated in the region in 1962, as can be seen from the available topographic maps (Yong et al., 2023). This area was one of the main sources of paddy before the land was converted for housing, infrastructure and industrial purposes. The Wasan rice cultivation scheme is located here. The alluvial-gley soil here are recent river deposits that have been chemically reduced due to waterlogging (Hunting Technical Services, 1969).

In 2008, a team from the Commonwealth Scientific and Industrial Research Organisation (CSIRO) found that potentially acidic sulphate soils occur in the low-lying areas of Brunei-Muara (at Betumpu, Si Tukak Limau Manis, Si Bongkok Parit, Masin, Lumapas, Limpaki and Pengkalan Batu) and to a lesser extent in the Belait district. These areas are dominated by organic soils, which usually require intensive treatment. Small areas of sulphur-rich soils require more intensive treatment than sulphide-rich soils, which require moderate treatment (Grealish et al., 2008). There are extensive areas of acid sulphate soils in Wasan, but as these are 'cracking clay soils', they only require low to moderate treatment (ibid.).

In the past, the Limbang River contributed a large amount of freshwater when it flowed into this area before its mouth was diverted to the east due to natural geomorphic processes (Yong & Hassan, 2022). Currently, small rivers draining from the western ridge (Ladan Hills) supply the lowlands with freshwater. At Wasan, the Imang River, with a catchment area of only 14 km^2 is the main source of water. In 1995, the Imang Dam, with a reservoir of 10 m^3, was built to provide the Wasan scheme with irrigation.

Paddy Cultivation Program

The programme for rice cultivation in Brunei has been in place since the 5th NDP (1986–1990). Under the DAA's 'Skim Intensif Pertanian dan Agrimakanan' (SIPA, or 'Agriculture and Agrifood Incentive Scheme' in English), rice farmers are offered various incentives and subsidies. These include basic agricultural supplies such as fertilisers, pesticides and plant protection products at half price, as well as agricultural mechanisation, the provision of basic agricultural infrastructure, marketing support and a buy-back programme, in which the government buys the harvest from rice farmers at a fixed price of USD 1.60 per kilogram. In its 6th NDP (1991–1995), the Government of Brunei focussed on strategic agricultural development with the aim of improving food security by promoting self-sufficiency in staple foods, specifically rice. High-yielding rice varieties as well as new agricultural technologies and practices were introduced. Collaboration with leading research institutes yielded the Laila rice, high-yielding varieties (HYV) (with the Philippine's International Rice Research Institute (IRRI)) and high-yielding hybrids, Sembada 188 and Titih (with Indonesian Institute of Rice Research (IIRR), or 'Balai Penelitian Tanaman Pangan') in 1994.

In 2007, the DAA published a manual for swamp rice or 'Sistem Semakan Padi Bagi Padi Paya' in Malay, which is aimed at rice farmers/entrepreneurs. It is a systematic guide for cultivation to achieve production targets based on the Palaycheck system developed by IRRI (1999). This is an innovative rice cropping system developed to optimise irrigation management and increase the efficiency of rice production. It also included field assessment, nutrient management, pest control, variety selection and farmer training. However, after more than 15 years since its revitalisation in 2008, rice production is still plagued by persistent problems. According to the DAA, they include: (i) acidic sulphate soils, which the agency says are characteristic of soils in Brunei (mean pH 3.6; recommended 5.5–6.5) as well as high iron levels (mean 553 ppm; recommended 2.5–4.5 ppm); (ii) insufficient water supply at critical stages of the cropping cycle; (iii) inadequate irrigation infrastructure; (iv) intense climate (hot weather and intense rainfall); and (v) numerous pests and diseases that are difficult to control. There is also a lack of commercial operators as 99% of rice farmers are smallholders (<5 hectares) who cultivate on government land, DAA land or land under Temporary Occupancy Licence (TOL).

The DAA's long-term solution are as follows (Department of Agriculture, 2009b) are as follows:

1. Open new rice farming areas, covering 1,430 ha in the Brunei-Muara District (Wasan, Bebuloh and Limau Manis), 2,860 ha in the Belait District (Buau, Kandol, Labi Baru, Lot Sengkuang), 955 ha in Tutong District (Supon Kecil) and 980 ha in the Temburong District (Lekiun/Perdayan, Selapon and Senukoh) (see, Fig. 5.2).
2. Upgrade existing field infrastructure, such as roads, irrigation and drainage, and flattening areas.
3. Switch from traditional to HYV to facilitate double or triple cropping.
4. Organise courses for DAA staff and farmers, e.g. Rice Farmers Field School (RFFS), offered to farmers to upgrade agricultural practices and farm technology.
5. Establish a Rice Research Centre.

Wasan Case Study

The Wasan programme was launched in the 1970s to improve food security through the use of HYV paddy. However, the project did not made much progress until 2006, when it was revived by a co-operative of retired army personnel called 'Koperasi Setia Kawan Sdn Bhd', or 'Koseka'. In 2008, following the global food security crisis, the DAA made a push towards rice security. The following year, His Majesty the Sultan officially launched a revitalised rice system in Wasan, promoting Laila as the national rice variety. However, the DAA also encouraged local farmers to continue growing traditional rice varieties such as pusu, adan and bario. As part of the rice self-sufficiency programme, the DAA established a buy-back programme for traditional rice and HYV rice.

Two groups participated in the Wasan programme: (a) Koseka; and (b) farmers from seven villages in the Pengkalan Batu sub-district ('mukim'), namely (i) Junjungan; (ii) Limau Manis; (iii) Pengkalan Batu; (iv) Masin; (v) Wasan; (vi) Panchur Murai; and (vii) Batong. They are referred to as the 'Mukim' group. During the 2010 fieldwork, 98 army retirees were cultivating the Koseka plots and 68 farmers were cultivating the Mukim section. Figure 5.3 shows the agricultural plots of the two groups. The aim of the fieldwork conducted between May 2010 and January 2011 was to assess the effectiveness of the Wasan programme in meeting the Brunei government's goals of economic diversification and food security (Kamarulzaman, 2017). The study utilised a mixed methods approach and involved a sample of 52 farmers from the two groups (36 Koseka and 16 Mukim). A number of DAA officials were consulted in the study. Semi-structured interviews, observations and archival methods were used to collect data. A questionnaire was also deployed.

Although 98 farmers were registered with Koseka in 2010, only 90 reported yields and 67 stated that they sold part of their yield to the DAA. Farm sizes ranged from 0.7 ha to 5.1 ha, with an average of 1.8 ha. Although the majority of farms were small (<2 ha), 33% were 2–3 ha, 10% 3–4 ha, 9% 4–5 ha and 7% larger than 5 ha. Paddy was grown twice a year. The peak season (October to February) coincided with the generally wetter northeast monsoon, while the 'off-season' (April to July) coincided with the slightly less wet southwest monsoon. The study found that the 52 farms with a total area of just over 92 ha achieved a total yield of 164,172 kg in the main season, with a mean of 2,180 kg/ha. A closer look revealed a wide range of yields. Three farms produced nothing at all, while the most productive farms produced more than 8,000 kg/ha, more than four times the mean yield. This indicates that a quadrupling of the total yield is possible if all farms were to operate like the highly productive ones. There are no significant differences in yields between Koseka and Mukim. As expected, there is a broad correlation between yield and farm size, i.e., yield increases with farm size. However, there are large differences between the larger farms, some of which produce about the same as farms of 1.5 ha to some that are 3–4 times larger. This reflects the condition and management of the farm.

In terms of the percentage of yields sold to the DAA, the data shows a weak trend suggesting that farmers with higher yields tend to sell a greater proportion to the

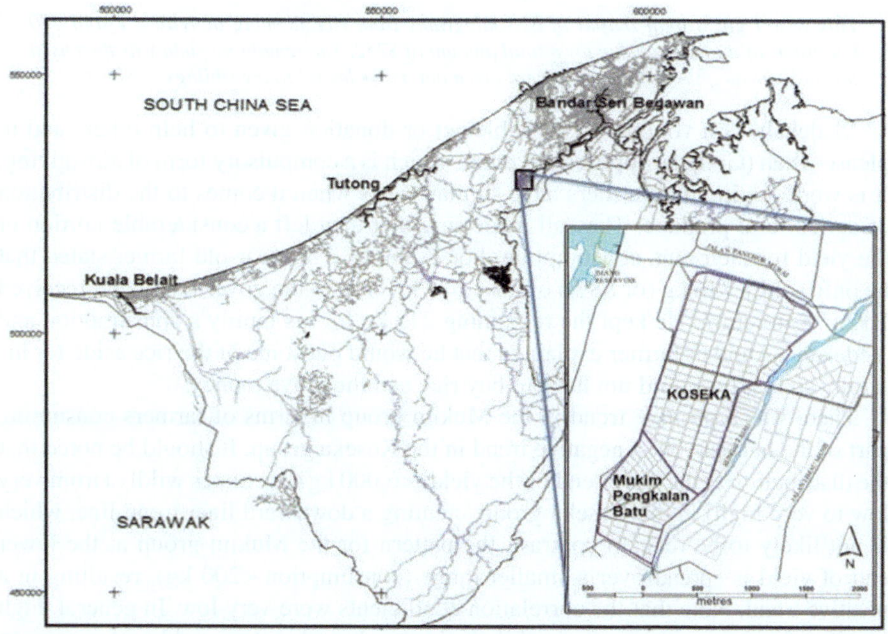

Fig. 5.3 The Wasan scheme showing Koseka and Mukim plots

government scheme. However, four farmers sold all their rice to the local market (i.e., 0% to the buy-back programme), while two farmers sold all their produce to the DAA (100%). Farmer #4 justified his decision to sell all his rice to the local market by saying that he needed the money to maintain his farm and the government scheme was slow to turn over the money.

> *I sell rice so that I can receive immediate cash to pay my workers, to buy fertilisers and other inputs. If you sell to the government, it will take four months. You need money for farming.*

Some farmers said that local market gave them a better deal than the government because there is a demand for Laila rice. As a result, they do not even want to entertain the thought of selling their produce to the government. Some farmers, however, have the opposite view. For example, farmer #11 said,

> *I prefer to sell my rice to the government purchasing scheme because it does not involve much work, such as winnowing and drying the rice under the hot sun, and then sending it to a private milling centre to get unhusked rice.*

This farmer appears to prioritise convenience over higher profits and faster returns. In the example below, the attitude of farmer #8 is typical of Bruneians, where culture is strongly influenced by religion (Islam) and allegiance to the monarchy. He readily participated in the program set up by the government, and also abided by the Islamic practice of 'sedekah' ('Sadaqah' in Arabic).

This year I got a total output of 828 kilograms from rice farming activities. I sold 470 kilograms of it to the scheme for a total amount of $752. The remaining yield was then used for own consumption while some were given out as 'sedekah' to my siblings.

'Sedekah' is a voluntary charitable act or donation given to help others and to please Allah (God), as opposed to 'zakat' which is a compulsory form of almsgiving. It is worth noting that farmers have similar views when it comes to the distribution of agricultural produce. They all made sure that they left a considerable portion of the yield for their own consumption. For example, a 52-year-old farmer stated that he only sold 1,736 kg (or 88%) of his 1,981 kg crop to the government and received B$2,778 in return. He kept the remaining 245 kg for his family's consumption and 'sedekah'. Another farmer explained that he would put some of the rice aside for his family so that he would not have to buy rice and thus save money.

There was a positive trend in the Mukim group in terms of farmers consuming part of the harvest, but a negative trend in the Koseka group. It should be noted that the distribution at the lower end of the yield (<6,000 kg) fluctuates wildly (from very low to very high) in the Koseka group, creating a downward liner trend line, which is not likely to be real. In contrast, the pattern for the Mukim group at the lower end of yield is spread over a smaller range (consumption <200 kg), resulting in a positive trend. Note that the correlation coefficients were very low. In general, high yielding farmers consume more of their rice produce than low yielding farmers, on average 7% of yield. However, as mentioned above, the proportion for the Koseka group shows a much wider range of rice produce consumed.

The study found that, apart from keeping a portion for themselves, a subsistence farming trait, 'sedekah' is an important feature of rice cultivation in Wasan. For the farmers, who are Muslims, this represents a non-financial gain, which, from a religious perspective, will bring them a reward in the hereafter. In the words of one farmer,

The sedekah we give out to people are not only meant for our family members, but also we give out to people who do not have enough. This is the way we can also contribute to the society.

Another farmer echoed this sentiment, saying,

giving out sedekah is something that cannot be counted and valued. How much I gave should not be spoken about. I must do it sincerely to get my rewards in the afterlife.

The data for the 'off season', the yield was similarly positively skewed, with a mean of 2,050 kg/ha, slightly less than the main season. There is, however, a statistically significant difference ($p = 0.00$) in average yields per hectare between the Koseka farmers (1,590 kg/ha) and the Mukim farmers (3,090 kg/ha). This could be attributed to some Mukim farmers planting the Titih variety, which has a higher yield than Laila. There is a slightly stronger relationship between the amount of rice that is sold to the government purchasing scheme and total yield from each farm, such that farmers who have lower yields tend to sell a smaller percentage of their rice to the purchasing scheme. A typical response was,

Since my yield is low, I would rather keep the rice for my family's consumption and give some as sedakah.

As pointed out earlier, this is the characteristic of traditional subsistence farming, which is still the dominant agriculture practice in Brunei. Modern agriculture is still at an incipient stage. The differences in average yields between the two growing seasons are negligible. This is probably due to the wide usage of irrigation, fertilizers and pesticides which allow for round-the-year growing. If agricultural support become more dependable and affordable, higher yields could be achieved because the seasonality appears not to be a controlling factor. Finally, it is interesting that the Mukim farmers with smaller yields tend to have larger off-season yields compared to their main season yields. In contrast, the Koseka farmers with large yields tend to have considerably larger yields during the main season compared to their off-season. However, this might be a feature of the scheme in 2010.

Reflections on the Findings

Based on study conducted (Kamarulzaman, 2017), fundamental contradictions appear to be hindering progress in rice production. For example, a competitive and efficient rice production system would require large commercial farms run by knowledgeable farmers with access to the necessary machinery and labour. However, in the Wasan scheme, retired military personnel with no farming experience were given small plots of land and were not allowed to merge into larger farms to achieve economies of scale. In addition, the experienced Mukim farmers were not familiar with the Laila and hybrid varieties, because these were foreign to Brunei; they were also not familiar with wet paddy cultivation. In addition, the law on migrant workers is too restrictive, which resulted in labour shortage. Wasan was partially successful in that it has promoted economic diversification and created economically viable communities where more farmers sell in open markets and earn profits, albeit because of subsidy and a buyback programme. However, it has not improved food security in the way that the DAA had hope because many farmers are not achieving full productivity. Instead, they were achieving yields well below the potential and earned lower incomes than they could. The challenges faced by farmers need to be addressed in order to improve the sustainability of food production and food security. However, the widespread practice of 'sedakah' contributes to food security at the local level.

The field survey revealed that the main problems are: (i) inadequate water supply (Koseka 29%, Mukim 44%); poor drainage (Koseka 13%, Mukim 16%); diseases and pests (Koseka 28%, Mukim 19%); unreliable DAA deliveries (Koseka 13%, Mukim 0%); inadequate infrastructure and technical support (Koseka 12%, Mukim 19%); and late DAA payments (Koseka 5%, Mukim 3%). The main problems identified in the Wasan study ranged from land sizes too small for commercial rice cultivation, poor soil quality, high acidity and low nutrient content, to inadequate machinery and poor drainage system. These factors are categorised as follows: (i) water problems,

(ii) poor farm management, (iii) pests and diseases, (iv) inability to pay, (v) farm equipment and (vi) labour issues.

(i) Water Issues

During the rainy season, farmers faced difficulty removing excess water from the rice fields as the open drainage network is also flooded with water. The problem is compounded by unevenness of the rice plots, which created ponds in certain parts. The bearing capacity of the soil is weakened in these patches causing farm machinery to be stuck in the mud. During drier periods, some farms have insufficient water supply. The problem appears to stem from the management of the irrigation system. Water was turned on/off manually following a fixed schedule, rather than based on needs for specific plots and at specific times. Furthermore, as the irrigation network is fixed, farms farther away in the network were not adequately supplied, while farms closer to the source suffer from excessive water.

(ii) Poor Farm Management

Farmers lack the necessary knowledge and skills to cultivate Laila rice. The more experienced farmers employed traditional knowledge, which were not suitable for Laila, a wet paddy variety. Moreover, although there were courses offered on cultivating Laila rice, the experts from the Philippines did not adapt the training to the local context. This problem was further compounded by the inexperience of local agricultural extension officers. Farmers also lack knowledge on managing drainage in their fields as well as proper fertilizer/pesticide use, resulting in the occurrence of land toxicity caused by the concentrated accumulation of minerals in various parts of the fields. Consequently, plant growth was stunted and yields low.

(iii) Pests and Diseases

The most important pests in Wasan were: rice bugs ('kekisang'), snails, birds and rats. Birds attacked during the day, while rats at night. Some farmers used a traditional bird repellent called 'telinting', where metal cans are attached to a string to drive the birds away by the sound they make when the string is pulled. However, this requires farmers to remain vigilant, especially in the early morning (around 8.30 am) and in the late afternoon (3.30–6.30 pm). As far as rats are concerned, black plastic fencing erected along the outer perimeter of the field has proven effective. Leaf moth is the most common disease. It is caused by the larvae of the rice leaf moth (Cnaphalocrocis medinalis), which damages and wrinkles the leaves of rice plants as they feed, hindering photosynthesis, resulting in reduced plant vigour and ultimately lower yields. Another serious infestation is the brown leafhopper. This insect leaves burn-like spots ('hopper burn') on the leaves. It can cause yield losses of up to 60%. Dealing with diseases was the biggest challenge, as most farmers lack the necessary knowledge.

(iv) Inability to Pay

It was found that many of the farmers interviewed were unwilling to buy additional herbicides, pesticides or fertilisers, unless such items were not supplied by DAA.

Some farmers, however, believe that the products offered by agricultural retailers were more effective than those supplied by DAA. They disclosed that they had bought products that cost over $100 to enhance their farm's productivity. This included weed control, suitable fertilisers, systemic pesticides, and integrated plant protection systems.

(v) Farm Equipment

The government faces challenges in renting out mechanical equipment like transplanters and combine harvesters due to their limited number. This short-coming in the rental process necessitates a review of government policies regarding the procurement of appropriate machinery. Delays in accessing proper equipment would lead to setbacks, such as late harvesting, which can result in overriped gains being consumed by birds. Consequently, farmers are forced to harvest manually using hand tools like sickles, knives, or grass cutters. This manual process requires additional labour for threshing, incurring extra costs and increasing the risk of grain deterioration and losses. While renting combine harvesters from the DAA could save time and labour, they could also become stuck in muddy conditions, especially during the rainy season. Additionally, many machine operators are unwilling to work outside their usual schedule, leading to longer waiting time for farmers who came later.

(vi) Labour Issues

Most farmers at Wasan are middle-aged (45 and older) and have little family support. They therefore relied on foreign migrant labour, primarily from Indonesia. However, as there is no labor quota specific to paddy farming, many must turn to domestic or seasonal workers. Domestic labour costs about B$300 or USD 220–225 per month, while seasonal workers demand higher pay and refuse to start until their wage expectations are met. This situation creates financial strain for many farmers.

Wasan Revisited

A field study was conducted in Wasan from September to November 2020. Paddy farmers were selected using stratified random sampling. The aim of this follow-up study was to further investigate the current water management and cropping practice of farmers in the area through a rapid field survey to collect information on water issues, crop health, pest information and farm management. Soil pH and moisture were measured at various locations using a portable garden probe (convenience sampling). In addition, soil samples were collected for laboratory testing under the supervision of a PhD student investigating soil properties for agriculture.

The first survey, conducted towards the end of the growing season, revealed that the most acidic areas (pH < 5) were located in the southern part of Wasan, which was categorised as gley soil. In other locations, the pH range was suitable for cultivation. A few cultivation areas were categorised as slightly alkaline (pH 8). Samples taken (after tillage) for laboratory testing showed that 5 of the 6 soil samples had a pH of

5.3–6.3 (good). Sample 8A, taken in the acid sulphate area in the north-east corner, had a pH of 4.5. In a number of plots (e.g., samples 43F and 40C), the soil pH was acidic at this late stage of the growing season. The acidity appears to have developed as a result of agricultural practise, as the soil pH was normal on the second survey after tillage.

According to the DAA, the soil is predominantly acidic, as assessed by CISRO (Grealish et al., 2008). There appears to be soil and farm management problems and weak support from extension services, perhaps due to insufficient manpower and/or technical knowledge/experience. Low pH contributes to low yields as roots cannot take up essential nutrients in an acidic medium. However, the survey showed that some farms were able to achieve quite high yields (e.g., 10 MT/ha). In terms of water supply, it was found that the inadequate supply was largely due to the DAA's recommendation to flush the fields every few days due to the high acidity. As farmers can switch the pump on and off at will, those closest to the irrigation outlet used the most water, while fields further away suffered from inadequate supply. Field observation also revealed that the farms were at different stages of growth, indicating poor management.

Discussion: Food Security

The Food and Agriculture Organisation of the United Nations (FAO, 1996) defines food security as a state in which all people at all times have physical, social and economic access to sufficient, safe and nutritious food that meets their dietary needs and preferences for an active and healthy life. The four main aspects are: (i) availability, i.e., food through domestic production, imports or food aid; (ii) access (economic and physical) to sufficient food; (iii) utilisation, i.e., proper and effective use to ensure that the food supplied is safe and nutritious; and (iv) stability of the food security system so that it is robust and resilient to external shocks such as economic downturns and disasters. This is a pragmatic definition that helps authorities to ensure the availability of and access to safe and nutritious food for specific communities. However, from an academic perspective, it might be instructive to consider the different dimensions of the concept.

The 'safety' part of the term obviously refers to the easily accessible supply of food, which can be 'secured' by growing one's own food, operating a food production system, importing and storing. However, while the agencies responsible for distributing food consider it to be solid commodities such as rice, maize, sugarcane and cassava, as in the case of AFSIS, it is much more diverse in form and condition. For example, a variety of foods can be grown, raised or caught/gathered in the wild. A community with such capabilities, characteristic of remote tribes, would be food secure even if the state, local authority or market are unable to provide rice, maize and other staple foods. In addition, raw foods could be processed for later consumption, e.g., in fermented, dried or preserved form. These are foods in different states that can be used to produce new forms of food. In mechanised systems, a bumper crop of an arable crop could be processed into dry powder (flour), which extends its use value for a much longer period of time. It is also easier to store. With mechanised systems, even food waste (which is rejected because of its appearance but still contains considerable food value) can be processed into packaged food. If these are taken into account, they contribute significantly to food security.

The term 'food' itself appears differently to different people and individuals in different cultures. A hunter-gatherer, for example, may consume a wide variety of animals and plants, while a city dweller may consider only chicken legs, wings and breasts and other typical offerings in the modern environment as edible food. The latter would react with disgust when they see the former preparing and eating their food. In this chapter, we have focussed on rice because it is a staple food in Brunei and throughout the region. However, this is a recent development. The environment in Brunei is better suited to the cultivation of the rhumbia palm (Cycas revoluta) for the production of sago, from which sago flour ('ambulong') is obtained through traditional processing. Sago was a staple food in Borneo and an important food in the Sultanate of Brunei. Rice was brought to Brunei in the fifteenth century by Sultan Bolkiah following a visit to Java (Sharifuddin, 1973). Even then, the brackish, swampy environment of Brunei was not conducive to wet paddy cultivation. The paddy cultivated by the Kedayans, which is believed to have originated from Java (ibid.), probably evolved from varieties brought to Brunei and adapted to the local conditions. In any case, the tropical environment is known for its great biodiversity, i.e., a wide range of potential foods, be it plants, fruits, animals or mushrooms. Many land dwellers in the region enjoy insects and worms, which are an excellent food due to their nutritional content. From an ecosystem perspective, food is abundant in a tropical environment. From this perspective, urbanisation (deforestation) is therefore a major threat to food security.

Regional Situation

ASEAN has several documents and initiatives related to food security. The most current and comprehensive one is The ASEAN Integrated Food Security (AIFS) Framework and Strategic Plan of Action on Food Security in the ASEAN Region (SPA-FS) 2021–2025. This is the latest iteration of ASEAN's food security strategy, building upon previous versions. The AIFS Framework and SPA-FS 2021–2025 were adopted by ASEAN member states in 2020 and came into effect in 2021. This plan outlines ASEAN's approach to addressing food security challenges in the region for the five-year period from 2021 to 2025. The key components of this framework are:

1. Enhancing food security and nutrition
2. Increasing sustainable food production and productivity
3. Enhancing trade facilitation and market integration
4. Ensuring food stability and safety
5. Improving farmers' income, livelihoods, and welfare

6. Empowering smallholders for food and nutrition security

ASEAN has a number of related initiatives and documents that complement this main framework, such as: its Guidelines for Promoting Responsible Investment in Food, Agriculture and Forestry; the Regional Framework for Public–Private Partnerships for Technology Development in the Food, Agriculture and Forestry Sector; and a Roadmap for Strengthening the Role of Agricultural Co-operatives in Global Agricultural Value Chains 2018–2025.

ASEAN has launched its own programme specifically for rice supply in cooperation with China, Japan and South Korea. This ASEAN Plus Three Emergency Rice Reserve (APTERR) was established in 2011 and has been fully operational since 2013. The main objectives of APTERR are: (i) to provide food aid and strengthen food security in emergencies caused by natural or man-made disasters; (ii) to maintain price stability for rice in the region; and (iii) to improve farmers' income and welfare. To date, APTERR has established a stockpiling programme whereby member states contribute to a physical stockpile of rice for the region in case of emergency. Each member state has committed to setting aside a certain amount of rice for emergencies. Brunei, for example, has pledged 3,000 MT, while China, Japan and Korea have pledged 300, 250 and 150 thousand MT, respectively. APTERR has also set up a phased release system to manage the distribution of rice depending on the nature and scale of the emergency.

An ASEAN Food Security Information System (AFSIS) has been established to promote regional and national food security development. Table 5.2 shows the food security status of member states based on the data available in AFSIS. Although most of the data represent the mean for the years 2008 to 2021, a few extend to 2018 (for C. Rate). There were also gaps in the data set for Thailand (for consumption). The values for production, import, consumption, opening/closing stocks and supply–demand is given in 1,000 metric tonnes. The beginning stock refers to the amount of rice available at the beginning of the year, mainly from carryover stocks from the previous year, while the ending stock is the amount of rice remaining at the end of the year after adding new stocks from production and deducting the total amount consumed, exported or lost (e.g., damage). It gives an indication of the degree of food security in the event of an interruption in supply. The value in Table 5.2 is the average over 14 years of the beginning and ending stocks. This is an attempt to derive a more reliable estimate for a highly variable parameter. DCC rate refers to the daily per capita consumption rate, i.e., the amount of rice in grams (g) that each person consumes per day. It is calculated by dividing the consumption ($\times$ 1,000,000 to convert to grams (g)) by the population $\times$365 days. The difference between supply and demand is calculated by subtracting the consumption from the sum of production, imports, and the mean of opening stock and closing stock. Buffer refers to the number of days that the opening/closing stock would last. It is calculated by dividing the average opening/closing stock by consumption and multiplying 365 days.

It is clear that Indonesia is the leading rice producer. Together with Vietnam, Thailand, Myanmar and the Philippines, they are the rice producers in the region.

Table 5.2 ASEAN rice security, 2008–2021 mean values

Country	Production (1,000 MT)	Import (1,000 MT)	Consumption (1,000 MT)	DCC (g/day/ Capita)	Beginning ending (1,000 MT)	Supply -demand (1,000 MT)	Buffer (days)
Brunei	1.6	34.3	33.2	227	17	37	190
Cambodia	5,860.0	0.0	2,420.7	410	2,420	6,818	367
Indonesia	38,959.7	853.0	36,539.1	400	5,094	12,993	53
Lao PDR	2,200.8	20.6	1,666.6	696	349	678	80
Malaysia	2,116.3	822.0	2,542.9	235	425	768	68
Myanmar	18,098.7	0.0	13,823.3	743	5,081	11,944	145
Philippines	11,836.8	1,578.9	12,151.3	331	2,626	5,273	80
Singapore	N/A	362.8	273.6	141	N/A	106	N/A
Thailand	21,731.4	21.8	8,945.7	375	12,831	39,925	702
Vietnam	27,481.4	203.6	11,942.0	346	3,351	17,777	92

Source AFSIS Paddy Balance Sheet. https://aptfsis.org/statistics

However, Indonesia is also the leading rice consumer and its DCC rate is similarly high. Myanmar and the Lao People's Democratic Republic have excessive DCC rates. Unsurprisingly, DCC rates are high in all rice-producing member states. However, supply is sufficient to meet demand in all member states because they have healthy levels of ending and beginning stocks. Singapore is the exception with no details of beginning or ending stock. It relies entirely on imports, and exports 20% of the amount import on average. It therefore appears to be most vulnerable, as it does not have a stockpile reserved for rainy days. The Buffer column shows that Thailand is the most securely supplied with rice and can feed its population for 702 days (or 2 years). In general, all member states except Singapore seem ready to absorb a disruption in rice supply. Indonesia, the largest producer and consumer, is somewhat vulnerable, probably because of its large population.

Sustainability and Sustainable Food Security

Ironically, the current view of food security is somewhat rigid, which poses a threat to its sustainability. As mentioned earlier, deforestation is systematically reducing natural resources and life support systems. Urbanisation is displacing nature and diminishing the richness of food resources on land and water in the process. The creatures and plants that have adapted to the built environment are seen by many as pests and chemical products (pesticides) are being developed to kill them. Gardening in the built environment requires the use of industrial products that release pollutants into the air, soil and water, which are eventually absorbed into the natural environment. Modern agriculture, promoted as part of food security programmes to increase food production, destroys the ecosystem in the same way as described above, but on a larger scale and with greater intensity. As mechanised agriculture expands, its deleterious impacts increases in range and scale via systemic feedback processes. This threatens not only food security, but also people's health and quality of life. This is the ironic outcome of food security projects that are based on mechanised agriculture of a narrow range of food. Ecosystem-orientated agriculture is not new. Ewel (1986) provides a guide for the humid tropical environment, which is applicable to most SEA countries.

The AIFS framework aims to promote better nutrition in the food supply and sustainable production. These cannot be achieved by conventional means, as mechanistic industrial approaches are the main cause of environmental problems and the reason for the emergence of the sustainability movement. Our common future (WCED, 1987) depends on balanced developments in the interrelated areas of economic growth, environmental protection and social development to create a more equitable society. The last area is addressed in the AIFS framework for empowering smallholder farmers and improving farmers' incomes. The hallmark of sustainability is interconnectedness, i.e., an ecosystem perspective. Therefore, the AIFS framework, like any food security programme at local and regional level, must not be implemented in isolated piecemeal manner, but considered and implemented as an integrated whole. A coherent mosaic of home-based permaculture farms would provide a wide range of food, maintain high biodiversity and natural ecosystem functions, and resemble a natural area, even if it is part of the built environment. This is a more ecological alternative to achieve food security. Salleh et al. (2018) present an interesting approach that combines permaculture with local knowledge in land use planning, which would be very appropriate given the strong entrenchment of culture and traditions in Brunei.

Returning to the Brunei case, the limited success in rice production and the low contribution of agriculture to GDP does not necessarily mean that the population is vulnerable to shocks in the food supply chain. There is plenty of food in the area. Sweet potatoes, tapioca (cassava), maize and a variety of fruits, vegetables and other food crops are/could be grown in gardens and on farms. They are simply not considered in studies and discussions on food security. Although rice has a much higher energy content than other food sources, there is a wide range of potential food available in the tropical environment, which has nourished people and ecosystems since historic times. Currently, many of these sources are being neglected as modern society becomes increasingly disconnected from the natural world and lives almost entirely in a constructed and increasingly artificial reality, sustained by manufactured goods. This includes food, which we only see in its fully processed, ready-to-eat state. Unwanted parts of the food are simply thrown away, with no regard for the living plants and animals that were used to produce it. In the case of packaged food, even the workers in the industrial plants or in the fields are quite disconnected from the fruits, vegetables, animals, etc., because they are harvested en masse by monstrous machines capable of processing large quantities, even on site. There is a need for more research and investment in agroecology (Andres & Bhullar, 2016) for a more sustainable approach to food security and sustainable development.

Conclusion

The chapter examined the issue of food security using the example of rice cultivation in Brunei. At first glance, Brunei appears to be secure as it has a small population and the means to import food and develop domestic food production. However, although it has the same humid tropical environment as the region, it is a coastal state with small drainage basins, which means it is not suitable for wet paddy cultivation. The soil is infertile, soil erosion is severe and there are many pests. Although Brunei has achieved some success with its efforts to develop modern agriculture, it still falls short of its goals. However, the country has taken the necessary steps to stockpile imported rice to ensure a six-month supply. A broader look at food security revealed that there is plenty of food in the environment. In a world increasingly built and sustained by artificial systems, society tends to view food in a ready-to-consume form. Authorities focus on ensuring that all members of society are fed, rather than considering their food needs according to culture and environment. As modern means of food production are inextricably linked to environmental impacts, including climate change, it is necessary to seriously consider agroecology and integrate permaculture into agriculture and also into urban development.

Acknowledgements The author is grateful for the continued support of DAA in her on-going research on agriculture in Brunei.

References

Andres, C., & Bhullar, G.S. (2016). Sustainable intensification of tropical agro-ecosystems: Need and potentials. *Frontiers in Environmental Science, Section: Agroecology, 4*. https://doi.org/10.3389/fenvs.2016.00005

DAA (Department of Agriculture and Agrifood). (2024). *Brunei Darussalam agriculture and agrifoods statistics 2023*. Ministry of Primary resources and Tourism, Government of Brunei.

Department of Agriculture. (2009a). *Towards self-sufficiency in paddy production Negara Brunei Darussalam*. Department of Agriculture.

Department of Agriculture. (2009b). *Strategic plan: Medium term implementation programmes (2008–2013)*. Ministry of Industry and Primary Resources, Brunei: Government Printing

DEPS (Department of Economic Panning and Statistics). (2023). *Brunei Darussalam key indicators*. Ministry of Finance and Economy, Government of Brunei.

Ewel, J. J. (1986). Designing agricultural ecosystems for the humid tropics. *Annual Review of Ecolological Systems, 17*, 245–71. https://people.clas.ufl.edu/jackewel/files/ewel-ann-rev-ecol-syst-1986.pdf

FAO. (1996). *World food summit: Rome declaration on world food security and plan of action*. Food and Agriculture Organization of the United Nations.

Grealish, G., Fitzpatrick, R., Ringrose-Voase, A., & Hicks, W. (2008). Brunei: Summary of acid sulfate soils. In R. Fitzpatrick & P. Shand (Eds.), *Inland acid sulfate soil systems across Australia* (pp. 301–309). CRC LEME Open File Report No. 249. (Thematic Volume) CRC LEME, Perth, Australia.

Hajah Aidah binti Haji Hanifah. (2011, April). The 25th ASEAN ASWGL. *Paper presented at the department of agriculture and agribusiness, Ministry of Primary Resources and Tourism, Brunei.* Retrieved from http://www.pelitabrunei.gov.bn

Hunting technical services. (1969). *Brunei land capability survey* (Vol. 3). Bandar Seri Begawan.

International Rice Research Institute. (1999). *The PalayCheck system: An integrated approach to rice pest management in the Philippines.* IRRI.

Integrated Environmental Consultants. (2011). *Tutong district plan.* Department of town and country planning, ministry of development, government of Brunei.

Kamarulzaman, N. (2017). *Economic diversification in Brunei Darussalam* (206 pp.). Uni of Sussex.

NDP (National Development Plans)

Salleh, A. M., Rosli, F. M., Esa, N., & Ibrahim, M. K. (2018). Permaculture design: Linking local knowledge in land use. *SHS Web of Conferences, 45*, 03003. https://doi.org/10.1051/shsconf/20184503003

Planning for House Compound

Sandal, T. S. (ed.) (1996). *The geology and hydrocarbon resources of Negara Brunei Darussalam.* Brunei Shell Petroleum Company & Museums Brunei.

Sharifuddin, P. M. (1973). Museum development in Brunei, Borneo. *Brunei Museum Journal, 3*(1), 51–61.

Tham, A., & Chin, W. L. (2024). Food, glorious food! The intersection of food cultures and creative tourism in Brunei. *Southeast Asia: A Multidisciplinary Journal, 24*(2), 121–133. https://doi.org/10.1108/SEAMJ-10-2023-0076

WCED (World Commission of Environment and Development). (1987). *Our common future.* Oxford University Press.

Yong, G. Y. V. (2010). Brunei (Negara Brunei Darussalam). In E. Bird (Ed.), *Encyclopedia of the world's coastal landforms* (pp. 1142–1143). Springer.

Yong, G. Y. V., & Hassan, N. H. (2022). Historical geography of the Limau Manis archaeological site, Brunei Darussalam. *Brunei Museum Journal, 2022*, 53–72.

Yong, G. Y. V., Azalie, I. A., & Hassan, N. H. (2023). Riverine lifelines of the Brunei Sultanate. *Paper presented at the Fluid Phenomena: Re-engaging Rivers in the Anthropocene, Universiti Brunei Darussalam, Brunei, 27–28 February.*

Norzurianie Hj Kamarulzaman is a lecturer in Geography, Environment and Development Studies Programme, FASS at Universiti Brunei Darussalam. A human geographer concerned with the study of people and their communities, cultures, economies, and interactions with the environment by studying their relations with and across space and place. Her current research focuses on agriculture in Brunei and social and economic issues in Brunei.

Chapter 6
ASEAN's Halal Production Networks: Towards Regional Integration and Effective Industrial Governing: A Brunei Case Study

Izni Azrein Noor Azalie

Abstract The demand for halal products should have taken a larger market share due to wider acceptance, expansion of the halal brand to include broader sustainability and health safety concerns, and a growing population. How would the halal network, including its supply chains, connect the SEA region and the world? This chapter highlights the importance of taking a regional and global approach to halal production by strategising how to overcome the barriers to industrial production that are still dictated by national borders, protectionism and other non-tariff barriers such as differing halal standards. For smaller economies such as Brunei Darussalam, maintaining the sustainability of the halal food industry requires a number of measures that go beyond the conventional halal hub models that mainly examine the national agglomeration of halal production activities. Incorporating the theory of production networks, both regional and global, this chapter has utilised the case studies of halal food producers in Brunei to demonstrate the convergence of local–regional-global networks as an alternative pathway for the development of the halal industry in the region, as ASEAN celebrates the tenth anniversary of the establishment of the ASEAN Economic Community (AEC) in 2025.

Keywords Regional integration · Industrial reorganisation · Multi-scalar · Halal food networks · Sustainability

Introduction

Historically, the Sultanate of Brunei, which included the coastal areas of present-day East Malaysian states of Sabah and Sarawak to the present-day territories of the central and southern Philippines (Luzon, Visayas and Mindanao) during its golden age in the fifteenth century. With the arrival of the Spanish in the sixteenth century

I. A. N. Azalie (✉)
Faculty of Arts and Social Science, Universiti Brunei Darussalam, Gadong, Brunei Darussalam
e-mail: azrein.azalie@ubd.edu.bn

G. Y. V. Yong and A.K.M. A. Ullah (eds.), *Brunei and the Future Transformation of Southeast Asia*, https://doi.org/10.1007/978-981-96-7901-0_6

and the British in the nineteenth century, the sultanate lost the Philippines to Spain, Sabah to British capitalists and Sarawak to Brooke's White Rajah dynasty. The loss of the Baram and Limbang basins to Sarawak in 1882 and 1890, respectively, dealt a blow to the sultanate, as they were major sources of food and granary. It began the dependence on foreign supply of food. This reliance, coupled with the increasing need for sustainable and culturally aligned food options, prompted a focus on leveraging Brunei's Islamic identity to establish a halal food industry. Brunei Darussalam's path to developing its halal food sector was never smooth. All new initiatives that were to be launched and implemented were often met with new problems and the re-emergence of old challenges. The reasons for this lie in the small geographical and demographic size of the country and the deep-rooted rentier economy that has sustained itself since the discovery of oil in 1929.

The wealth generated by the oil and gas industry created a consumer society that transformed communities, which had relied on subsistence farming for food. Faced with the influx of food into the domestic market, the Brunei government, on the orders of His Majesty the Sultan, established a control and monitoring agency, the Halal Food Control Division (HFCD), to liaise with the Royal Customs and Excise Department on issues relating to the importation of halal meat and retail trade. The establishment of this agency was a result of the Brunei Halal Meat Act 1998, which was enhanced by the Brunei Halal Certification and Halal Label Order (BHCHLO) in 2005. The process to halalise the food sector occurred in two stages. The first stage in 2011 involved the need for halal certification and approval for food businesses as part of the renewal of their Miscellaneous or Business Licence. The second, in 2017, all businesses, formal and informal, must have halal certification and authorisation in order to continue operating in the country. Despite numerous studies, the rationale behind the development of halal in Brunei, particularly in relation to regional production linkages, has never been articulated. The royal decree issued by His Majesty the Sultan during his visit to Malaysia in 2005, and its follow-up in 2016 on the occasion of Nuzul Al-Quran celebrations, show that a forward-looking perspective needs to be adopted. The 2005 decree contained elements calling for Brunei and Malaysia to work together to develop sustainably and increase economic competitiveness, as both have the advantage and speciality of developing as centres for Islamic banking and the halal food industry. The 2016 decree reaffirmed the development potentials of the halal industry and questioned the current halal nature of the global food industry. Due to its expansive nature, the Sultan emphasised the need to seize the opportunities presented to fulfil one's role as a Muslim.

Brunei Certification Guidelines (BCG) 1–4 focussed on different aspects, ranging from general certification requirements, internal auditing and surveillance, post-surveillance requirements, as well as appointment of auditors and inspectors and the required code of conduct during checks and audits (refer to Azalie (2023, p. 178) for details). Apart from technical guidelines, the centrality of the State's religious institutions is crucial in streamlining halal development in the country. The Brunei State Mufti's Office (2007) has published religious fatwas (rulings) on emerging contemporary issues relating to halal food production and consumption. The Ministry of Religious Affairs (MORA), through the HFCD, is responsible for governing and regulating halal industrial activities, as well as the socialisation and certification processes.

Halal Certification and Industrial Development in Southeast Asia

There has been active development of the halal industry in ASEAN over the past 30 years. However, individual ASEAN Member States (AMS) and their industries and companies have implemented different strategies. The signing of the ASEAN Free Trade Agreement (AFTA) in 1992 offered new regional development opportunities. The global halal industry, which excludes Islamic finance, is estimated to be worth USD 2.3 trillion per year, with the halal food industry taking up 36% of the total halal market according to a 2016 estimate (Azam & Abdullah, 2020). The global halal industry is projected to reach USD 3.2 trillion by 2024 (Dinar Standard, 2019, cited in Ab Talib, 2020). The development of the halal industry in ASEAN is highly competitive, with Malaysia leading the way. In 2022, the Indonesian government declared its vision to further develop its halal food industry with the relocation of its capital to Kalimantan. Vietnam organised its first halal conference in 2024 to raise awareness, enrich knowledge through knowledge transfer and share industrial insights with scholars inside and outside Southeast Asia. Meanwhile, the Chinese investment in ASEAN is catalysing new industrial developments in the region in line with the ASEAN Economic Blueprint (Bi, 2021).

Similarly, the Regional Comprehensive Economic Partnership (RCEP) in 2022 and sub-regional growth triangles of BIMPEAGA and the Greater Mekong Subregion (GMS) offer potential for growth despite ongoing challenges in terms of financing and connectivity issues (Bi, 2021). As for the halal industry, despite efforts towards regionalism, protectionist issues persist. The current situation could be characterised as a collection of national systems with limited focus on effectively integrating production-related activities across borders to enhance regional productivity and food safety. Instead, current halal production activities still take place in silos and follow the competitive rules akin to a zero-sum game, which is unsustainable, unfair and unjust to smaller players in the halal network. In the case of Brunei, deeper integration with the regional network is essential to ensure sustainability. Therefore, it is imperative to understand Brunei's status within the regional networks in order to address challenges and shortcomings in relation to the network linkages between businesses and the role of state as enabler and facilitator.

A Shift from Global Value Chains (GVC), Global Production Networks (GPN) to Regional-Localised Networks of Production?

The globalisation process since the end of the twentieth century has been characterised by the relocation of industries, decline of former world production centres in the West (Storper et al., 2015), and the emergence of new economic regions in Asia, led by China (Li, 2018), which created new growth potentials that are increasingly founded on sustainability. Wallerstein's World Systems Theory (1974) has undoubtedly transformed current approaches to commodity production. With the emergence of Global Value Chains (GVCs) and Global Production Networks (GPNs) in discourse, current debates focussed on the importance of understanding similarities and differences in the radix scientiae (scientific roots). This gained prominence in the aftermath of the 2008 global financial crisis and the post-COVID-19 changed the functioning of GVC and GPN as right-wing populism and nationalism came to the fore (Hess, 2021).

Gereffi's (2001) seminal work on GVC increased the complexity of the already multi-layered, multi-faceted aspects of contemporary industrial production by going beyond simply tracing the origin of traded goods and analysing where value-creating activities take place in a value chain that has become global. Subsequent work by Gereffi et al. (2005), Nadvi (2008) and Ponte and Gibbon (2005), to name but a few, have emphasised the existence of an invisible governance or hierarchical structure that determines how firms operate within GVC activities. Mudambi (2008) also pointed out that such governance structures lead to geographically uneven development, especially for those mainly engaged in lower value-added activities. Campling and Quentin (2021) postulated that current GVC has further exacerbated uneven development when combined with Global Wealth Chain (GWC). The levels of wealth creation, from high to medium to low, refers to both the tangible, such as income, technology and assets, and the intangible, such as knowledge, skills and influence along these chains. GVCs have led to the reproduction of inequalities worldwide across the global inequality chain, regardless of whether they are located in the West or in emerging economies (ibid).

The post-COVID-19 world is characterised by a different dynamic within GVCs, of onshoring or nearshoring activities carried out mainly by small and medium-sized enterprises to mitigate the inherent systemic risk they have experienced during the pandemic (Canello et al., 2022). Two years after the reopening of national borders, the region has resumed its pre-COVID-19 economic boom aided by the rise of China. ASEAN is expected to become the fourth largest regional economic bloc by 2030. This monumental rise was further cemented by the accession of four AMS (Malaysia, Indonesia, Thailand and Vietnam) to BRICS in its Kazan conference in October 2024. This will undoubtedly impact the future regional and global value chains, which has implications on smaller economies, such as Brunei and Timor Leste, as economies continue to integrate. The GVC theory is useful in this study as it allows for multilinear tracking of the activities of Bruneian producers linking their production chain to regional and global networks. However, a simplified treatment could also contribute to shortcomings if it fails to account for the effects of biopower or biopolitics in the functioning of GVCs. Therefore, GPN theory is used to complement the gaps in the GVC theory to analyse the status of halal production activities in relation to their global regional partners in Brunei.

In contrast to the GVC, the GPN theory (Henderson et al., 2002) deals with the deeper interconnectedness in the local, regional and global networks, and postulates the role of other non-economic actors and institutions in their operationalisation and functioning. GPN 1.0 has been instrumental in changing the current approach to industrial production, particularly because it incorporates spatial realities. Coe et al. (2008) refined the theory by defining GPN as the "nexus of interrelated functions, operations and transactions through which a particular product or service is produced, distributed and consumed" (p. 274). Yeung (2009, 2018) emphasised the central importance of understanding and analysing intra- and inter-firm linkages, coordination, partnerships, collaborations and control through coupling and decoupling mechanisms as crucial features of the theory. In his study of the geographical shifts of semiconductor production to East Asia, Yeung (2022) examined how spatial proximity, changing market dynamics, customer proximity in both intermediate and final markets, and consumer demand, helped shape the types of linkage strategies and coping mechanisms that firms adopted during production processes. While Frohm and Gunnella (2021) considered the existence of key suppliers in global nodes as crucial determinants for the emergence of spillover effects of GPN activities, the author was of the view that such overarching universalism was not applicable to small economies, such as Brunei. There has also been a clear resurgence of Marshallian industrial districts or manufacturing clusters within the current GPN, for example, in the Chinese industrial districts of Dongzhen and Wenzhou compared to the Italian industrial districts of Brenta and Manzano, which produce consumer goods such as electronics (Dongzhen), footwear (Wenzhou), luxury and fashion shoes (Brenta) and chairs (Manzano) (Bretznitz et al., 2024). Another study by Kuroiwa et al. (2024) also showed the clear shifts in production networks towards more localised development in their analysis of the Thai automotive industry.

Despite this development, Yeung's (2021, 2024) critique of GPN provided fertile ground for more in-depth studies of the inner workings of the GPN. In his view, the theory was never about making assertions about how all industrial transactions and linkages must work. Instead, it should be treated as a black box in which processes and outcomes derived from the application of such theory in an applied study are dependent on the circumstances in which they occur (Yeung, 2021). Any GPN phenomenon observed is bounded by its spatial–temporal context, including events such as politics, changes in regulation, and pandemics, to name a few. This is especially true based in Arias-Loyola's (2024) study, where both generic and specialised firms often use their bargaining power to negotiate with the more prominent actors or leading firms within three-dimensional environments, which may be (i) episodic, i.e. times when bargaining is not possible but must be conducted, (ii) non-decisive, where firms control the political agenda, and (iii) ideological, a period when firms view bargaining as natural or inherent. Yeung (2021) also argued that despite the re-emergence of the state as a critical actor in industrial governance, their functions are not new as GPNs have always operated or been embedded in the state system. Their functionalities are also coming into focus due to the success of China's state-led socialist industrialisation, which is moving closer to capitalism without alienating or compromising the well-being of small economic actors and communities, as part of a global civilisation initiative declared.

Hence, this chapter used the GPN approach to identify the strategies used by Brunei's halal food producers in producing their halal products, their linkages and their firm's role in these network linkages. Apart from tracing the material supplies, the theory is also used to address potential issues surrounding the dependencies of such firms over materials outsourced from beyond the borders while identifying the industrial capacity status or condition relating to the general standard requirements. Towards the end of the chapter, functionality of the State towards the operationalisation of the GPN will also be discussed to understand the close relationship between the industrial players within the GPN and the State.

Study Data Collection and Methods

This study used a case study approach in which fifteen (15) halal food companies were surveyed and interviewed. The data sets used were collected between 2016 and 2023. Where there were data gaps or omissions, the researcher utilised information from secondary sources such as newspaper reports from physical and online publications to produce a near-complete picture of the current development and status of the actors. Where information on the digital platforms is accessible to third parties, the direct names are used. However, where there is a need to protect the competitiveness and confidentiality of the companies, pseudo names or coded short names are used for these companies. Therefore, mixed methods are used in this study.

Brunei Halal Brand and Certification

The Brunei Halal brand was a project initiated by the Ministry of Industry and Primary Resources (now the Ministry of Primary Resources and Tourism) in collaboration with the Islamic Religious Council of Brunei, the Ministry of Religious Affairs, and the Ministry of Health. The establishment of the Brunei Halal Brand is an outcome of the country's commitment to develop a diversified, competitive and sustainable economy. In July 2009, His Majesty the Sultan and Yang Di-Pertuan of Brunei Darussalam unveiled the Brunei Halal brand during the fourth International Halal Products Expo (IHPE) (Masli, 2009). A new state-owned company, Brunei Wafirah Holdings, was established as the sole owner of the Brunei Halal brand to provide halal food and beverages worldwide and to facilitate access to the international market for Brunei halal producers.

When it was first launched in 2010, the Brunei Halal range comprised 50 products, mainly ready meals (Temporal, 2011). By 2013, the Brunei Halal brand had introduced 23 new products. Currently, the Brunei Halal range, which is labelled with a green logo, is manufactured in various countries including Australia, China, Spain, Malaysia and India for Original Equipment Merchandise (OEM) products, as well as by some local companies in Brunei, such as Sahamada Corporation and Marimin Enterprise. In addition to strict religious accreditation, all manufacturers must meet food safety standards, including Hazard Analysis and Critical Control Points (HACCP) certification and International Standards Organisation (ISO) certification. Currently, most local manufacturers do not have these certifications, with a few exceptions, including large companies such as Sabli Food Industries, PDS Food Group, Brunei Meat Company, Hana Soy and World, Sahamada Corporation, Superwater Marketing, McBrumby, Simpor Pharma and Marimin Enterprise.

Brunei Halal is keen to partner with local SMEs to help them achieve their export goals by advising them on product development, market potential, packaging design and halal and commercial accreditation requirements. In terms of distribution, Brunei halal products are sold in Singapore through NTUC Fairprice, while in the United Kingdom (UK), they are available in major retail outlets and supermarkets (Shahminan, 2015). Wafirah has established an office, warehouse and distribution centre at Saltley Business Park in Birmingham, UK, which has been renamed Ghanim International. The company is responsible for the sale of halal products from Brunei. It also operates an abattoir in partnership with Blakemore Wholesale UK to supply the UK market. This initiative aims to supply halal meat and poultry products to retailers in the UK who are looking for non-stunned halal options from a recognised brand, such as Brunei Halal (Brunei Halal Press, 2015).

BIRC/MUIB accredited companies or firms can use the Brunei Halal purple logo on their products. In 2017, ASEAN countries such as Brunei, Malaysia, Indonesia and Singapore (MABIMS) issued a joint declaration to harmonise their respective halal standards to strengthen the regional halal industry (MABIMS, 2017). In recent years, Brunei's halal accreditation procedures have been in demand internationally, and countries, such as Russia, the Philippines and South Korea have requested assistance in developing their halal standards. It should be noted that Brunei's halal certification is not exclusive to the meat industry. It also includes the food and beverage processing, pharmaceutical and cosmetic industries, as well as logistics. The concept will be extended to the tourism, Islamic banking and finance sectors. As for the Brunei Halal brand, the certification process includes a halal compliance audit by MORA. Once the audit is passed, manufacturers must undergo a second audit, which is conducted by Ghanim International through international audit specialist STR and covers areas such as health and safety, environmental and social (Temporal, 2011).

Brunei Halal Agrifood Sector: A Localised Development?

The agri-food industry comprises sectors involved in the production of food and beverages. Within the halal agri-food industry, there are currently 64 halal-accredited domestic companies (as of 2014) (HFCD, n.d.), an increase from 46 companies in 2011 (Al-Haadi, 2011). These companies range from meat processors and canned food manufacturers to bakeries and dry food manufacturers. There is still no historical record of when the halal agri-food industry, especially animal processing, started. This could be related to the informal nature of the sector. According to the official report from the Ministry of Agriculture and Food in Brunei, the processing industry only started in 2005. There were 106 food processors in 2006, of which 24 were involved in livestock processing, 77 in crop processing and 5 in fisheries (Anjah, 2007). In the 2010 Economic Census Report on Small and Medium Enterprises, 234 enterprises are engaged in crop, animal and related services, 66 in food and product manufacturing and 5 in beverage manufacturing (Department of Economic Planning and Development, 2011). Presently, there are 344 entrepreneurs operating in agri-food processing, with an estimated total number of 4,541 employees, of which 2,744 are locals and 1,797 are foreigners (Department of Agriculture and Agri-Food, 2024).

Figure 6.1 shows the locations of businesses across in (a) the Brunei-Muara and (b) Tutong districts. In 2014, 634 restaurants were issued halal certificates (Halal Food Control Division, 2014; Azalie, 2016). It should be noted that the approval of halal certification is undertaken by the Brunei Islamic Religious Council (BIRC, or in Malay, MUIB), a stand-alone entity, and administered by HFCD under MORA. Each halal certificate or label is only valid for three years (BIRC, n.d.).

Currently, there is no data on the value of the halal and food industry in Brunei. However, the information on Brunei's food and beverage industry can serve as a guide as technically almost all food and beverages produced in Brunei are halal. Based on the latest statistics from the MPRT, the development of the agri-food industry in Brunei shows significant growth between 2014 and 2023, slowing down between in 2021–2023. Figure 6.2 shows that the local agri-food industry was worth BND 170.18 (about USD 127) million in 2023 and the total contribution of the agriculture sector to the national gross domestic product (GDP) was less than 1% of the total GDP.

The rise of Brunei's agrifood industry is closely linked to the change in government policy on food security by venturing into both traditional and newer industrial fields. There has been a paucity of research on Brunei's food production, particularly on the multi-scalar linkages beyond the national boundaries. The GPN theory is used to analyse the pioneers of the halal food industry and evaluate their status in relation to sustainability and competitiveness. This will be discussed in the following findings and analysis subsections.

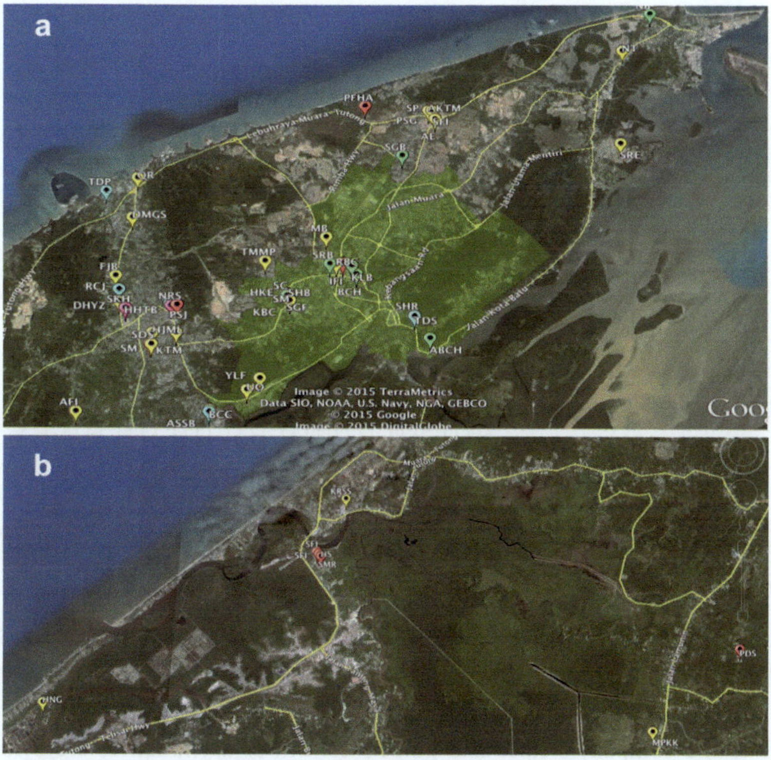

Fig. 6.1 Images showing the location of Halal food producers in **a** Brunei Muara and **b** Tutong districts

Global-Regional Production Networks of Brunei Food Producers: Analysis on the Halal Pioneers

Four of the thirteen halal pioneers interviewed will be used to illustrate the typologies of regional linkages or networks utilised in growth and expansion strategies. Previous studies (Azalie & Samad, 2022; Azalie & Kamarulzaman, 2023) examined regional networks of local coffee and tea producers, as well as of food processers, revealing an interesting pattern of development.

The results and analyses of the survey show four different strategies used by the companies concerned. Company 1, a Government-Linked Company (GLC), opted for outsourcing and offshoring its manufacturing activities to overtake incumbents by rebranding, repackaging and manufacturing OEM products. The GLC acted as the lead company responsible for most of the decision making, branding and marketing, capital raising and investment strategies, while the foreign partners engaged in manufacturing and other production-related activities. These companies usually consist of general suppliers that can be hired or fired depending on the success of the lead company (tying and decoupling). However, the partners who join their brand must

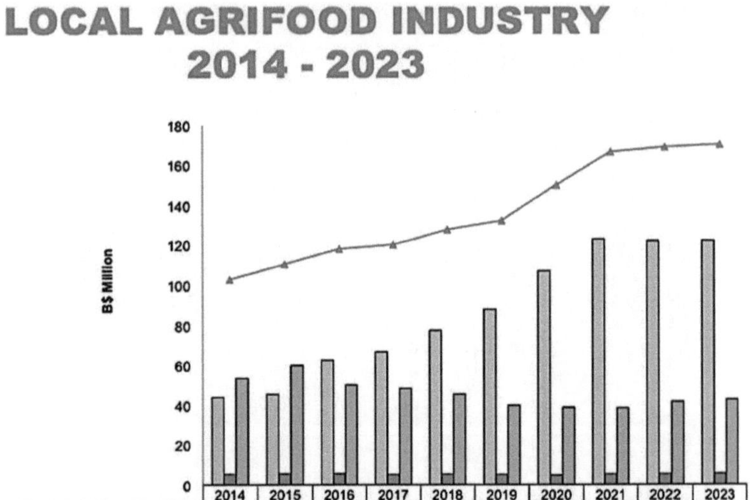

	2014	2015	2016	2017	2018	2019	2020	2021	2022	2023
Livestock Processed and Value Added Products	43.87	45.27	62.63	66.66	77.22	87.83	107.08	123.08	122.03	122.18
Crop Processed and Value Added Products	5.34	5.53	5.59	5.19	5.42	5.09	4.63	5.40	5.41	5.44
Others (Food and Drinks)	53.47	59.69	49.90	48.34	45.33	39.55	38.51	38.21	41.48	42.56
Total	102.68	110.48	118.11	120.19	127.98	132.47	150.23	166.69	168.92	170.18

Fig. 6.2 Growth trends of local agri-food industry between 2014 and 2023 based on retail value (Department of Agriculture & Agrifood, 2024)

adhere to the company's rules and guidelines. Power asymmetries are evident in the relationships between the companies, resulting in some international companies, particularly from China and Spain, as well as some domestic manufacturers, to decouple from the lead company due to the weak influence and reception of the GLC's brand name. For the GLC's offshore company, the plan to enter the European market was also halted in 2018 as it was uncompetitive, poorly managed and not innovative enough despite solid customer support in its host country. The company exited the market to avoid further losses. It then decided to build its production facility and reduce the scope of its current repackaging and rebranding activities, as the return on investment (ROI) from such activities is negligible and uptake of export-orientated activities was slow. Figure 6.3 shows the production linkages between the GLC (parent company) and its offshore company.

Company 2 developed from a small firm engaging metallurgical activity to an industrial-sized company. It was established in 2003 and accredited with Brunei Halal certification, Good Manufacturing Practises (GMP) and HACCP and had a production capacity of 20,000 bottles per day. The plant was operating below its capacity, producing 40,000 bottles per month in 2018. The company has however since tapped into the Singapore, Malaysia and Vietnam markets. In contrast to Company 1's strategy of branding and packaging, Company 2 mainly focused on outsourcing some of the raw materials for the production of chilli sauces to AMS (Thailand, Philippines and Malaysia). The company took this step due to uncertainties in local supply, which is affected by weather and price fluctuation of local chillies. The company also

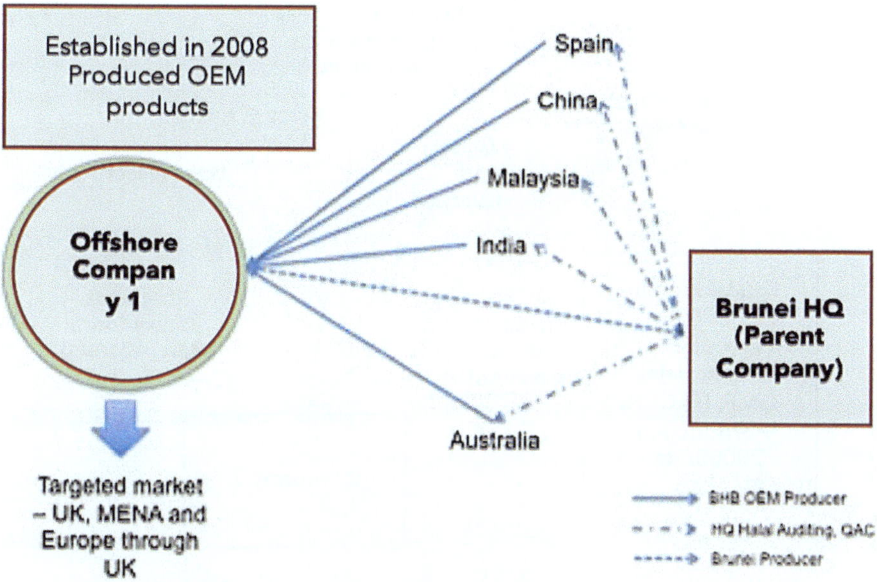

Fig. 6.3 Strategic coupling and rebranding: Offshoring and outsourcing activities of Company 1 (Government Linked Company)

took the initiative to conduct research on chilli varieties to differentiate their product from the others. This included 'right' flavour and the level of heat, measured using the Scoville scale, to cater to the East and Southeast Asian communities. They also utilised their family ethnic diaspora connections in Malaysia and the Philippines to build linkage with regional partners and leverage them for market entry. In this case, Company 2 established deeper connections with its specialised suppliers than the previously analysed Company 1, leveraging on personal connections (familial and kinship ties) in contrast to the 'hire and fire' strategy used by company 1 (Fig. 6.4).

In contrast to the two previous cases, Company 3 is an offshoot of a company with experience in the oil and gas and dairy sectors. It identified as a start-up and with seed funding from its parent company, established a joint venture with another start-up from Japan in 2014. Company 3 utilised its parent company's network in the dairy industry to develop a new venture. It utilised the existing network of distributors and retailers to promote its products, eliminating the need and long waiting period for agreement and contracts, allowing access to the market within a short time. In addition to market access and capital, the tacit and codifiable knowledge provided by its parent company, acquired through many years, played a significant role. As a result, they were able to obtain industrial accreditations: HACCP, GMP, ISO and Brunei Halal certification. Beside its Japanese partner, company 3 also invested in research and development (R&D) in South Korea. Its joint venture opened up export opportunities in Singapore and Hong Kong. Interestingly, Company 3 is linked to Company 1 and participated in the same market access abroad with products

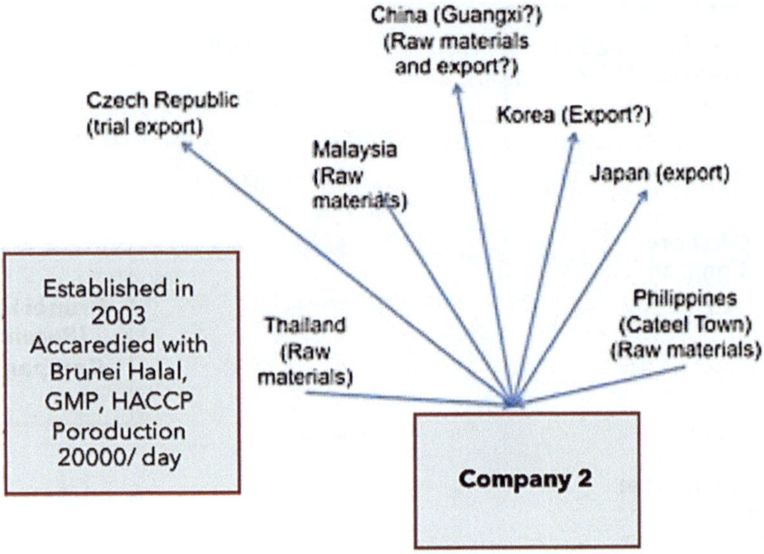

Fig. 6.4 Company 2 (transformed cottage-to-industrial business) outsourcing of raw materials and utilisation of B2B networks

of similar weight and typology, where one of its products has been renamed and repackaged under Company 1's flagship product. Network analysis (Fig. 6.5) shows that company 3 is the leading company in the production arrangement, as it made most of the decisions and has the capital. It is linked to specialised suppliers and companies responsible for high value-added activities such as research, branding and distribution. Its relationship with Company 1 is mainly about market access and B2B negotiations and can therefore be considered generalised. Company 3 dictates the policy direction using its partner to gain further leverage, but not necessarily in the medium or longer term. Company 3's rise was abruptly curtailed with the unexpected demise of its founder-CEO, after which it declined and eventually faded from the market between 2018–2019. Under the leadership of its CEO, it could be regarded as an exemplary case of a functioning regional production network.

Company 4 is an experienced company with 26 years of experience in the industry and follows a similar strategy to Company 2, but with a larger number of suppliers from abroad, mainly from China, India, Malaysia, Australia and Japan. However, unlike the other three companies, Company 4 has linkages with experienced players in the industry, especially those with higher industry standards such as GMP, HACCP, and some, with Halal certification from other recognised institutions. This association helped the company to obtain domestic certifications, such as the Piawaian Brunei Darussalam (PBD), also known as the Brunei Darussalam National Standards and the Brunei Halal Standard. This was made possible as its supply chain already possess the required quality standards from the beginning to the end. Due to the sheer size of the production (about 70,000 units per month for each product),

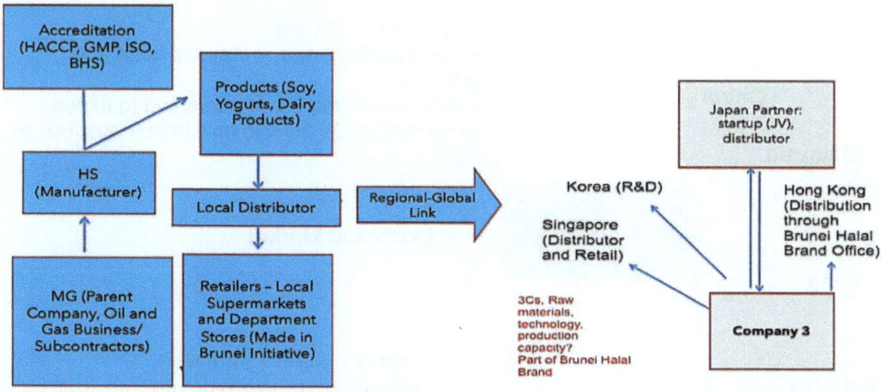

Fig. 6.5 Company 3 (Start-up) strategic coupling through Joint Ventures (JV) with promising foreign start-ups

Company 4 is one of the few Bruneian manufacturers that have managed to grow cumulatively to gain a foothold overseas. To date, the company has more than a hundred products on offer, ranging from canned food, frozen meat-based products and ready-to-eat products to bottled sauces. Company 4 has also expanded its capabilities through technology transfers from other partners in Malaysia, China and Japan in the areas of canning, printing, embossing and packaging. When data collection began in 2017, the company lamented the difficulties in exporting since the mid-2000s. However, it made a breakthrough and was able to export to the ASEAN market (Malaysia, Indonesia and Singapore) and found opportunities in East Asia and the Middle East. In the former, it secured production licences for Japanese kitchens and a major Japanese food manufacturer. Company 4 therefore has succeeded as a leading company where production activities and other forward industrial linkages such as marketing, branding, research and development are internalised, but has also developed networks with specialised suppliers, which enhanced its production capabilities (Fig. 6.6).

Industrial Standard Conformance: Experiences of the Bruneian Food Producers

Although Brunei was among the first to express the need for a global halal food brand in 2008, Brunei's ambition to become the leading player in global halal food production came to a halt when reality set in, especially among the pioneers of the domestic halal food sector. The fact that they are able to produce their own branded items (OBM) and increase their production capacity only partly explains the reality of being part of the regional or global halal food network (see, Azalie & Kamarulzaman, 2023; Sulaiman et al., 2023). Extensive industrial arrangements

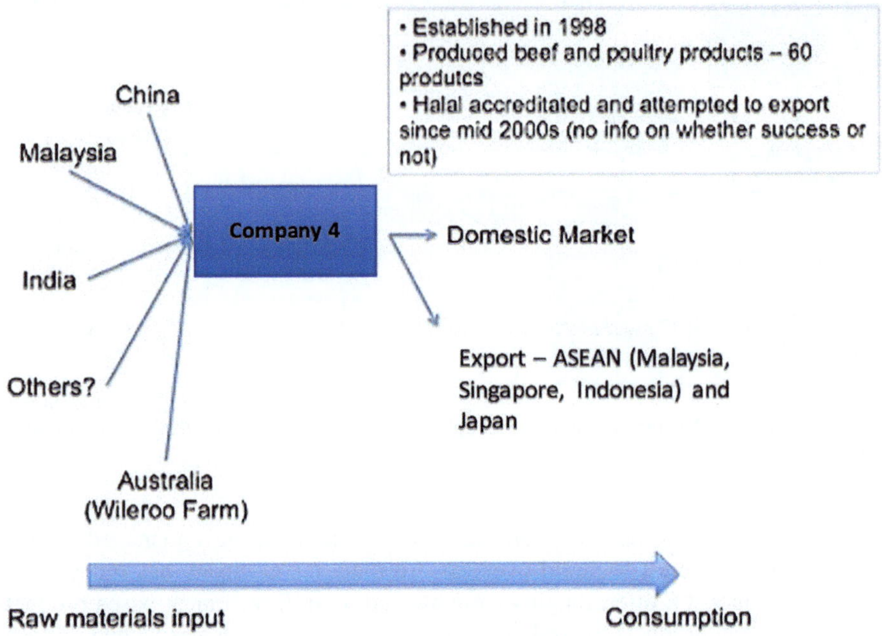

Fig. 6.6 Company 4 (seasoned) producer coupling with established foreign players

and procedures are required to participate in external markets. Only five of the 15 companies surveyed have managed to achieve this, including the four companies analysed above. Additionally, there is the need to fulfil the minimum requirements for industrial compliance.

Two companies, coded as IDFD and SMC, are used to illustrate this challenge: HACCP for IDFD and ISO 22000 for SMC, as part of their halalisation strategies and as required by BCG-1 at the time of data collection (2016–2019). Both companies argued that the government agency convinced them to undertake these international standardisations as part of the halalisation programme after receiving support from the then MIPR and Company 1 to expand their products domestically and abroad. According to the two companies, the industry standards were required for overseas expansion (under Company 1). While waiting for government subsidies to be granted, these companies must commit to paying a certain amount upfront to acquire certification for these standards, including application cost, renovating the premises, reorganising the entire production system and adapting process flows to comply with the required criteria. However, the subsidies never came, and according to IDFD, the ISO certification programme supported by the government was withdrawn, leaving the companies with substantial losses. IDFD waited at least 4 ½ years, during which the company had spent over B$ 1 million (about USD 760,000) on facility renovations, new equipment, and internal and external audits. SMC lost around B$ 80–90,000 (USD 60–68 thousand). This does not include the operating costs incurred in the certification process. As of 2021, the subsidies never came, and neither company has received any of these certifications. These examples thwarted the very principle of Islamic ethics in governance, particularly in relation to the wealth creation necessary for the achievement of communal posterity through fair and just action promoted in Islamic economy.

As halal certification and approval in Brunei involves the need to either follow the submission of other recognised industry standards or apply for such standards as part of the halal requirements (see item 3.1.2.(o) of BCG-1), this indirectly increases the existing overheads or costs once the certificates or approvals are renewed. The survey of the 13 formal companies revealed that the cost of renewal varies depending on the number of products for which such certifications are required and the geographical location where the renewal was undertaken. The two tables below show the location of renewals and the overall average cost that companies spent on renewals.

Tables 6.1 and 6.2 show that in addition to the existing halal standards, local producers must also comply with other existing national standards as part of their annual renewal. Apart from the outsourcing of raw materials from global-regional players, the results of the survey of the 15 companies also revealed the need to outsource some of the advanced laboratory and trade-related services beyond Brunei. It can be argued that high-end activities related to the future, such as branding and export product registration, are located in some offshore partner agencies, creating a different kind of dependency based on advanced services. Most renewals were on a three-year cycle. Other compliance requirements relating to health and safety, hygiene and sanitation requirements in the business premises or production facilities were met locally either monthly or annually. The application of GS1 barcode (7 digits), one-time basic nutritional analysis of food and Brunei Halal certification are all done on site. Apart from these, other advanced industrial requirements need to be applied overseas. For example, the advanced nutritional analysis is usually done in Malaysia, and the 10-digit GS1 barcode, which is mainly used for import and export purposes, in Vienna, Austria, as in the case of AST and NORB. For higher-level industry standards such as ISO 22000, HACCP and GMP, renewals have been carried out every three years, and the success of the renewals depends on site audits carried out at various stages of the applications. The local partner for these important industry standards manages the standards applications. The fees charged by companies are slightly higher compared to other parts of the world, perhaps because this company acts as an intermediary. The table shows that the costs also vary depending on the industry standard. The 'miscellaneous' category includes certification for pest control, fire hazards and safety, as well as food safety and quality control testing conducted on-site, as well as travel cost of exerts involved in the process, where their physical presence are required. Costs range from a minimum of BND 3,000 to a maximum of BND 10,000 (approximately between USD 2,268 and USD 7,560 annually). For micro and small producers, numerous standards, such as those required under Brunei's halal certification, mean that their businesses have to enter into more costly operational agreements, which affects their competitiveness.

Table 6.1 Number of firms renewing their industrial certification locally (under government scheme) in contrast to foreign (privately)

Location	(i) Brunei Halal	(ii) HACCP	(iii) ISO	(iv) GMP	(v) GAP	(vi) ASEAN-GAP	OTH
Local government	13	1	0	0	0	0	10
Foreign-private	0	2	1	1	0	0	2

Table 6.2 Detailed breakdown of certification type based on the number of firms, the cost of each certification as well as its frequency of renewals (Prices as of 2019)

Certificate type	No. Of Firms	Renewal cost (min)	Renewal cost (max)	Frequency of renewals (years)
Brunei Halal	13	<BND1,000	>BND2,000	3
HACCP	3	BND1,000	BND5,000	3
ISO	1	BND2,000–5000	BND15,000	3
GMP	1	BND1,000	BND5,000	3
Basic food analysis (Local)	6	BND30/raw material		One off
Food analysis (Foreign)	2	BND600/raw material		1
GS1 Barcode	8	BND70/product	BND210/product	1
Miscellaneous	1	BND3,000	BND10,000	1

As illustrated in the above cases, other international standards are also required in the Brunei's halal certification programme. The HFCD and the Brunei National Standard Committee (BNSC) claimed that incorporating them with halal requirements, such as the halalan tayyibban principles, would strengthen the brand and prestige of halal products, thus enabling them to better capture the market. The research carried out however found numerous halal-certified products from Brunei unable to gain foothold in overseas markets. However more studies are needed to shed light on the impact of Brunei's halal standard on producers.

Recent studies by Sulaiman and Hashim (2022) and Ab Talib (2021) on the impact of COVID-19 on the supply chains of food producers in Brunei have revealed that Bruneian producers are susceptible or vulnerable to supply uncertainties in the event of significant disruptions. According to Kopytov et al. (2024), in a study of relationships between firms in production networks with high vulnerability to supply insecurity, a positive relationship was found between supply disruptions and decoupling of production links. They also found a positive relationship in the decision to connect with all available suppliers, despite the potential increase in production costs due to the higher price of supplies or materials in the event of major disruptions. Of the 15 Bruneian halal producers interviewed, ten stated that they would have to relocate or reorganise their production in the event of an unprecedented disruption. Figures 6.7 and 6.8 show the details of coupling from production network analysis.

Figure 6.7 shows the strategic linkages of the companies surveyed based on the origin of raw materials or ingredients. Apart from eggs, which are mainly sourced domestically, the most important tying activities were found to be in food labelling and packaging materials (14 out of 15 companies). This is followed by food additives, colourings and sugar with six manufacturers. For industrial oil and meat-based materials, five manufacturers have tied up with players from meat-producing countries such as Australia, India, Indonesia and China for meat supply, while industrial oil was imported from Malaysia, Indonesia and China. Five companies have close links with foreign traders in countries to which they exported their products, four of which were discussed in detail in the cases presented. Four producers imported industrial salt from countries such as China, India, Pakistan and Indonesia. Three producers imported the majority of these materials from Argentina, India and Pakistan for

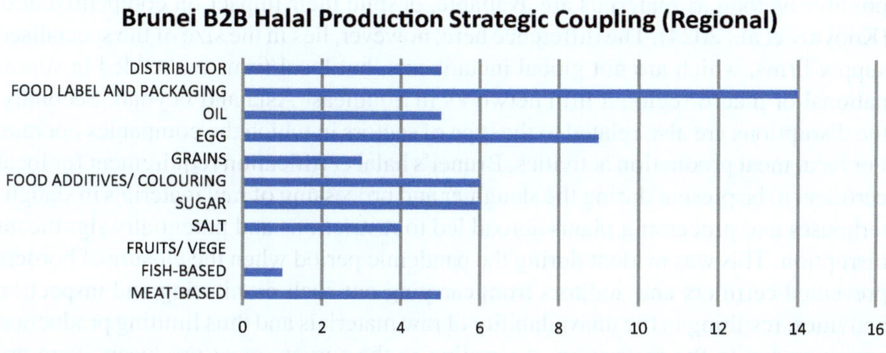

Fig. 6.7 Brunei Halal production strategic coupling

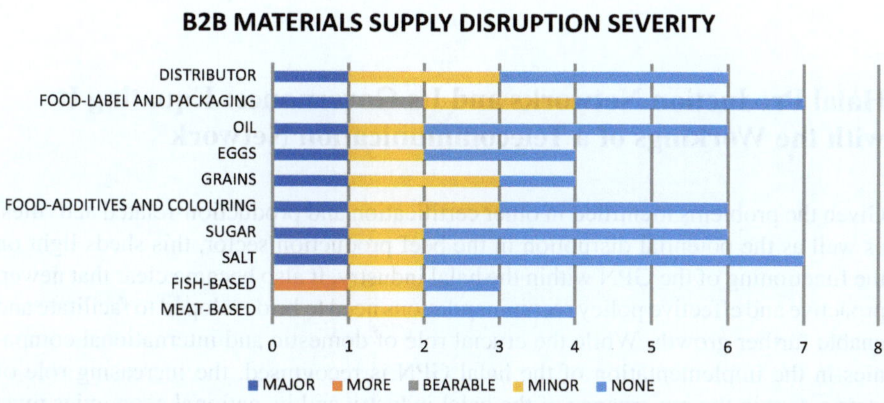

Fig. 6.8 Business-to-business (B2B) strategic coupling of Brunei and regional suppliers based on production materials (top). Materials supply disruption severity based on scale

grains and spices used in canned food. One producer, which uses fish or seafood as the basis for its products, imported mainly from the Malaysian states of Sabah and Sarawak in Borneo.

Interestingly, the question on supply disruption (Fig. 6.8) shows interesting patterns that reflect the resilience of local producers. The question used a Likert scale to measure the severity of interruptions, with 'Major' referring to interruptions of more than two months, 'More' referring to interruptions of one to two months, 'Bearable' referring to interruptions of 3 to 4 weeks, 'Minor' referring to interruptions of two weeks or less and 'None' referring to no interruptions at all. Most of the companies surveyed stated that they had 'minor to no interruption'. Only one company involved in meat processing indicated serious disruptions, as most raw materials are sourced from abroad and overall disruptions are expected to exceed two months. One of the reasons why other producers have managed to reduce the severity of disruptions is the availability of local replacement suppliers, despite the higher costs involved. The localisation of supplies in the face of severe disruptions is possible as long as materials are available, despite their impact on competitiveness (Kopytov et al., 2024). The difference here, however, lies in the size of the specialised supply firms, which are not global incumbents, but local firms embedded in supranational or macro-regional firm networks in Southeast Asia and beyond. Secondly, the disruptions are also related to the type of sectors in which the companies operate. For halal meat production activities, Brunei's halal certification requirement for local certifiers to be present during the slaughter and processing of raw materials in slaughterhouses and processing plants abroad led to restrictions and potentially significant disruption. This was evident during the pandemic period when the closure of borders prevented certifiers and auditors from carrying out such monitoring and inspection activities, resulting in the unavailability of raw materials and thus limiting production activities due to the restriction. According to the survey, most producers were not involved in the production of beef or red meat, so the impact of a serious disruption was less.

Halal Production Networks and Its Governance: Equating It with the Workings of a Telecommunication Network

Given the problems identified in other certification and production-related activities, as well as the potential disruption in the beef production sector, this sheds light on the functioning of the GPN within the halal industry. It also became clear that newer, proactive and effective policy recommendations need to be developed to facilitate and enable further growth. While the crucial role of domestic and international companies in the implementation of the halal GPN is recognised, the increasing role of state actors in the governance of the halal industry and its national economies must be considered. As outlined in the previous sections, the role of the state is crucial in supporting the growth of fledgling industries. In the case of halal food manufacturers in Brunei, support from the relevant government agencies or parties have not been consistent, which leads to costly uncertainties. To assess how the state's competency and capacity in the industry affect production activities, it is essential to look at this from a developmental perspective. An analysis using Foucault's (1991) governmentality perspective could shed light on the limitations and weaknesses of the current state governance of Brunei's Halal GPN. The analogy of a telecommunication service or network is used to illustrate how current governance works in order to make it easier to understand (Fig. 6.9).

Foucauldian governmentality, of which Dean's analytics of government is used as the anchor, was used as the primary theoretical framework to examine the halalisation programme, from the problematisation of the halal problem into strategising the mechanisms and selecting the tools to be used towards achievement of the envisioned growth and compliance as well as in actual identification and manifestation of the real effects of halal governing and standardisation had towards the halal businesses and industry in general as experienced.

The telecommunications analogy used the problematisation of halal as a vision made by the executive authority (CEO), supported by members of the Board of Directors (in the case of halal, the BIRC/MUIB), which makes the final decisions. The top management utilised the network operators and the supporting structure (HFCD and its sister agencies) as the primary mechanism for providing services to the clients. The HFCD not only acts as the 'provider of broadband services', but is also responsible for operating and maintaining the servers, information carriers and technical infrastructure providers, as well as client identity protection (in the case of halal certification, as the technical committee responsible for halal certification of applications, knowledge dissemination and enforcement), marketing and retail promotion (such as the promotion of halal applications). For the provision of efficient broadband services, broadband connectivity (wireless (Wi-Fi) or wired via LAN), is essential to meet the clients' demands. In the case of the HFCD, both direct and indirect forms of regulation have been used, directly through the use of tools such as halal workshops, lectures, briefings, clinics and other practical programmes, and indirectly through knowledge sharing and information dissemination via online platforms on halal guidelines and requirements and other technical halal information. However, as with any telecommunication service, problems can arise. In this case, despite the use of numerous tools for halalisation, it is not easy to comply with standards and achieve the required industry code of conduct. Problems such as loss of signal and dead-ends due to inaccessibility, exclusion of certain groups and throttling of network speed and misuse of the network disrupt the efficiency of the system.

Despite the information sharing, capacity building and halal initiatives implemented in halal governance, weaknesses have led to unwarranted outcomes. Imbalance in certification management was characterised by mistrust and unfairness, confusion over halal requirements and a lack of community feedback, which prevented the desired outcome from being achieved. Despite improvements or bug fixes and enhancements to the network, similar results still occurred. The feedback mechanism for the halal standard and the improvements in governance followed similar pattern, with some improvements made in the recruitment of government

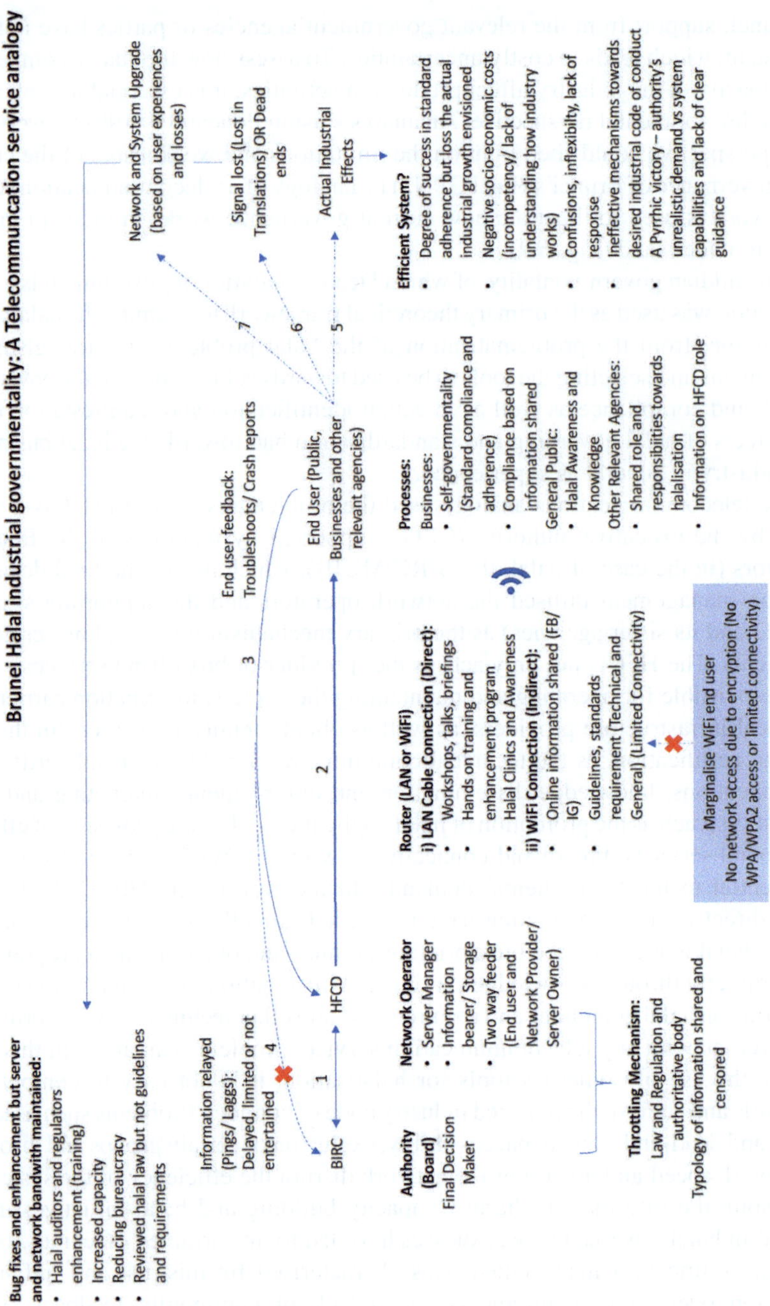

Fig. 6.9 Effective Halal Governance using a telecommunication service analogy: final representation

officials, capacity and capability building, and changes in permit fees, among others. However, the root of the confusion, the halal guidelines and laws, remained unchanged or unreviewed. Instead of understanding how industrial production works in the region and the costs associated with the certifications (see, Tables 6.1 and 6.2), in addition to the vulnerabilities of the current GPN to shocks and disruptions (Figs. 6.7 and 6.8), the government continues to focus only on improving the programme, training of regulators and auditors on GMP and GHP certification, instead of the companies' situations. The increase in production and processing costs and the lack of studies on the impact of the halal logo on product competitiveness would only compound the unjustified costs to the industry. Further research is needed to illuminate the problem and prevent losses suffered by the clients.

Conclusion and Recommendation

The different strategies employed by Bruneian producers to connect with supranational and actors have resulted in different outcomes. Some have managed to maintain production activities and become the leading companies with connections to specialised suppliers. Others became over-dependent on external suppliers and became 'unproductive' because they focussed only on marketing and rebranding, and relegated production to others, wasting resources and financial capital in the process. There are companies that developed their capacities and competences in the 'right' way by building from the bottom up and growing from there. This is much more sustainable in the long term. However, a continuous and positive sum supportive policy is required to ensure longevity. Young companies need significant financial support. There is a need to improve the current value chain ecology of halal production by enabling value addition and co-creation, and this requires more than just supportive policies. Instead, innovation that can help increase industrial productivity is critical. The halal industry in Brunei is currently at a crossroads. With dwindling funding, the authority must take decisive action on whether to risk investing potentially productive sectors or continue the status quo. The success of Silicon Valley in the United States and Shenzhen in China as global centres for digital technologies was an outcome of taking risks. Success is achieved after numerous failures by pioneers and the lessons learned by all involved. This is the only way to achieve success, no matter how elusive it may seem.

Rome was not built in a day, and similarly, the halal food industry needs to go through such a heuristic process in order to develop successfully and sustainably. There is also a need to create an integrated halal production network that integrates and harmonises standards beyond the current MABIMS framework. Fixed political boundaries should not impede the continuous flow of resources. Although a framework for standard compliance has been established under AFTA, compliance criteria or requirements are still tied to different national requirements. In 2022, the Halal Development Corporation (HDC) Malaysia and the Standards and Metrology Institute for Islamic Countries (SMIIC) conducted a comprehensive study to standardise the various halal standards by creating a single global halal standard. The results of the study have not yet been published. However, it is hoped that the introduction of this standard will solve the problems associated with harmonisation and facilitate the further integration of production networks. It also enable member states to trace flows in the network to ensure safety, support decision making, enable smooth transfer of resources, reduce border impermeability and promote the growth of the regional halal industry, which bodes well for small fledgling systems, such as Brunei Halal.

References

Ab Talib, M. S. (2020). Halal logistics in Brunei. *Journal of Emerging Economies and Islamic Research, 8*(1). 69–76. https://doi.org/10.24191/jeeir.v8i1.7179

Ab Talib, M. S. (2021). Identifying halal logistics constraints in Brunei Darussalam. *Journal of Islamic Marketing, 12*(6), 1145–1158.

Al-Haadi, A. B. (2011, May 19). 46 Businesses now certified Halal. Brunei Times. Retrieved from http://www.bt.com.bn/business-national/2011/05/19/46-food-businesses-now-certified-halal

Anjah, R. (2007 December). Food processing in Brunei Darussalam: case study on chicken-based and beef-based processing industries. In: *Symposium on market liberalization and its relationship with market structure conduct and performance of selected food processing industries of APEC member economies.* Retrieved from publications.apec.org/file-download.php?filename=07_atc_mkt...

Arias-Loyola, M. (2024). Negotiating (under) development? Expanding bargaining power within globalised production networks. *Political Geography, 115*, Article 103233.

Azalie, I. A., & Samad, N. (2022). Brunei as a global halal hub: Production network and strategic relations in halal food production. *Journal of Contemporary Islamic Studies, 8*(3), 1–22.

Azalie, I. A. B. N. (2016). Exploration on the effects of Brunei halal standard and its implementation on agri-food business development in Brunei Darussalam in the context of Islamic governance. *International Journal of Business, Economics and Law, 10*(5), 79–90.

Azalie, I. A. N. (2023). Constructing an integrated sustainable Halal Policy (ISHP) in Brunei Darussalam: Refiguring the domains of conventional policy multiverses. In *Public policy's role in achieving sustainable development goals* (pp. 167–189). IGI Global.

Azalie, I. A. N., & Kamarulzaman, N. H. (2023). Development linkages of Brunei Darussalam's coffee and tea sector: Global and regional value chain perspectives. *Journal of Business and Economic Analysis, 5*(3), 255–279.

Azam, M. S. E., & Abdullah, M. A. (2020). Global halal industry: Realities and opportunities. *IJIBE (International Journal of Islamic Business Ethics), 5*(1), 47–59.

Bi, S. (2021). Cooperation between China and ASEAN under the building of ASEAN economic community. *Journal of Contemporary East Asia Studies, 10*(1), 83–107. https://doi.org/10.1080/24761028.2021.1888410

Breznitz, D., Buciuni, G., & Murphree, M. (2024). Still sticky after all those years? The resurgence of Marshallian districts in a world of global production networks. *Regional Studies, 58*(3), 522–536.

Brunei Halal Press. (2015, July 3). Blakemore wholesale launch new range of Halal products with Brunei Halal. Retrieved from http://www.muslimnews.co.uk/news/uk/blakemore-wholesale-launch-new-range-of-halal-products-with-bruneihalal/

Brunei State Mufti's Office. (2007). *Issues on halal products: Compilation of State Mufti's Fatwa on issues on Halal products 1994–2006 Brunei Darussalam.* State Mufti's Office, Prime Minister's Office.

Brunei's Islamic Religious Council (n.d.). *Brunei Halal guideline for certification*. Retrieved from https://www.brunei.gov.bn/en/SiteAssets/eForms/Business%20e-Forms/Halal/Senarai%20Restoran%20Halal.pdf

Campling, L., & Quentin, C. (2021). Global inequality chains: How global value chains and wealth chains (re) produce inequalities of wealth. In *Rethinking value chains* (pp. 36–55). Policy Press.

Canello, J., Buciuni, G., & Gereffi, G. (2022). Reshoring by small firms: Dual sourcing strategies and local subcontracting in value chains. *Cambridge Journal of Regions, Economy and Society, 15*(2), 237–259.

Coe, N. M., Dicken, P., & Hess, M. (2008). Global production networks: Realizing the potential. *Journal of Economic Geography, 8*(3), 271–295.

Department of Agriculture and Agrifood. (2024). *Brunei Darussalam agriculture and agrifood statistics 2023*. Ministry of Primary Resources and Tourism.

Department of Economic Planning and Development. (2011). *Final report of economic census of business enterprises*.

Foucault, M. (1991). *Discipline and punish: The birth of a prison*. Penguin.

Frohm, E., & Gunnella, V. (2021). Spillovers in global production networks. *Review of International Economics, 29*(3), 663–680.

Gereffi, G. (2001). Beyond the producer-driven/ buyer-driven dichotomy: The evolution of global value chains in the internet era. *IDS Bulletin, 32*(3), 30–40.

Gereffi, G., Humphrey, J., & Sturgeon, T. (2005). The governance of global value chains. *Review of International Political Economy, 12*(1), 78–104.

Halal Food Control Division. (n.d.). Senarai Kilang Dan Pembuat Makanan Yang Telah Mendapat Sijil Halal. Retrieved from https://www.brunei.gov.bn/en/SiteAssets/eForms/Business%20e-Forms/Halal/Senarai%20Permit%20Halal.pdf

Halal Food Control Division. (2014, August 26). *Senarai Restoran Dan Kedai Makan Yang Telah Mendapat Sijil Halal*. Retrieved from https://www.brunei.gov.bn/en/SiteAssets/eForms/Business%20e-Forms/Halal/Senarai%20Restoran%20Halal.pdf

Henderson, J., Dicken, P., Hess, M., Coe, N., & Yeung, H. W. C. (2002). Global production networks and the analysis of economic development. *Review of International Political Economy, 9*(3), 436–464.

Hess, M. (2021). Global production networks: The state, power and politics. In *Rethinking value chains* (pp. 17–35). Policy Press.

Kopytov, A., Mishra, B., Nimark, K., & Taschereau-Dumouchel, M. (2024). Endogenous production networks under supply chain uncertainty. *Econometrica, 92*(5), 1621–1659.

Kuroiwa, I., Techakanont, K., & Keola, S. (2024). Evolution of production networks and the localisation of firms: Evidence from the Thai automotive industry. *Journal of the Asia Pacific Economy, 29*(1), 260–281.

Li, L. (2018). China's manufacturing locus in 2025: With a comparison of "Made-in-China 2025" and "Industry 4.0." *Technological Forecasting and Social Change, 135*, 66–74.

Masli, U. (2009, July 31). Brunei Halal Brand Unveiled. *Brunei times*. Retrieved from http://www.bt.com.bn/home_news/2009/07/31/brunei_halal_brand_unveiled

Ministers of Religious Affairs of Brunei, Indonesia, Malaysia and Singapore. (2017). *MABIMS harmonisation of halal standards: Strengthening the halal standards*. Retrieved from https://www.isdb.org/sites/default/files/media/documents/2022-04/MABIMS-Harmonisation%20of%20Halal%20Standards%20May17.pdf

Mudambi, R. (2008). Location, control and innovation in knowledge-intensive industries. *Journal of Economic Geography, 8*, 699–725.

Nadvi, K. (2008). Global standards, global governance and the organization of global value chains. *Journal of Economic Geography, 8*, 323–343.

Ponte, S., & Gibbon, P. (2005). Quality standards, the conventions and the governance of global value chains. *Economy and Society, 34*(1), 1–31.

Shahminan, F. (2015, August 27). Russia seeks Brunei expertise in halal industry development. *Brunei times*. Retrieved from http://www.bt.com.bn/business-national/2015/08/27/russia-seeks-brunei-expertise-halal-industry-development

Storper, M., Kemeny, T., Makarem, N., & Osman, T. (2015). *The rise and fall of urban economies: Lessons from San Francisco and Los Angeles*. Stanford University Press.

Sulaiman, N. S., & Hashim, N. (2022). The emergence of Halal food crimes amidst COVID-19: An analysis of illegal and frozen meat issues in Brunei Darussalam. *Journal of Contemporary Islamic Studies, 8*(3).

Sulaiman, N. S., Abdullah, R., & Hashim, N. (2023). Halal industry development in Brunei Darussalam: Realities and challenges. *KnE Social Sciences*, 1013–1029.

Temporal, P. (2011). *Islamic branding and marketing: Creating a global islamic Business*. Wiley.

Wallerstein, I. (1974). *The modern world-system: Capitalist agriculture and the origins of the European world economy in the sixteenth century* (p. 410). Academic.

Yeung, H. W. (2009). Transnational corporations, global production networks and urban and regional development: A geographer's perspective on multinational enterprises and the global economy. *Growth and Change, 40*, 197–226.

Yeung, H. W. C. (2018). The logic of production networks. In G. Clark, M. Feldman, M. Gertler & D. Wojcik (Eds.), *The new Oxford handbook of economic geography*. Oxford University Press.

Yeung, H. W. C. (2021). The trouble with global production networks. *Environment and Planning a: Economy and Space, 53*(2), 428–438.

Yeung, H. W. C. (2022). Explaining geographic shifts of chip making toward East Asia and market dynamics in semiconductor global production networks. *Economic Geography, 98*(3), 272–298. https://doi.org/10.1080/00130095.2021.2019010

Yeung, H. W. C. (2024). From regional to global and back again? A future agenda for regional evolution and (de) globalised production networks in regional studies. *Regional Studies, 58*(7), 1–12.

Izni Azrein Noor Azalie is a human geographer who graduated with an MSc in Global Transformations from Loughborough University UK and recently completed his PhD at University College London, researching halal industrial governing and its effects on Brunei agri-food producers. He has been a lecturer at Universiti Brunei Darussalam (UBD) for 14 years. He has been involved in professional consultancies involving environmental and communal advisories relating to oil and gas explorations, environmental monitoring and impact assessments on urban development projects, and socio-economic costs of industrial development towards coastal fisheries and communities in Brunei Darussalam. His research specialisation covers industrial production networks, urban economy and development and industrial development in general. His research has expanded to include studies on creative industrial development, heritage industry conservation and poverty alleviation within the old urban centres. As Brunei embarked on a path of leveraging the development of the blue economy as part of its economic diversification strategies, it became imperative to understand the dynamics between policymakers and governmental strategies and how they are translated into actual practices. This sphere is currently of interest to the researcher.

Chapter 7
Young Consumers Jumping on the Bandwagon of Sustainability

Noor Hasharina Hassan

Abstract The young generation understands 'sustainable consumption' on the basis of green consumer guides. This includes using bamboo straws, taking water-filled bottles or cups to restaurants, buying items or participating in activities with emotional slogans to save the planet or wildlife. These young consumers often ignore the environmental or socio-cultural aspects of the urban lifestyle and services they demand and desire because they believe they belong to the environmentally conscious community or civil society. They often blindly join the sustainable bandwagon via social media without necessarily understanding the environmental and urbanisation connections to sustainable development and consumption. Their urban lifestyles include regularly spending time in air-conditioned spaces connected to the digital world. Younger consumers have a preference for imported trendy items and consumption spaces such as cafés with a large environmental and carbon footprint. The aim of this study is to get some insights on the thoughts behind their consumption behaviour to determine the depth or superficiality of their understanding and commitment to sustainability and sustainable consumption. This study uses interviews and thematic analyses to address the research aim involving 55 first-year university students. It examines whether they practise what they preach, what intermediaries shape their consumption behaviour, and whether they recognise the links between their consumption and sustainable urban development.

Keywords Sustainable consumption · Urbanisation · Intermediaries · Young consumers · Superficial sustainable

N. H. Hassan (✉)
Faculty of Arts and Social Science, Universiti Brunei Darussalam, Gadong, Brunei Darussalam
e-mail: hasharina.hassan@ubd.edu.bn

G. Y. V. Yong and A.K.M. A. Ullah (eds.), *Brunei and the Future Transformation of Southeast Asia*, https://doi.org/10.1007/978-981-96-7901-0_7

Introduction

The consumption-based economy is a product of the Industrial Revolution, which emerged in Europe in the late nineteenth century. It is linked to modern economic models, social and technological development, and urbanisation. Preindustrial societies were often producers of their daily needs and exchanged the surplus production they possessed with others who had surplus of what they needed. This is described as subsistence consumption. What they consumed was simple and basic. As Europe became more urbanized and its economy industrialised, people's lifestyles change radically and the landscape transformed from predominantly rural to increasingly urban. This period is characterized by mass production and the proliferation of consumer goods sold at increasingly affordable prices. Cities were established, as more people moved to urban and industrial centres in pursuit of paid employment. Many worked in factories and no longer have the time to make their daily necessities. These city dwellers relied on buying what they needed and had access to a wide range of goods, including many that are beyond their needs. The marketing of consumer goods through advertisements shaped consumer behaviour, motivating people to consume what they wanted rather than just what they needed.

This gave rise to inflation, which was exacerbated by the emergence of the affluent middle class in the nineteenth century, eventually becoming a significant social-economic force by the turn of the century. In the twentieth century, the world witnessed the consolidation of the consumer culture, which stratified into different classes according to consumption patterns and nature/quality of products (clothes, furniture, houses, cars, food, etc.) they consume. In the period of economic boom following the Second World War, access to credit and financing, such as, through credit cards and bank loans, created the culture of hyper-consumerism that is characteristic of the present-day economy and a way of life, especially in mega-urban centres, such as London, New York, Tokyo and Singapore.

Consumption is more than a process by which artefacts are simply bought and consumed. It includes adding meaning to people's lives through their active incorporation (Jackson, 1993, p. 209). Consumption leads to identity formation, where the commodities are identity signifiers and social markers (Hassan, 2023). People could consume more than they could afford to be associated with identities beyond their ascribed status. The actual use-value of an object, though important, thus becomes secondary to the object's social symbolism and signifying effects. Hebdige (1993) calls this the "theology of appearances" (1993, p.89).

Consumption in Southeast Asia has undergone significant changes in tandem with urbanization, rising incomes, technology adoption, globalization, and shifting social values. From the 1960s to 1995, Southeast Asia underwent remarkable economic growth and industrialization resulting in an urban middle class with improved purchasing power and a consumption-based lifestyle (Chua, 2000). The affluent urban middle-class consumers acquired status, or positional goods, such as imported fashion, cars and foreign education for their children to promote social distinctions. The expansion of consumption coincided with the rapid globalisation in the 1990's, where luxury West designer brands, which carry socially symbolic meanings flooded the retail stores of the more developed Southeast Asian cities such as Singapore, Kuala Lumpur and Bangkok (Hassan, 2017).

However, rampant urbanisation, production and consumption rates have major environmental and social impacts, and therefore undermines sustainable development (Ash et al., 2010; Jackson & Michaelis, 2003; Michaelis & Lorek, 2004; Watkins & Aitken, 2020; Whitmee et al., 2015). Rampant globalization facilitated global production and distribution of goods from one part of the world to be consumed in another. Online retailing and social media further drove and facilitated consumption and consumerism. The rise of social influencers aided in the dissemination of culture and information, shaping consumption and influencing lifestyle choices, particularly the younger generation.

The combination of natural resource depletion and disruption of the planet's ecosystem functions are key sustainability concerns in the field of consumption. Presently, there is noticeable climate change, occurring in the backdrop of urban growth and increasing transition to urban lifestyle or urbanism, which is contributing to overconsumption. The international community's response, which was initiated at the 1992 Earth Summit and guided by Agenda 21 and its follow-up targets Responsible Consumption and Production under Sustainable Development Goal (SDG) 12 of its current programme. Associated issues are addressed in SDG 11 on Sustainable Cities and Communities and SDG 13 on Climate Action. They demand transformational changes to promote sustainable lifestyle, which necessitates a shift in the mindset of consumers, requiring them to be more conscious and aware of consumption patterns as well as how urban design and development affects our environment and climate (Hassan & Yong, 2023). However, greenwashing in advertisements misleading consumers by claiming to be eco-friendly, organic, ethical or sustainable, when the products or services offered do not follows sustainability ideals.

This study aims to highlights different levels of sustainable consumption, contrast between superficial or shallow, to "green" and a comprehensive understanding. It involves a survey of first year university students.

Sustainable Consumption Approaches

The Brundtland Report (WCED, 1987) defines sustainable development as development that meets the needs of the present without compromising the ability of future generations to meet their own needs. Sustainable consumption is guided by the triple goals of this definition, which is to promote sustainable economy growth, protect the natural environment from impacts, and promote social justice and equity. Among its guides is the ecological footprint concept, which should not be greater than the biocapacity of the natural systems needed to produce the goods. Sustainable consumption therefore promotes consumption habits with small ecological footprints. This is also extended to include the environment's capacity to absorb waste/emissions (Fischer et al., 2023, p. 108). Advocates of sustainable consumption include choosing items that employs more efficient production processes and renewable or recycled materials (Jackson, 2006. p. 4), which reduces and/or optimizes resource use.

"Sustainable Consumption" is an umbrella term that encompasses numerous aspects, including meeting the needs of consumers and improving their quality of life, resource use, efficiency of production system, use of renewable energy, waste minimization, taking a life-cycle perspective of product development, and the equity dimension in production (Fischer et al., 2023, p. 109). The UN (2002) contends that focusing solely on product-oriented approaches is inadequate to achieve sustainable development and that consumer lifestyle and behaviour that could result in over-consumption should be curbed (Liu et al., 2016). When addressing sustainability in consumption, a more holistic approach is needed to consider the social, environmental, economic and long-term aspects, instead of focusing only on the environment. Environmental well-being includes reducing the ecological footprint, waste reduction and resource conservation. The social component of sustainable consumption encompasses social justice and employee well-being involving ethical practices throughout the commodity chain (production, consumption and disposal); consuming organic food; fair trade; and free-range animals for healthier food and support ethical farming. Economic prosperity can include new forms of economic activities such as green economy, circular economy and fair trade, but economic activities that could result in overconsumption should be curbed (Liu et al., 2016).

Although sustainable consumption is often perceived as originating from Western Environmentalism, consumers outside the Global North or West have been practicing sustainable consumption through indigenous practices, culture and religious values (Bakar et al., 2022; Hassan et al., 2022, 2024; Zainuddin, 2021). Sustainable consumption and green consumption are often used interchangeably, despite differences between them. However, green consumption can be considered a subset of sustainable consumption. Although sustainable consumption is a growing trend, it has not yet manifested itself in the typologies and range of purchases. The decision to consume sustainably often ignores ethical products or boycotts unethical products (Šálková et al., 2023).

Green consumption involves the acquisition and use of products that are promoted as environmentally friendly, energy-efficient, and having low ecological or environmental impacts, and low carbon footprint. Several methods to enhance green consumption includes the use of eco or green label certifications such as Energy Star, Fair Trade, or organic. According to Haba et al. (2023, p. 1), green consumers engage in behaviours that improve social and environmental outcomes while boosting consumer well-being. Ethical consumption involves knowing the impact or being conscientious of the impacts of one's consumption practices and choices on the environment and society (Arman & Herbert, 2024). It involves abiding with certain ethical issues and moral values such as the support of fair wages, worker's rights and safety, and observing the protection of the ecosystem (Berki-Kiss & Menrad, 2022). It involves refraining from purchasing or boycotting products sold by irresponsible companies, such as, exploiting poor producers, farmers and workers (Berki-Kiss & Menrad, 2022).

Islamic consumption aligns well with sustainable consumption. It emphasizes consumption in moderation, promoting resource conservation, and social responsibility for the well-being of people, animals and the environment. These values and concerns echo the growing global distress relating to sustainable consumption and production. In addressing the need for controlled consumption and avoiding over-consumption, the Qurʾān (45:23) talks about those who worship their desires, and that this act makes them blind to their welfare. Muslims are taught and urged to control their desires for instant gratification and to reflect on what truly makes us happy. In a modern consumerist society, this is often socially constructed or fashioned by retailers through marketing and advertisements telling us what we need to consume. Consuming beyond our means and giving in to our desires could result in damage to oneself, such as debts, not controlling what we eat or consume that could lead to health effects, or laziness, are not permissible. In Islam, aside from consuming what is permissible (e.g., Halal food and modesty in fashion), Muslims are also taught to respect, protect and preserve natural gifts or resources from God, whether it is regarding sustainable use of water or the destruction of flora and fauna (Zaman & Qadir, 2021). Islam therefore encourages the practice of ethical consumption. Therefore, the *Maqasid* or objectives of *Shari'ah* in Islam are in harmony with sustainable development preached by the West and the promoted in the UN's sustainable development programme.

Shari'ah alone provides the natural law for mankind to regulate social behaviour (Hasan, 2017, p. 16). Moderation and balance in worldly pursuits that the verses of the Qur'an repeatedly emphasize are intended to support this basic Islamic concept of sustainable development (Hasan, 2017). Religion such as Islam is an important intermediary and shaper of consumption. "O Children of Adam! Wear your beautiful apparel at every time and place of prayer, eat and drink, but waste not by excess, for Allah loveth not the wasters" (Quran Al-A'raf: 31–32).

According to Quoquab and Mohammad (2016, p. 603), the barriers to sustainable consumption include: (a) lack of adequate knowledge or ignorance of the effects of extravagance and excess consumption; (b) having a "don't care" mentality; (c) pressure of globalization; (d) wanton electronic media aggression for consumerism (e) profit maximization motive.

Young Consumers

The definition of young consumers may vary but it refers to those between the ages of 18–34 years old in this study. This age group would include part of the Generation Y (Millennials) and Gen Zs. Both generations have been characterized as the age group whose lives revolve around digital technology, are tech-savvy, and are often influenced by global culture and media (Chen et al., 2023). Gen Z consumers are generally more environmentally and socially conscious than the previous generation and have a greater tendency to participate in sustainable movement, including consumption practices (Zhang et al., 2023). Forbes reported in 2019 that 62% of Gen

Z prefer to buy from sustainable brands, with 54% willing to pay up to 10% more for sustainable products (Campbell, 2022).

The younger generation of consumers is identified to be "hedonistic" consumers, and they consume to conform to the social group (Chen et al., 2023; Hassan, 2017). Being digital natives and highly integrated into social media, these consumers are up to date, have a strong awareness of brands, especially for ethical and sustainable brands, particularly with better access to different recommendations and reviews of products, are price-conscious, and always try to influence others' consumption behaviour through their social media (Chen et al., 2023). Being the second youngest generation (the youngest generation currently is Gen Alpha), Gen Z consumers' worldview is mainly shaped by the Internet particularly social media (Adamson et al., 2018). These consumers are often termed the TikTok Generation and are typically known to be involved in the "woke" culture. These young consumers are people whose knowledge about the environment is often related to the "climate crisis" (Chen et al., 2023). Williams & Hodges (2020) believes that the younger generation's movements are influenced by greenwashing. Chen et al. (2023) highlighted that Bruneian Gen Zs appreciate nature, are interested in learning more, are exposed to vast amounts of information via the cloud, but religion still plays an important factor in their lives including consumption. In woke culture, the genuineness and perseverance of their actions to fight for sustainability I important.

However, it is pertinent to question whether this generation is merely virtue signalling and their actions are part of performative activities, or whether they truly belief in what they claim to stand for. This is relevant particularly in relation to the recent surge in popularity of sustainability globally, making it a buzzword across industries from development programmes, production to consumption. In education and research, questions arise on the authenticity versus superficial adoption of sustainability (Young, 2024). Is there a bandwagon effect in sustainable consumption? Identical to the collectivist consumption culture and the self-interdependent concept (Wong & Ahuvia, 1998), the bandwagon consumption behaviour relates to consumers who buy stuff to establish or maintain social relations within a social group and to "keep up". When a product becomes popular, it pushes people to "get on the bandwagon" and signals a person's aspiration towards obtaining the commodity or brand to conform and be accepted into the perceived mainstream (Kang & Ma, 2020). Young Gen Zs consumers often call this the Fear of Missing Out or FoMo. Perceptions and attitudes toward sustainability are generally positive, but few incorporate sustainable consumption habits in their everyday lives (Buvár et al., 2023).

Apart from religion being an important intermediary as discussed in the previous section, family plays a significant role as a cultural intermediary that has shaped consumption of Southeast Asian consumers including Bruneians (Hassan, 2017). Consumers often conform to the norms and values of the family and is still an important social group that shapes consumers' perceptions and behaviour. This is mainly because of religious beliefs stating that family, especially parents, should be respected, and many offspring live with their parents until they start a family of their own. Additionally, maintaining social relations is an important attribute for

Southeast Asian communities (Wong & Ahuvia, 1998), especially in Brunei where the community is small and know one another (Hassan, 2017).

Anglo-American consumers are often identified as hedonistic and individualistic consumers, highly materialistic facilitated by less regulated finances (lending and borrowing) with capitalist institutions such as banks and subprime lenders having more control or influence. On the other hand, consumers in Southeast Asia especially Brunei's consumption and access to finance is regulated by the government through the Brunei Darussalam Central Bank (See Hassan, 2017). Unlike claims of some scholars offering distinctions between Western Consumers as hedonistic and individualistic and Asian consumers as collectivists (Wong & Ahuvia, 1998), Hassan (2017) and Chua (2000) argue this binary is not accurate, rather consumers including Bruneians may practice both collective and individualistic traits.

The government plays a significant role in consumers' everyday consumption in Brunei. The discovery of offshore oil and gas fields in the 1960s enabled rapid development in Brunei. As one of the richest countries in Southeast Asia, Brunei has one of the most comprehensive welfare programmes including providing subsidized housing for its citizens, quality infrastructure and amenities (Hassan & Yong, 2019). Young Bruneian consumers have access to welfare programmes offered to universally to Bruneians particularly subsidized education and healthcare (Hassan et al., 2022, 2024; Hassan, 2017, 2023). This also included scholarship provisions to study in any local public universities that included a monthly stipend of BN $300–400. Most items imported are not taxed, and high tax is 20%. Food commodities, namely, rice, salt and sugar, are subsidized. Brunei's welfare provision is, therefore, a significant contributing factor to its consumption pattern.

With regards to sustainability, the government has initiated measures to curb unnecessary consumption, including controlling loans and debts of Bruneians through their financial loan directives (Hassan, 2017), no plastic bag week, where people had to bring their own shopping bags, as well as its commitment to achieving UN SDGs. (Hamdan et al., 2023). In a government report on SDG in 2023, Brunei has much to do to achieve SDG 12 on Responsible Consumption and Production but is "profoundly devoted" to attaining it through various initiatives including waste management, waste minimization, sustainable tourism, e.g., staycations to increase domestic tourism to reduce carbon footprint and green consumerism (Prime Minister's Office, 2023). Green consumerism initiatives include circular economy and recycling packages to avoid single-use packaging and introducing more sustainable products into the local market. To combat issues of climate change (SDG 13), the government's initiative that curb the negative effect of consumption includes reviewing plans to introduce electric vehicles (EVs) or energy efficient vehicles into the local market, better public transport and smart urban planning that would address consumers' carbon footprint and improve ridership. Brunei can also manage its consumption of resources and pollution from consumption by implementing a fully integrated smart city approach that would support achieving SDG 11 to make sustainable urban development and consumption. An initiative by the government for sustainable city development is reducing the allocation or consumption of land by increasing density and promoting green living spaces; it recommends a 10%

green space for all types of development as well as improved pedestrianisation and walkability.

The UN SDG 12 on responsible consumption and production asks us to consider how we use materials, energy and resources to minimise our impact on the environment while meeting the needs of future generations (WCED, 1987). The Oslo Symposium (1994) promotes the use of services and related products to meet our basic needs and improve quality of life, while minimising the use of resources and toxic materials. It also promotes the reduction of emissions, pollution and waste throughout the life cycle of the product or service, all with minimal impact on the needs of future generations. Both consumers and manufacturers have an important role to play here. Awareness of issues related to global and excessive consumption and disposal of discarded and toxic materials during production and after consumption has increased worldwide. As people's environmental awareness of the impact of unsustainable consumption has increased, consumers' perceptions and values are often incorporated into the global value chain, impacting the commodity chain. This is because the behaviour of consumers who choose to make ethical and sustainable choices have an impact on demand and production, particularly for producers and retailers. More common advice and sustainable consumption practices are the 3Rs: reduce, reuse and recycle. In addition, sustainability has changed from focussing solely on the environment to combining the importance of the environment and the economy. Today, there is a triangular relationship between environment, economy, well-being and sustainability. New entrepreneurs are developing green and affordable technologies and the rise of the circular economy. These products and services should be affordable so as not to socially exclude vulnerable groups, especially low-income consumers. Therefore, this chapter shows that SDG12 can only work if the other SDGs such as SDG 11 and 13 are included. Hassan and Yong (2023) also proposed similar approach to address SDGs in Kampong Ayer, Brunei, a historic settlement exposed to climate change risk, where a significant portion of the residents rely on welfare.

Methods

This research takes employs an open-ended questionnaire survey to gain insights into the sustainable consumption motivations of young consumers in Brunei. It involved 55 first-year university students between the ages of 19–29; 83% concentrated in the 19–21 range. These are Gen Z, comprising 56% female and 44% male. For 74% of these young consumers, monthly income range from BN $300–399 (mostly from their student allowance).

The questionnaire used included considerations on changing realities due to climate change. Respondents were asked about what they understand as "sustainable consumption", "green consumption" and "climate change". Understanding the three keywords would determine their consumption practices, which was also questioned in the survey. The questionnaire asked the young consumers to:

(a) Identify the main drivers of sustainable consumption, to get insights on their understanding of the term, concept and practices;
(b) relate sustainability to urbanization, to determine their understanding of the connection.

Open-ended survey questions were used because they allow the respondents to answer in their own words, rather than selecting from predefined options. Respondents' details were collected. They include age, gender, discipline and income. A selective sampling method was used to distribute the questionnaire to first-year students in the premier university. The frequency of keywords used by the young consumers provides insights into the respondents' understanding of sustainability and consumption decisions. The data was analysed thematically to determine the level of knowledge on various aspects of sustainable consumption, practices and influences. Likert scales were also used to determine levels of knowledge by self-rating. The respondents were also asked to list their urban comforts vis-a-vis sustainable practices to gauge their comparative levels of importance.

Findings and Discussion

The section discusses the results of the survey. It includes their understanding of the different types of consumption, climate change, the main sources of their knowledge about sustainable consumption, followed by their motivations for sustainable consumption.

Understanding of Consumption

The young consumers surveyed have a common understanding of consumption: consumption is the use, spending or purchase of resources that satisfy their or other people's needs and desires.

(i) In contrast, they indicated time as a factor in sustainable consumption, e.g., when using resources or materials, they consider the future and use by other generations. Most of the respondents describe sustainable consumption as involving the use of resources that do not affect future generations or that materials are used for longer duration. This aligns with the inter-generational equity aspect of the Brundtland Report's definition (WCED, 1987).
(ii) They understood green consumption as the use of materials, including resources, that do not harm the environment or nature. It includes the 3Rs (reduce, reuse and recycle), environmentally friendly, energy efficient, low impact and biodegradable items. Green consumption still falls under the umbrella of sustainable consumption (Fischer et al., 2023), where it encompasses the management of resource use, but the young consumers placed more

emphasis on green approaches and care for the environment when it comes to green consumption. This is similar to the general existing discussions on green consumption (Haba et al., 2023; Arman and Herbert, 2024).

(iii) Although they revealed a good level of understanding, the majority did not mention the ethical aspect, such as social justice, equality, and well-being.

(iv) They did not make any connection to Islamic consumption, even though the majority were young Muslims.

Sources of Knowledge

The main source of their knowledge on sustainable consumption is the media (N = 25, or 45%), especially social media such as TikTok or social media influencers. The results are consistent with the general discussions about social media or the internet as an important source of knowledge for Gen Z and internet connoisseurs (Chen et al., 2023, Adamson et al., 2018). Government policies, events and actions rank second (N = 8, or 15%), followed by educational institutions (N = 5, or 9%). A previous study on Bruneians' everyday consumption show how excessive consumption led to high levels of debt (Hassan, 2017, 2023).

While the younger generation has access to a buffet of knowledge, it is unclear how they sieve and process the information gleaned from social media and the internet. Although young consumers were taught and often reminded while in college to look for credible sources to avoid misinformation, more often, Gen Zs worldviews are shaped by sources from the West and they are susceptible to be draw by the bandwagon effect in sustainable consumption (see Kang & Ma, 2020), a FoMo reaction (Buvár et al., 2023). This is a typical trait of many collectivists Asian consumption pattern (Chua, 2000; Hassan, 2017, 2023; Wong & Ahuvia, 1998). Their knowledge on consumption, sustainability and climate change determines their motivations, reasons and engagement in sustainable or green consumption. This often leads to a general change in urban lifestyles.

Climate Change

Regarding climate change, the most common keywords used were: (a) temperature rise (hotter) (N = 32, or 58%), (b) weather changes (due to temperature changes) (N = 27, or 49%), (c) sea level rise/changes with (N = 8, or 15%), (d) greenhouse effect and thinning of the ozone layer with (15%), (e) carbon dioxide (CO2) (N = 4, 0r 7%), and (f) others (N = 5, or 9%). The last could be broken down to harmful environmental impacts (N = 2), environmental change (N = 2), and waste produced/emitted (N = 1). Only a minority of young consumers do not know what climate change is, with one saying that it only occurs during the monsoon season. Temperature and weather changes are the most common associations with climate change, as identified by the young consumers. This is quite consistent with the UN definition, which is the long-term shifts in temperatures and weather patterns (UN, n.d.). The UN attributed climate change to human activities, such as the burning of fossil fuels, deforestation and the release of methane from agriculture.

Sustainable Consumption Motivation

The key motivation for sustainable and green consumption for the young consumers surveyed are to reduce negative effects on the environment, such as climate change and environmental degradation (N = 17, or 31%); influence from friends and family (N = 10, or 18%); influence from social media (N = 4, or 7%); to save the earth (N = 4); and to reduce water or electricity bills, particularly from air conditioner usage due to the hot weather. This aligns with reasons cited to practice sustainable consumption: firstly, to reduce impacts on the environment, other people and animals due to climate change and disasters associated (N = 27, or 49%); secondly, for a greener, safer, better quality of life in Brunei (N = 9, or 16%); thirdly, to avoid waste (N = 5, or 9%), and to understand environmental problems and finding solutions (N = 5).

Islamic Consumption

Religion shapes many Southeast Asian and Asian consumers' consumption, particularly within the debates of collectivist and individual consumption patterns (Hassan, 2017 and Wong & Ahuvia, 1998). Though not mentioned directly as their sources for sustainable consumption, the survey found that religion shape young Muslim consumers' consumption practices (N = 45, or 82%). Only a minority are not affected by religion (N = 5) or unsure (N = 5). More than half of the consumers noted that Islam teaches them to be more mindful, to consume in moderation, to be more appreciative of resources or materials owned and not to be wasteful of resources or materials including food (N = 47, or 85%), to not harm objects (animals, environment and plants) (N = 5), promote cleanliness (N = 3), conform to norms and expectations (N = 2), to think or reflect how their actions or consumption affect their din and sin (N = 1). These are teachings in the Quran and have been promoted by scholars on Islamic consumption (Hasan, 2017; Quoquab & Mohammad, 2016; Zaman & Qadir, 2021).

Interestingly, none of the consumers see halal as aligning to sustainable consumption. This is more likely due to young consumers relating sustainable consumption mainly about conserving the environment and resources for their and future generations' consumption. Again, there is not much element on social justice such as fair trade. Islam has always emphasized green and ethical values including protecting the environment, social justice including fair trade, as well as welfare, cleanliness, health and the methods of slaughtering of animals and high standard of ingredients and cleanliness meal preparation are part of ensuring the halal standards for safe and clean Muslim consumption. Halal is ethical consumption as it safeguards the welfare of animal and people from the rearing of animals, the humane slaughtering of animals as part of production and processing of the product, the production process will involve certifying the product is safe to be consumed with no intoxicants to the product that could be harmful to people or the environment (Teng et al. 2017; Rezai et al., 2015). However, young consumers do not see these associations.

You Are What You Consume: Joining the Bandwagon of Sustainability

The provided insights into how young local first-year university student consumers view sustainable consumption and climate change, their sources of knowledge, and motivations and reasons for engaging in sustainable consumption. This section discusses the findings further. There are several ways in which young consumers can help conserve and prolong the use of resources and materials for the next generation as part of their sustainable consumption: (1) Energy efficiency and energy saving strategies (N = 22, or 40%), e.g., switching off appliances when not in use, followed by reducing the use of air conditioning during the day or opting for an electric fan (N = 9), using solar panels, electricity and appliances, and drying clothes in the sun (N = 2). With regards to 3Rs (N = 21, or 38%), recycling paper, cans and plastic bottles are the most popular, followed by reusable carrier bags, bottles or cups, reducing wastage of water, and buying food in plastic bags.

The most important environmentally friendly consumption habits of the young consumers surveyed also show similar patterns to those of sustainable consumption, but with a stronger emphasis on the 3Rs. As with sustainable consumption, two main themes emerged from the thematic analysis of the responses: (a) environmental protection and (b) energy saving. However, sustainable consumption places more emphasis on responsible and alternative energy use, while green consumption places more emphasis on preserving the environment. Almost half of the respondents apply the 3R approach (N = 27, or 49%) and use reusable bags, especially personalised carrier bags, which are fashionable, as well as reusing plastic bags (N = 13, or 24%). Contrary to widespread concerns about pollution from single-use plastic bags, some of the young consumers surveyed use plastic bags multiple times, i.e., not treating them as single-use items. A third important approach used by young consumers is to minimise fuel consumption (n4) by carpooling, using fuel-efficient cars, public transport and walking to nearby shops or destinations. One consumer mentioned "thrifting" as a method of eco-friendly consumption, indicating the importance of fashion in the practise of eco-friendly and sustainable consumption.

Most young consumers admitted that their consumption has changed due to climate change (N = 47, or 85%). Only five young consumers admitted that their consumption had not changed, while three others said they were unsure or did not answer. The most noticeable changes in consumption were due to changing temperatures (warmer) or hotter weather and the need to keep cool. Such changes consist of using more air conditioning and fans (N = 25, or 45%). This is followed by higher water consumption for showering, drinking and watering plants (N = 8). Young consumers are also consuming more sustainable or environmentally friendly goods and services (N = 44, or 80%). Five consumers believe that climate change is having an impact on their cost of living and their economy, as they feel that food and drink has become more expensive.

The study found consumption behaviour to range from individualistic and hedonistic (use of air-conditioner to cool themselves) to collective (e.g., caring for the environment). The latter is evident in their adherence to sustainable consumption bandwagon's practices, e.g., bring their own reusable cups and bags. Their individualistic character however is revealed in the preference for customising items or choosing fashionable products. This means that their consumption is neither only individualistic nor only collective. A person's consumption behaviour, including in sustainable consumption, could be a combination both individualistic and collective traits (see Hassan, 2017).

Urban Consumption

As urban dwellers, the students were asked to list down their urban comforts. The few leading urban comforts selected are: (a) Wi-Fi or the Internet (N = 12, or 22%), (b) access to water for drinking, showering and watering plants (N = 10, or 18%) and (c) air conditioning (N = 7, or 13%). The demonstrations of the identity of these consumers were drawn from what is important to these young urban consumers and compared to their consumption patterns. Based on the urban comforts data, the young consumers surveyed matched the general identity and characteristics of Gen Zs where they are digital technology savvy and influenced by global culture and ideologies due to their access to social media (Chen et al., 2023). This includes riding on the bandwagon of sustainability (Zhang et al., 2023). Social media, therefore, becomes a significant cultural intermediary affecting their consumption and urban lifestyle.

The intermediaries that shape their consumption decisions also contribute towards their identity formation. All the consumers surveyed are Asians, mostly Malay Muslim Bruneians living in Brunei. Much of consumption is collective, where conforming to social group expectations is a common feature. This includes family and peers as well as religious expectations. However, they also have individualistic and hedonistic tendencies. Observations conducted in universities and in retail stores and cafes found young consumers (Gen Zs) typically carry around with them branded tumblers or flasks (e.g., Stanley, Hydro Flask, Corkcicle and Owala) with prices ranging between B$40–100 (USD 30–74) depending on the size. Some have accessories to go with their flasks or tumblers such as Water Bottle Boot or Anti-Slip Protective Sleeve for Water Bottle that can cost between an extra B$15–25 (USD

11–19). Many engraved their names or initials onto their water bottles, flasks or tumblers.

When asked whether sustainable consumption is fashionable, many said "yes" and agreed that it is a fad (N = 29, or 53%). Seven consumers thought that advertising and influencers in the media and social media make sustainable consumption cool or fashionable (hedonistic feeling). A few (N = 7) found the growing new thrift culture of buying second-hand or used clothes with brand names, especially from designers, at a cheaper price both fashionable and sustainable, although they do not think about the carbon footprint associated with flying the used items in from abroad. Five said that reusable bags and bottles are fashionable and sustainable products because they can be personalised. Four agreed that sustainable consumption is fashionable because there are many eco-friendly brands, while three said that it is a part of their urban lifestyle. Two respondents thought sustainable consumption is an affordable practice while another two thought it was part of the growing circular economy.

Many relate its fashionability to social status. The colourful designs of recycling bins are fashionable, and it reduces harm, educates people, and uses new technologies to improve our lives. However, 15 (27%) of the young consumers do not consider sustainable consumption to be a fad or fashionable. Their reasons include the products are not appealing, attractive (e.g., old clothes, ugly wind turbines that ruin the aesthetics of a place), or expensive. A few believes that sustainable consumption is difficult to practise due to the car culture in Brunei and this makes it unfashionable. Instead, they said that sustainable consumption should be a long-term practice, rather than part of fashion or a fad. The remaining 11 (20%) were either unsure or had no answer. Interestingly, none of the respondents commented on greenwashing.

The various fashion trends and brands (e.g., eco-certified gadgets and devices, designer clothing, reusable customised branded bags and cups) identify consumers as those on the sustainability bandwagon. At the same time, the bottles or cups fulfil their basic function as water bottles that store drinks at the desired temperature. The brand names are a social symbol of the socio-economic status of consumers. Sustainable consumers can be both adherent to the norms of social groups and hedonistic.

Interestingly, half of the young consumers surveyed rated their sustainable consumption as satisfactory or passable (3 in the 5-level scale, where 1 indicates the highest and 5, the lowest). Based on rating, the distribution are as follows: 2% at level 1; 16% at level 2; 55% at level 3; 18% at level 4; and 4% at level 5. Three did not rate themselves. There were more males that gave lower ratings than females. The opposite pattern is observed among those with higher ratings.

While much of the sustainable consumption covers personal or micro level consumption performed by consumers, the survey also asked the Gen Z consumers how sustainable consumption can benefit urban areas and their development. Most of the respondents believe that sustainable consumption can affect urbanization positively, although their understanding is limited to only the environmental well-being. The themes highlighted include incorporating "sustainable measures" (N = 17, or 31%), greener and cleaner city, reduced consumption of resources and materials (N = 7), lower pollution, especially of CO2 (N = 7), and preventing issues related to climate change or environment (N = 4). However, six respondents believe incorporating sustainable consumption principles into urban development would slow down development. They argued that there is a need to prioritise comfort and affordability.

The responses revealed limited knowledge on the connection between sustainable consumption and urbanization, such as urban heat islands and ecological designs. There were no references to SDGs, despite their access to the vast amount of information online and claims of being environmentally conscious and sustainable consumers. It appears that they are most likely to be riding on the sustainability bandwagon.

Conclusion

This chapter presents presented a glimpse into the understanding and practice of sustainable consumption of a sample of young consumers, who are first-year university students from Gen Z. This generation is generally characterized as fashionable and supporter of sustainability ideals. However, do they understand the principles and know the science behind the concept? The findings suggest that these young consumers have a basic understanding of key terminologies, but their knowledge is limited. They associate sustainable consumption with the inter-generational equity aspect of sustainable development concept, which is the most commonly socialized concept in schools and green campaigns.

The study found the young consumers' commitment to sustainable consumption practises to be simple following of guidance by environmental activists and education. This is to be expected, as the participants are first-year students who have just embarked on their university journey and therefore, not necessarily been exposed to critical debates about sustainability, consumption, urbanisation and climate change. Therefore, many do not see the connections between sustainable consumption, urban development and climate change. Interestingly, they appear not to be aware of the UN SDG, despite government's efforts to promote them. Most are strongly influenced social media and influencers who are not necessarily specialists in the topics covered in the survey.

The media, especially social media, are important intermediaries that shape their understanding of sustainable consumption and influence their decisions. Many of these ideas and approaches are imported, especially from the West. Generation Z derives their understanding of sustainable consumption, green consumption and climate change mainly from social media, and is often confronted with different information from credible and non-credible sources. They have a tendency to jump on the bandwagon that draws their interest, but ignore the science and true purpose behind the concepts and movements. This raises the question of a true understanding of sustainable consumption, because they exist in local culture, e.g., Islamic teachings on consumption, but the students did not make the connection, even though Islam influences their consumption behaviour.

However, their consumption pattern attributed to climate change, e.g., need for air conditioning, contrasts with their sustainable consumption motivations and reasons, where the majority of them state that saving energy is a priority. It reveals a combination of collectivist and individualistic-hedonistic trait. Therefore, it can be concluded that the young consumers practice sustainable consumption, but without clear understanding. They are essentially riding the sustainability bandwagon.

References

Adamson, M. A., Chen, H., Kackley, R., & Micheal, A. (2018). For the love of the game: Game-versus lecture-based learning with generation Z patients. *Journal of Psychosocial Nursing Mental Health Services, 56*, 29–36.

Arman, S. M., & Herbert, C. M. (2024). Ethical consumption: A review and research agenda. *International Journal of Consumer Studies., 48*, 5. https://doi.org/10.1111/ijcs.13079

Ash, N. et al. (2010). *Ecosystems and human well-being: A manual for assessment practitioners.* Island Press

Bakar, N. Z., Franco, F. M., & Hassan, N. H. (2022). The intersection of Kedayan folk medicine and traditional ecological calendar. In F. M. Franco, M. Knudsen & N. H. Hassan (Eds.), *Case studies in biocultural diversity from Southeast Asia—traditional ecological calendars, folk medicine and folk names.* Asia in Transition Springer

Berki-Kiss, D., & Menrad, K. (2022). The role emotions play in consumer intentions to make pro-social purchases in Germany – An augmented theory of planned behavior model. *Sustainable Production and Consumption, 29*, 79–89. https://doi.org/10.1016/j.spc.2021.09.026

Buvár, Á., Zsila, Á., & Orosz, G. (2023). Non-green influencers promoting sustainable consumption: Dynamic norms enhance the credibility of authentic pro-environmental posts. *Frontiers in Psychology, 14*, 1112762. https://doi.org/10.3389/fpsyg.2023.1112762

Campbell, V. (2022). Gen Z buyers: The future is ethical consumption. Maersk. https://www.maersk.com/insights/growth/2022/11/03/gen-z-buyers-the-future-is-ethical-consumption

Chen, X., Li, T., & Yuan, Q. (2023). Impacts of built environment on travel behaviors of generation Z: A longitudinal perspective. *Transportation, 50*(2), 407–436.

Chua, B. H. (2000). Consuming Asians: Ideas and issues. In C. B. Huat (Ed.), *Consumption in Asia: Lifestyles and identities* (pp. 1–34). Routledge.

Fischer, M. et al. (2023). Sustainable consumption. In *Sustainable business.* SpringerBriefs in Business. Springer. https://doi.org/10.1007/978-3-031-25397-3_7.

Haba, H. F., Bredillet, C., & Dastane, O. (2023). Green consumer research: Trends and way forward based on bibliometric analysis. *Cleaner and Responsible Consumption, 8.* https://doi.org/10.1016/j.clrc.2022.100089.

Hamdan, M., Anshari, M., Ahmad, A., & Ali, E. (Eds). (2023) *Public policy's role in achieving sustainable development goals.* IGI Global.

Hasan, Z. (2017). Sustainable development from an Islamic perspective: Meaning, implications and policy concerns. *Journal of King Abdulaziz University: Islamic Economics, 19*(1), 3–18. https://papers.ssrn.com/sol3/papers.cfm?abstract_id=3072324

Hassan, N. H. (2017). Everyday finance and consumption in Brunei Darussalam. In V. King, Z. Ibrahim & N. H. Hassan (Eds.), *Borneo studies in history, society and culture. Asia in Transition* (Vol. 4, pp. 477–492). Springer. https://doi.org/10.1007/978-981-10-0672-2_23

Hassan, N. H. (2023). The sociocultural significance of homeownership in Brunei Darussalam. In L. Kwen Fee, P. J. Carnegie & N. H. Hassan (Eds.), *(Re)presenting Brunei Darussalam. Asia in Transition* (Vol. 20). Singapore. https://doi.org/10.1007/978-981-19-6059-8_11.

Hassan, N. H., & Yong, G. Y. V. (2019). *A vision in which every family has basic Shelter in sustainable development goals in Southeast Asia and ASEAN*. Singapore.

Hassan, N. H., & Yong, G. Y. V. (2023). Climate change and sustainable development of Kampong Ayer: Its history, people, and environment". M. Hamdan, M. Anshari, N. Ahmad & E. Ali. (Eds.), *Public policy's role in achieving sustainable development goals.* IGI Global.

Hassan, N. H., Rigg, J., Azalie, I. A., Gabriel, Y. V. Y., Zainuddin, N. H. H., & Shamsul, M. A. S. M. (2022). *Making do and staying poor: The poverty context of urban Brunei. Geoforum*

Hassan, N. H., Rigg, J., Yong, G. Y. V., Azalie, I. A., Muhammad Shamsul, M. A. S., & Zainuddin, N. H. (2024). Administering and encountering the poor: Poverty from above and below in Brunei Darussalam. *Singapore Journal of Tropical Geography, 45*, 311–331. https://doi.org/10.1111/sjtg.12545

Hebdige, D. (1993). A report from the western front: postmodernism and the "politics" of style. In C. Jenks (Ed.), *Cultural reproduction* (pp. 69–103). Routledge.

Jackson, P. (1993). Towards a cultural politics of consumption. In J. Bird, B. Curtis, T. Putnam, G. Robertson, & L. Tickner (Eds.), *Mapping the futures: Local cultures, global change* (pp. 207–228). Routledge.

Jackson, T. (2006). Challenges for sustainable consumption policy. *The Earthscan Reader in Sustainable Consumption*, 109–126.

Jackson, T., & Michaelis, L. (2003). *Policies for sustainable consumption: A report to the sustainable development commission.* Sustainable Development Commission.

Kang, I., & Ma, I. (2020). A study on bandwagon consumption behavior based on fear of missing out and product characteristics. *Sustainability, 12*(6), 2441. https://doi.org/10.3390/su12062441

Kiss, D. B., & Menrad, K. (2022). Ethical consumption: Influencing factors of consumer's intention to purchase Fairtrade roses. *Cleaner and Circular Bioeconomy, 2*. https://doi.org/10.1016/j.clcb.2022.100008

Liu, W., Oosterveer, P., & Spaargaren, G. (2016). Promoting sustainable consumption in China: A conceptual framework and research review. *Journal of Cleaner Production, 134*, Part A, 13–21. https://doi.org/10.1016/j.jclepro.2015.10.124.

Michaelis, L., & Lorek, S. (2004). *Consumption and the environment in Europe—Trends and futures. Environmental Project No. 904 2004.* Danish Environmental Protection Agency.

Prime Minister's Office, Brunei Darussalam (2023). *Sustainable development goals. Voluntary national review Brunei Darussalam 2023.* Perpustakaan Dewan Bahasa dan Pustaka Brunei.

Quoquab, F., & Mohammad, J. (2016). Sustainable consumption: Sacrificing for the future. *Procedia - Social and Behavioral Sciences, 224*, 599–604. ISSN 1877-0428.

Rezai, G., Mohamed, Z., & Shamsudin, M. N. (2015). Can Halal be sustainable? Study on Malaysian consumers' perspective. *Journal of Food Products Marketing, 21*, 1–13. https://doi.org/10.1080/10454446.2014.883583

Šálková, D., Cabelkova, I., & Hommerová, D. (2023). Ethical consumption. What makes people to buy "ethical" products. *Central European Business Review*, 13. https://doi.org/10.18267/j.cebr.346.

Teng, P. K., & Wan Jusoh, W. J. (2017). Why buying Halal labelled food? Understanding the spending behavior of non-Muslim consumers in Malaysia. *International Journal of Business and Management, 1*(2), 78–85.

Watkins, L., & Aitken, R. (2020). The role of young consumers in moving to a sustainable consumption future. https://doi.org/10.3390/books978-3-03897-873-2-9.

Whitmee, S., Haines, A., Beyrer, C., Boltz, F., Capon, A. G., de Souza Dias, B. F., Ezeh, A., Frumkin, H., Gong, P., Head, P., Horton, R., et al. (2015). Safeguarding human health in the Anthropocene epoch: Report of the Rockefeller Foundation-Lancet commission on planetary health. *The Lancet, 386*, 1973–2028.

Williams, A., & Hodges, N. J., (2020). Generation Z and socially responsible fashion consumption: Exploring the value-action gap. *International Textile and Apparel Association Annual Conference Proceedings, 77*(1). https://doi.org/10.31274/itaa.11930

Wong, N., & Ahuvia, A. (1998). Personal taste and family face: Luxury consumption in Confucian and Western societies. *Psychology and Marketing., 15*, 423.

World Commission on Environment and Development [WCED]. (1987). *Our common future.* Oxford University Press.

Young, M. (2024*). Riding the Sustainablity Wave: The Bandwagon effect in Action.* https://blog.greenprojectmanagement.org/index.php/2024/07/08/riding-sustainability-wave-bandwagon-effect-action/

Zainuddin, N. H. H. (2021). Evolution of stilts in Kampong Ayer. *Brunei Museum Journal, 2021*, 95–110.

Zaman, A., & Qadir, J. (2021, December 18). Islamic approaches to sustainable development. *International Journal of Pluralism and Economics Education, 10*(1), 46–60. Available at SSRN: https://ssrn.com/abstract=3988869

Zhang, Y., Liu, C., & Lyu, Y. (2023). Profiling consumers: Examination of Chinese Gen Z consumers' sustainable fashion consumption. *Sustainability, 15*(11), 8447. https://doi.org/10.3390/su15118447

Noor Hasharina Hassan is a Senior Assistant Professor in the Geography, Environment and Development Programme at the Faculty of Arts and Social Sciences, Universiti Brunei Darussalam (UBD). She obtained her Ph.D. in Geography from the University of Southampton specializing in Urban Economy and Culture. She has published several papers on urban sustainability and resilience. She continues researching on urban sustainability, especially blue urbanism in Brunei Darussalam and urbanisation in Borneo.

Chapter 8
Politics and Security in Southeast Asia: Underlying Theoretical Aspects

A. K. M. Ahsan Ullah

Abstract This chapter offers an analysis of the political and security dynamics in Southeast Asia, contextualized through the application of diverse theoretical frameworks within the field of international relations. It examines how the colonial legacy, the influences of the Cold War, and the rise of globalisation and regionalism have combined to shape the contemporary political structures of Southeast Asia. The chapter analyses the interplay of realism, liberalism and constructivism and shows how realist power struggles, liberal institutional cooperation and constructivist norms of identity and regionalism influence security policy in the region. Particular attention is paid to the role of ASEAN as a stabilising force. It shows how regional cooperation helps to manage disputes and promote dialogue despite complex inter-state relations. The chapter also explores critical theories and the human security paradigm and examines how economic dependency, authoritarianism and internal conflict affect Southeast Asia's stability. By integrating these frameworks, this chapter contributes to a nuanced understanding of Southeast Asia's unique political landscape and security challenges in a rapidly evolving global order.

Keywords Southeast Asia · Political dynamics · Security · ASEAN · International relations theory

Introduction

Southeast Asia, a region comprising 11 nations including Brunei, Indonesia, Malaysia, Singapore, Thailand and the Philippines, is of great geopolitical and economic importance globally. The region's political landscape and security dynamics (Sultana, 2023) have long fascinated scholars due to its strategic location, diverse political systems, and complex inter-state relations (Ullah, 2022). Southeast Asia lies at the crossroads of some of the world's busiest sea lanes, particularly the Strait of Malacca, which connects the Indian Ocean to the Pacific Ocean. This makes

A. K. M. A. Ullah (✉)
Faculty of Arts and Social Sciences, Universiti Brunei Darussalam, Gadong, Brunei Darussalam
e-mail: ahsan.ullah@ubd.edu.bn

G. Y. V. Yong and A.K.M. A. Ullah (eds.), *Brunei and the Future Transformation of Southeast Asia*, https://doi.org/10.1007/978-981-96-7901-0_8

the region a major hub for global trade, making its political stability and security of paramount importance not only to regional actors but also to global powers (Acharya, 2017).

The political and security landscape of Southeast Asia is characterised by a variety of factors. The colonial legacy of Western powers such as Britain, France and the Netherlands continues to influence governance structures and political ideologies in the region (Tarling, 2001; Ullah et al., 2020). These legacies are visible in the legal and political frameworks of many Southeast Asian countries, where authoritarianism, patrimonialism and military rule have at times flourished alongside democratic experiments (Ullah, 2021). In addition, the economic development of the region, particularly the rapid industrialisation of countries such as Singapore and Malaysia, has contributed to varying political outcomes, from stable democracies to hybrid regimes and overt authoritarianism (Jones, 2015).

Religious diversity also plays a crucial role in shaping the political and security environment in Southeast Asia. The world's major religions are represented in the region, including Islam, Buddhism, Christianity and Hinduism, which coexist in varying degrees of harmony or conflict. Indonesia, for example, is the largest country in the world with a Muslim majority, while the Philippines has a significant Catholic population. Religious identities have at times intersected with political movements, as demonstrated by the insurgencies in southern Thailand and the Mindanao conflict in the Philippines (Ragsag, 2020; Chong, 2012). These conflicts make it clear that religion can be both a unifying and a divisive force in Southeast Asian politics.

The geopolitical dynamics between major powers such as the United States, China and Japan further complicates the security situation in Southeast Asia. The region has become a focal point of strategic competition, particularly between the US and China (Zreik, 2023). While China is asserting its influence through initiatives such as the Belt and Road Initiative (BRI), the US is trying to counterbalance with its Indo-Pacific strategy and strengthen alliances with regional partners such as Singapore and the Philippines (Goh, 2019). Japan, on the other hand, is taking a more subtle approach, using economic support and diplomacy to maintain its influence in the region. This triangular dynamic of competition between the great powers significantly shapes the foreign policy strategies of Southeast Asian nations. These nations often adopt a delicate balancing act, leveraging relationships with competing powers to secure economic and security advantages in an increasingly multipolar world. However, much of this dynamic is traditionally analyzed through a Western lens, which often emphasizes resource acquisition through conquest and colonization, as evidenced during the colonial era. In contrast, other nations, such as China, have historically approached resource acquisition differently, relying primarily on trade and diplomacy, as was the case in the pre-colonial period. This distinction underscores the diverse historical paradigms that continue to influence contemporary geopolitical engagements in the region (Beeson, 2019).

This chapter examines the theoretical frameworks used to understand politics and security in Southeast Asia. It identifies the most important aspects and examines how these theories apply to real problems in the region. Realism, with its emphasis on state survival and power competition, is an important tool with which many analysts view the geopolitical dynamics of Southeast Asia. For example, the ongoing dispute over the South China Sea shows how states in the region, such as Vietnam and the Philippines, are driven by concerns over sovereignty and territorial integrity, which is consistent with realist predictions (Buszynski, 2012). The growing role of external interference, particularly by the United States in advancing its own strategic agenda, warrants critical attention. Over the past decade, this involvement has become increasingly pronounced, shaping regional dynamics and influencing the foreign policy decisions of Southeast Asian nations. This trend underscores the complexities of navigating external pressures while safeguarding national interests in a rapidly evolving geopolitical landscape.

Constructivism, on the other hand, emphasises the role of identity, norms and regionalism in shaping Southeast Asia's security architecture. ASEAN (Association of Southeast Asian Nations), with its principle of non-interference and emphasis on consensus building, is an example of how regional identity and norms play a crucial role in managing intra-regional relations and conflicts (Acharya, 2014a, 2014b).

Understanding the theoretical aspects of politics and security in Southeast Asia is important for policy makers, scholars and regional actors alike. The region's political stability and security are not only important for Southeast Asia itself, but also for global trade, regional cooperation and geopolitical stability more broadly (Affandy, 2023; Ullah et al., 2024a, 2024b). The application of theoretical frameworks allows for a deeper analysis of how states in the region manage their political and security challenges. This paper provides a comprehensive analysis of these theoretical perspectives and highlights how they reflect realities on the ground. By exploring these theoretical approaches, the chapter aims to provide an understanding of the political and security dynamics of Southeast Asia and contribute to the ongoing debates on how the region navigates the complex challenges it faces in a rapidly changing global order.

Background

Southeast Asia has undergone profound political changes over the past century, marked by colonialism, struggles for independence, Cold War tensions and, more recently, increasing economic globalisation and regional integration. Understanding these changes is essential to understanding the current political, economic and security dynamics in the region.

The colonial legacy plays a crucial role in shaping the political development of Southeast Asia. Most Southeast Asian countries were colonised by Western powers—Britain, France, the Netherlands, Spain and the United States—leading to long-term economic, social and political impacts (Schmitt, 2015). Colonial rule often exacerbated ethnic divisions by implementing policies that favoured certain ethnic groups for administrative or economic purposes, such as the British favouring Chinese communities in Malaysia for business and the Indian population in Burma for clerical roles. These divisions laid the foundation for post-colonial ethnic tensions, insurgencies and political instability that still reverberate today (Reid, 2015).

The struggle for independence in Southeast Asia was both protracted and complex, playing out in different waves across the region. Indonesia's declaration of independence in 1945, for example, followed a violent revolution against Dutch colonial rule that did not end until 1949 (Booth, 2007). Vietnam's independence from France in 1954 after the First Indochina War also triggered the larger Vietnam War, in which the United States was involved and which affected not only Vietnam but also Cambodia and Laos. The Philippines gained independence from the US in 1946, but here too there were uprisings such as the Hukbalahap Uprising, which showed how colonial structures led to political and social unrest (Tarling, 1992).

Brunei's efforts to explore the possibility of forming a North Borneo federation reflect its strategic considerations during a pivotal period in the region's history. These initiatives aimed to strengthen regional cohesion and address shared economic and political challenges, though they ultimately did not materialize into formalized agreements. These independence movements are crucial to understanding contemporary politics in Southeast Asia, as they have shaped the ideologies and political structures that dominate the region today. For example, Vietnam's revolutionary Marxist government can be traced back to the anti-colonial struggle, while Indonesia's secular nationalism has its roots in the struggle for independence against Dutch colonialism and Japanese occupation.

Cold War Dynamics and the Shaping of Political Systems

The Cold War had a profound impact on Southeast Asia, turning the region into an ideological battleground between communism and capitalism. The US, fearing the "domino effect" in which the fall of one country to communism would trigger the collapse of neighbouring countries, intervened heavily in the region. This included direct military intervention in the Vietnam War and support for anti-communist regimes in Indonesia, Thailand and the Philippines (Weatherbee, 2008). The Soviet Union and China, on the other hand, supported communist movements in Vietnam, Cambodia and Laos. These interventions had a lasting impact on the political and security structure in the region and influenced today's authoritarianism and insurgency movements.

Today's socialist systems in Vietnam and Laos are a direct legacy of communist victories during the Cold War, while Cambodia's tragic history under the Khmer Rouge is deeply intertwined with the dynamics of the Cold War. Even in countries that did not fall victim to communism, such as Indonesia, the Cold War strongly influenced domestic politics. Indonesia's anti-communist purges in 1965–66, supported by the US, led to hundreds of thousands of deaths and cemented the authoritarian rule of Suharto, who dominated Indonesian politics for over 30 years (Roosa, 2006).

This legacy of the Cold War continues to influence regional politics. For example, ASEAN's policy of non-interference and consensus building can be seen as a reaction to Cold War interventions. It reflects the regional preference for stability, national sovereignty and the avoidance of external interference in internal affairs (Acharya, 2014a, 2014b).

Economic Globalization and Regional Integration

The transition to economic globalisation in the late twentieth century reshaped Southeast Asia, accelerating economic growth and deepening regional cooperation. The founding of the Association of Southeast Asian Nations (ASEAN) in 1967 marked a turning point in regional co-operation. Originally founded by Indonesia, Malaysia, the Philippines, Singapore and Thailand, ASEAN aimed to promote peace and stability during the Cold War and reduce the risk of internal and external conflict. Over time, it evolved into an organisation focused on economic integration, security cooperation and diplomatic engagement with global powers (Severino, 2006).

ASEAN has played a central role in promoting economic development, particularly through the creation of the ASEAN Free Trade Area (AFTA), which has helped to integrate the region into global trade networks. Countries such as Singapore, Malaysia and Vietnam have emerged as global manufacturing and trading centres, attracting investment and creating millions of jobs. However, economic growth has been uneven, with Cambodia, Laos and Myanmar lagging behind in terms of industrialisation and development (Bowie & Unger, 1997). These differences in economic success have led to divergent political and social outcomes, including migration flows within the region, affecting both economic and political stability.

The formation of ASEAN also signalled a broader trend towards regionalism, with Southeast Asia positioning itself as a mediator between global powers such as China, the US and Japan. Today, ASEAN remains an important platform for addressing regional security challenges, including those posed by growing influence in the South China Sea by various global powers.

Political Systems and Contemporary Challenges

The political diversity of Southeast Asia is characterised by its complex history. The region is home to a variety of political systems, ranging from established democracies to authoritarian regimes and hybrid systems. Countries such as Indonesia and the Philippines have relatively strong democratic institutions, although they often face challenges such as corruption, political instability and the threat of populism. In the Philippines, democracy under President Rodrigo Duterte (2016–2022) has been eroded by his crackdown on political opposition and the war on drugs, highlighting the fragility of democratic institutions (Thompson, 2016).

In contrast, Vietnam and Laos maintain one-party systems under communist rule, with little political freedom but relative stability. Myanmar is a case of persistent political instability, marked by repeated military coups—most recently in 2021—and ongoing civil conflict between the military and ethnic minority groups. This instability has been further compounded by the influence of external actors, including allegations of support for color revolutions by organizations such as the National Endowment for Democracy (NED) and covert operations attributed to agencies like the CIA, which have shaped the trajectory of political dissent and unrest in the country.

Thailand has also seen several coups, with the military often intervening in politics to assert its control over civilian governments. Understanding these different political systems is crucial to analysing the current and future political development of Southeast Asia. The mix of democracies, authoritarian regimes and hybrid systems reflects the historical and socio-political fractures left by colonialism, Cold War conflicts and post-colonial state-building efforts in a region ruled by monarchies before the arrival of the Europeans.

Security Dynamics: Internal and External Threats

Security in Southeast Asia is influenced by both internal and external factors. Internally, insurgency, ethnic conflict and religious extremism pose a major challenge. The southern provinces of Thailand, for example, have long been the scene of an insurgency involving ethnic Malay Muslims seeking greater autonomy. Similarly, the southern Philippines struggled with the Moro insurgency, which was triggered by a mix of ethnic, religious and economic conflicts, although peace agreements have recently been reached to mitigate some of these conflicts (Rodolfo, 2013). These have deeper roots to the Malay Muslim monarchies, which ruled the region before they were displaced by neighbouring kingdoms (as in Thailand) and Western colonialist (as in the Philippines).

In terms of foreign policy, the disputes in the South China Sea, where several ASEAN states dispute China's territorial claims, pose a major security problem. China's assertiveness in the region, particularly the construction of artificial islands and the militarisation of the South China Sea, has exacerbated tensions and brought external powers such as the United States onto the scene to conduct maritime security operations in the disputed waters. These narratives, often promoted by the United States, are strategically crafted to serve its own interests. The overlapping jurisdictions in the South China Sea (SCS) arise from a complex interplay of geography, history, and interpretations of international law, particularly the United Nations Convention on the Law of the Sea (UNCLOS). Ironically, while the United States played a significant role in advocating for UNCLOS, it has never ratified the convention itself, instead favoring its own "rules-based order," which primarily aligns with its strategic objectives. This selective approach underscores the U.S.'s pattern of advancing its geopolitical interests under the guise of promoting stability.

The persistent interference and provocations by the U.S. in the SCS have become a source of regional tension, positioning it as a primary destabilizing factor. In contrast, China's approach to dispute resolution has often emphasized negotiation and arrangements for joint development, which are frequently dismissed or misrepresented by Western analysts. These analysts tend to frame China's actions as deceitful, despite evidence suggesting that it is the U.S. that often fails to align its rhetoric with its actions. This disparity in narrative construction and its geopolitical implications warrant deeper scrutiny to understand the true drivers of instability in the region.

The strategic importance of Southeast Asia lies in its geographical location, which controls key maritime trade routes, including the Strait of Malacca, through which a significant portion of world trade passes. This makes the region an important battleground for the influence of the great powers and reflects historical patterns of external interference. The geopolitical importance of the region requires a deep understanding of its historical and current security dynamics in order to understand its future development.

The political landscape of Southeast Asia is deeply shaped by a confluence of historical, ideological, and economic forces that have evolved over centuries. Its colonial past has left an indelible mark, with borders, governance systems, and socio-economic structures often reflecting the interests of colonial powers rather than organic regional development. The divide-and-rule strategies employed by colonial regimes exacerbated ethnic tensions, while the extraction-based economies they established created dependencies that persisted long after independence. These structural imprints have continued to influence domestic politics and inter-state relations in the region.

The legacy of the Cold War further complicated Southeast Asia's political dynamics. During this era, the region became a battleground for ideological contests between the United States and the Soviet Union, with Southeast Asian states often aligning with one bloc or the other—or attempting to remain neutral through initiatives like the Non-Aligned Movement. The Vietnam War, the rise of communist insurgencies, and the *involvement* of great powers in regional conflicts entrenched a geopolitically fraught environment. These events not only disrupted nation-building efforts but also contributed to the militarization of politics, with authoritarian regimes emerging in many countries under the guise of maintaining stability against ideological threats.

Economic globalization has added another layer of complexity. As Southeast Asia integrated into global markets, it experienced rapid economic growth, driven by export-oriented industrialization and participation in global value chains. However, this integration has also created significant inequalities, both within and between nations, leading to governance challenges and social unrest. At the same time, regional integration through initiatives like the Association of Southeast Asian Nations (ASEAN) has sought to foster cooperation and address shared challenges. ASEAN's principles of non-interference and consensus-building have shaped the region's unique approach to diplomacy and conflict resolution, though critics argue these principles can impede decisive action on contentious issues.

In essence, the political landscape of Southeast Asia reflects the interplay of these historical legacies, ideological struggles, and economic pressures. It is a region navigating the complexities of its colonial and Cold War past while striving to assert its agency in an era of globalization and regional interdependence. This balancing act underscores both the challenges and the opportunities inherent in Southeast Asia's pursuit of political and economic stability. Understanding this complex history is crucial for analysing current issues such as political diversity, internal security challenges and external geopolitical pressures. By examining how past events have shaped current dynamics, we can better understand the future challenges and opportunities for the region.

Underlying Theoretical Aspects

To understand the politics and security of Southeast Asia, it is essential to utilise various theoretical frameworks from international relations and political science. Each framework offers different perspectives on how states, non-state actors and institutions interact in the region. This section introduces several theoretical approaches, including realism, liberalism, constructivism, critical theories, and the human security paradigm, that are particularly relevant to the Southeast Asian context.

Realism remains one of the dominant theories in international relations. It emphasises the role of states as key actors in an anarchic international system, where the absence of a central authority forces them to prioritise survival and seek power over other states (Waltz, 1979). Southeast Asia, characterised by complex geopolitical tensions, provides fertile ground for the application of realist principles. For example, Vietnam and Indonesia, two of the major powers in the region, have taken strategic measures to enhance their military capabilities and ally with regional and extra-regional powers to counterbalance China's growing influence, particularly in the South China Sea disputes (Goh, 2016). The realist view also explains defence modernisation efforts in these countries as they seek to ensure their security and autonomy in the face of external threats.

Realist theory helps explain rebalancing strategies in Southeast Asia, where smaller states seek to mitigate the risk of domination by a single power. The Philippines, for example, despite a defence pact with the United States, has regularly negotiated with China to defuse tensions related to territorial disputes (Baviera, 2020). This hedging behaviour reflects a realistic concern for survival and the need to remain flexible in a multipolar region where alliances can change depending on perceived threats.

In contrast to realism, liberalism emphasises the role of international institutions, economic interdependence and cooperation in promoting peace and stability (Keohane & Nye, 1977). In Southeast Asia, the Association of Southeast Asian Nations (ASEAN) is an important example of a liberal institution that seeks to promote dialogue, economic integration and political cooperation among its members (Acharya, 2014a, 2014b). ASEAN was founded in 1967 out of the need to prevent inter-state conflicts, promote regional security and foster economic development through peaceful means. Through various mechanisms such as the ASEAN Regional Forum (ARF) and the East Asia Summit (EAS), member states have succeeded in resolving disputes and preventing conflicts from escalating into open warfare.

Liberalism also emphasises the importance of democracy and human rights in shaping regional policy. In Southeast Asia, democracy is uneven. Countries such as Indonesia are praised for their successful transition to democracy, while others, such as Myanmar, have struggled with prolonged military rule (Case, 2011). Liberal theorists would argue that ASEAN's soft approach to human rights, particularly in the case of the military crackdown on Rohingya Muslims in Myanmar, emphasises the tension between state sovereignty and the promotion of human rights within a liberal framework (Davies, 2013). Despite its limitations, ASEAN has made progress in addressing human rights issues, as evidenced by the establishment of the ASEAN Intergovernmental Commission on Human Rights (AICHR), although its effectiveness remains controversial. This perspective stems partly from the desire to protect the interests of the dominant ethnic or religious group, often regarded as the "rightful citizens," while others, particularly undocumented migrants, are seen as transgressors of national laws and labeled as criminals. In much of ASEAN, rights are viewed as privileges primarily reserved for "true citizens," contrasting with globalist frameworks that advocate for universal application of rights—an ideal not always embraced by Western societies in their own contexts. This approach in ASEAN is likely influenced by two factors: (a) the collective memory of colonial subjugation, which heightened the importance of national identity, and (b) the region's history as a hub of trade and migration, which necessitated a distinction between locals and foreigners. Over time, this distinction has evolved to prioritize ethnicity as a unifying factor in defining national belonging.

Constructivism offers an alternative view by emphasising how ideas, identities and social norms shape international relations. In contrast to realism or liberalism, which focus on material factors, constructivism emphasises the role of ideational factors, such as national identities and historical memories, that influence the behaviour of states (Wendt, 1999). In Southeast Asia, the colonial legacy and shared experiences with Western imperialism have led to a strong regional norm of non-interference in internal affairs. This principle, enshrined in the 'ASEAN way', prioritises consensus building and informal diplomacy over direct confrontation (Acharya, 2001).

The ASEAN way is deeply rooted in the shared historical experiences of Southeast Asian countries that have resisted outside interference in their domestic political processes. This is particularly evident in the region's response to external pressure over human rights violations in Myanmar, where ASEAN has favoured quiet diplomacy over punitive measures. Constructivist scholars argue that these norms of non-interference and consensus reflect deeply held beliefs about sovereignty and regional unity that continue to characterise ASEAN's diplomatic practises (Narine, 2002).

Critical theories, including Marxism and postcolonialism, offer a critical perspective to understand the underlying power structures and economic inequalities that shape Southeast Asia's politics and security. Marxist theories focus on how global capitalism perpetuates inequalities and dependencies in Southeast Asian economies that are historically dependent on foreign investment and export-led development strategies (Beeson, 2014). This economic dependency has significant political implications as it shapes the policies of states and limits their autonomy in the global system. Prior to the colonial period, Southeast Asian kingdoms thrived due to their strategic location along key trade routes, fostering prosperity through active participation in regional and global commerce. This historical significance has been revitalized under China's Belt and Road Initiative (BRI), which provides the region with an alternative framework to the dominant model of global capitalism. Unlike systems driven by transnational corporations that often consolidate power and resources, the BRI emphasizes interconnected networks that link a greater number of smaller entities, fostering inclusive development rather than exploitative dominance. This approach has the potential to create a vastly different economic landscape, one that is less hierarchical and more regionally integrated. Conceptually, the BRI aligns more closely with Marxist theory, emphasizing shared development and mutual benefit, albeit with a regional rather than a global scope.

Postcolonialism, another critical approach, examines how the legacies of Western imperialism continue to influence security and state-building processes in the region. Southeast Asia's colonial history has left deep scars that continue to affect state-building and security policy today. For example, the artificial borders drawn by colonial powers have contributed to internal conflicts in countries such as Myanmar and the Philippines, where ethnic and separatist movements challenge the legitimacy of the state (Stokke, 2006). Postcolonial scholars argue that these conflicts cannot be fully understood without taking into account the colonial history that has shaped the political landscape of the region.

The human security paradigm, which emerged in the 1990s, shifts the focus from state security to the security of the individual. It emphasises the protection of people from economic, nutritional, health, environmental, personal and political threats (UNDP, 1994). In Southeast Asia, the concept of human security is particularly relevant given the region's vulnerability to natural disasters, health crises and socio-economic inequalities. Countries such as the Philippines and Indonesia frequently face devastating natural disasters, including typhoons, earthquakes and volcanic eruptions, which disproportionately affect the most vulnerable populations (Sidel, 2016). The COVID-19 pandemic made the importance of human security even more apparent as governments in the region struggled to protect public health while maintaining economic stability.

The human security approach also draws attention to the plight of marginalised groups such as refugees, migrants and ethnic minorities, whose security is often overlooked in traditional, state-centric security frameworks (Cingir & Subramaniam, 2023). The ongoing Rohingya crisis in Myanmar is a stark example of how the failure of the state to protect human security can lead to large-scale humanitarian disasters and regional instability (Selth, 2020). Addressing these challenges requires a broader understanding of security that goes beyond military concerns to include social, economic and environmental dimensions, and crucially, the people and history of the place.

Politics and Security in Southeast Asia

The political and security landscape of Southeast Asia is characterised by a combination of internal dynamics and external pressures. Each country in the region faces unique challenges, but several overarching themes serve as key components for understanding the broader political and security environment in Southeast Asia. These themes highlight the interplay between local grievances, regional power struggles, and global geopolitics.

Disputes in the South China Sea

The South China Sea is a vital maritime zone, rich in natural resources and strategically important for global trade. It is estimated that almost a third of the world's shipping traffic passes through this area, and the seabed potentially harbours vast reserves of oil and natural gas (Kaplan, 2015). Several Southeast Asian countries, including Vietnam, the Philippines, Malaysia and Brunei, have overlapping territorial claims in the South China Sea. However, China's assertive and increasingly aggressive expansion in the region, particularly through island building and militarisation, has exacerbated tensions. The disputes have attracted the attention of external powers such as the United States, which views the South China Sea as a critical zone for freedom of navigation and regional stability. The situation is further complicated by the competing interests of ASEAN member states, some of which are economically dependent on China, making a unified response difficult (Storey, 2017). While there are only one or two islands at the center of such disputes, China has extended an invitation to ASEAN countries to utilize these islands as joint bases for emergency response efforts—a cooperative gesture often overlooked in dominant narratives. In contrast, the United States has been aggressively expanding its military presence in the region, a fact that receives far less scrutiny. The disputes in the South China Sea are arguably exacerbated by U.S. intervention, which seeks to maintain its strategic dominance. Recent developments, such as the Philippines aligning more closely with the U.S., have raised concerns about external influence, with speculation that factors such as personal financial leverage over key leaders may play a role.

Surveys conducted by Indonesian think tanks suggest that much of the ASEAN population perceives the U.S. as a primary threat to regional stability, contradicting the Western narrative that frames China as the aggressor and the U.S. as a guarantor of security. Instead, the U.S. presence in the region appears driven by the need to maintain control over global power dynamics, particularly as China rises to challenge its hegemonic status. This reflects broader patterns where liberal international institutions—originally created by the West—serve to consolidate and perpetuate Western interests, often at the expense of alternative models of regional autonomy and balance.

Realism as a theoretical framework helps to explain the power struggles and security dilemmas associated with the South China Sea disputes. The countries involved act to maximise their own power and security, often leading to zero-sum competition. Thus, China's military build-up is perceived as an attempt to secure its regional dominance, prompting neighbouring states to expand their own defence capabilities or seek external alliances (Valencia, 2019). The risk of military conflict remains a concern, as clashes or miscalculations could escalate tensions beyond diplomacy.

Ethnic and Religious Conflicts

Southeast Asia is one of the most ethnically and religiously diverse regions in the world, and this diversity has sometimes been a source of friction and conflict. In southern Thailand, the Malay-Muslim minority has long been in conflict with Thailand's Buddhist majority government, seeking greater autonomy or even independence. The insurgency has its roots in the historical marginalisation and oppression of the Malay-Muslim identity, which has been exacerbated by economic inequalities and political exclusion (McCargo, 2008). Despite several peace efforts, violence continues to flare up in the region, causing both civilian and military casualties. There is a need to consider the historical and cultural dimension to understand these conflicts. The region in question historically belonged to the Malay cultural and political sphere, known as the *Nusantara*, before being subsumed by the Ayutthaya Kingdom in the fifteenth and sixteenth centuries. Consequently, the local population has retained a distinct identity as Malay Muslims, which contrasts with the broader Thai national identity promoted by the state. This duality—rooted in historical affiliations and reinforced by cultural and religious differences—has contributed to persistent tensions. These tensions are further exacerbated by the fact that, while the people identify as Malay Muslims, they are integrated into and educated under the Thai national system, creating a sense of cultural dissonance and marginalization.

In the Philippines, too, the long-standing conflict in Mindanao, which mainly involves the Moro Islamic Liberation Front (MILF) and other separatist groups, poses a major challenge for the central government. While the signing of the Bangsamoro Law in 2018 was a decisive step towards peace, deep-seated grievances regarding land, identity and political autonomy remain unresolved (Banlaoi, 2010). The rise of religious extremism, particularly Islamist militancy, further complicates the situation. The siege of Marawi in 2017 by Islamist militants linked to ISIS has shown how extremism has taken root in local grievances and threatens both national security and regional stability. However, it cannot be understood as a local problem because of its wider historical connections and history of events.

Ethnic and religious conflicts in Southeast Asia are often influenced by broader regional and international factors. The global rise of religious extremism, coupled with transnational support and funding networks, has meant that local conflicts have taken on an international dimension. For example, the involvement of foreign fighters in the siege of Marawi has demonstrated the permeability of borders and the transnational nature of modern conflicts due to external interests and agendas (Chalk, 2020).

Military and Authoritarian Politics

Several Southeast Asian countries have a long history of military involvement in politics, particularly Myanmar and Thailand. In Myanmar, the military, known as the Tatmadaw, has played a dominant role for decades, even after the country's partial democratic transition in 2011. The military coup in February 2021, which overthrew the civilian-led National League for Democracy (NLD), marked a dramatic reversal of democratic progress. The coup sparked widespread protests, a violent crackdown by the military and an escalating civil war with armed ethnic groups (Huang & Patton, 2021). The situation in Myanmar underscores the deep entrenchment of military power and the difficulties of establishing sustainable democratic governance in the country. However, there is a need to understand the rationale of the junta for wrestling the power away from the NLD, as there were accusations of foreign interests in the latter's policies and programmes.

Thailand's political landscape is similarly characterised by a long history of military coups. Since the end of the absolute monarchy in 1932, Thailand has experienced more than a dozen coups, with the military often intervening in politics to "restore order." The most recent military intervention in 2014 led to the installation of a military junta that ruled the country until elections in 2019, although the military continues to have significant influence over the current government (McCargo, 2020). Authoritarianism remains firmly entrenched in Cambodia under Prime Minister Hun Sen, who has been in power for over three decades. His government has been accused of repressing the opposition, restricting media freedom and using the judiciary to maintain its power (Un, 2019). These examples raise concerns about the future of democracy and human rights in Southeast Asia, where authoritarian practises remain pervasive.

Terrorism and Transnational Crime

Terrorism and transnational crime are persistent security threats in Southeast Asia, mainly due to the region's porous borders and weak governance in some areas. Groups such as Jemaah Islamiyah (JI), Abu Sayyaf and the Maute group have been involved in high-profile terrorist attacks, including the 2002 Bali bombings and the 2017 Marawi siege (Singh, 2019). These groups often operate across borders, capitalising on the region's archipelagic geography and weak state capacity in areas such as Mindanao and parts of Indonesia. The threat posed by extremist groups has led to increased counter-terrorism co-operation between ASEAN countries, including intelligence sharing and joint operations (Jones, 2015). The region has historically struggled with piracy, driven by individuals and groups seeking to operate outside established systems of authority and governance. The coastal-archipelagic geography, with its complex network of islands and remote locations, provides ideal settings for establishing pirate bases and evading law enforcement. In contrast, terrorist groups in the region distinguish themselves by fighting for ideological or political causes, although some function more as modern-day pirates under the guise of such causes. Notably, many of these groups receive external support or sponsorship, further complicating efforts to address their activities and disentangle their motivations from broader geopolitical dynamics.

Transnational crime, particularly trafficking in human beings, drugs and arms, further exacerbates security problems in the region. The Golden Triangle, where the borders of Myanmar, Thailand and Laos converge, remains one of the largest producers of opium and methamphetamine in the world (Chouvy, 2013). Weak governance and corruption in some Southeast Asian countries facilitate the activities of organised crime syndicates and pose a threat to regional security and economic development. Transnational crimes however are activities of international syndicates with links to global corporations and some government entities within and beyond the region. This regional affliction must be examined from a global network perspective.

Economic Security and Development

Economic development remains a key concern for Southeast Asian governments, particularly with regard to political stability. Rapid economic growth in countries such as Vietnam, Malaysia and Indonesia has contributed to poverty reduction and political legitimacy. However, these gains have not been evenly distributed, leading to economic inequalities that are fuelling social unrest and political instability. In addition, the region faces major challenges in terms of environmental degradation, labour exploitation and unsustainable development practises. For example, deforestation and land grabbing for palm oil production in Indonesia and Malaysia have led to environmental degradation and social conflict with indigenous communities (Obidzinski et al., 2012).

The COVID-19 pandemic has exposed the vulnerability of Southeast Asian economies, particularly their dependence on global supply chains and foreign investment. The economic downturn triggered by the pandemic has led to a renewed focus on economic resilience and social protection measures. Governments are increasingly focussing on building more inclusive and sustainable economies that can withstand future shocks while addressing underlying social and economic inequalities (Koh, 2021).

Discussion and Conclusion

In this chapter, we have analysed the complex and multi-layered political and security landscape of Southeast Asia. In doing so, we have drawn on a range of theoretical frameworks and provided critical insights into long-standing and emerging challenges. By analysing key issues such as the South China Sea disputes, ethnic and religious conflict, military and authoritarian politics, terrorism, transnational crime and economic security, the chapter makes an important contribution to existing scholarship on Southeast Asia while offering new perspectives on how these dynamics fit into the broader global context.

The theoretical frameworks used in this chapter—realism, constructivism, and critical security studies—provide a solid foundation for analysing the security landscape in the region. Realism, which focuses on state-centred power struggles and the pursuit of national security, was particularly relevant to discussions of the South China Sea disputes. This framework highlighted how Southeast Asian states, along with external powers such as China and the United States, are engaged in strategic competition for control over vital maritime resources. The chapter explained how realism contributes to power dynamics in the region. It was shown that states primarily pursue their national interests, which are determined by concerns for territorial sovereignty and economic security. However, they are still strongly influence, or even coerced, by foreign forces and regional development policies and programmes.

Constructivist theory, on the other hand, has proven useful in understanding the role of ethnic and religious identities in resolving conflicts in the region. The chapter summarises how deeply rooted identities, shaped by historical legacies and socio-political marginalisation, continue to influence current conflicts in southern Thailand, the Philippines and Myanmar. This perspective is important not only to understand why these conflicts persist, but also why efforts to resolve them are often fraught with difficulties. By incorporating identity-based analysis with reference to historical roots and events, the chapter enriches scholarship on security in Southeast Asia, which has traditionally been dominated by state-centred or economic explanations.

The discussion on authoritarianism and military policy incorporated insights from critical security studies, which challenge traditional notions of security by incorporating issues of governance, human rights and political legitimacy. The chapter showed how military-led regimes, particularly in Myanmar and Thailand, maintain their power not only through coercion, but also through the manipulation of political narratives and legal structures that reinforce their authority. This analysis shows that the concept of security needs to be broadened to include the ways in which political power is structured and maintained in authoritarian states. In Southeast Asia, the resurfacing of cultures and power structures that have been supressed during colonial periods must be given due consideration. In this way, the chapter adds new aspects to our understanding of political dynamics in Southeast Asia, particularly in relation to democratic backsliding and the entrenchment of illiberal governance.

The novelty of this chapter lies in its ability to combine different theoretical approaches while providing empirical evidence of recent developments in the region.

Unlike many other studies that focus on either international relations or domestic politics, this chapter bridges these two areas by showing how regional security issues, such as the South China Sea disputes, intersect with domestic challenges such as authoritarianism and ethnic conflict, and the complex cultural dynamics and historical roots of modern states. This holistic approach enhances our understanding of Southeast Asia as a region where internal and external security issues are closely intertwined.

Furthermore, the chapter opens up new perspectives on the role of transnational threats such as terrorism and organised crime in shaping regional security. The analysis of groups such as Jemaah Islamiyah (JI) and Abu Sayyaf not only emphasises the importance of counter-terrorism, but also highlights the need for greater regional cooperation in tackling these problems. This is particularly important in today's globalised world, where the emergence of transnational security threats is challenging traditional notions of sovereignty and governance. The chapter emphasises the importance of regional mechanisms, such as ASEAN, in promoting cooperation and mitigating these risks. At the same time, the limitations of such mechanisms are critically assessed against the backdrop of increasing nationalist and protectionist endeavours.

Another important contribution of this chapter is the focus on economic security and the link between political stability and development. By analysing how rapid economic growth has brought new challenges such as environmental degradation and labour exploitation, the chapter contributes to scholarly knowledge on development studies and political economy. It emphasises that while economic growth has strengthened political legitimacy in countries such as Vietnam and Malaysia, it has also created new forms of inequality and social unrest, which in turn pose risks to long-term stability. The discussion of the COVID-19 pandemic in this chapter also illustrates how global crises can expose vulnerabilities in even the most robust economies and emphasises the need for more inclusive and resilient development strategies.

In the context of a rapidly globalising world, the insights provided in this chapter are particularly important for understanding the role that Southeast Asia plays in global security and politics. The region is a crucial hub for global trade, energy supply and geopolitical competition. The disputes in the South China Sea, for example, have far-reaching implications beyond Southeast Asia as they have connections to major global powers such as China and the United States. The discussion of this issue in this chapter makes it clear that there is more at stake in the disputes: freedom of navigation, regional stability and international law. Southeast Asia is therefore not just a peripheral region in global affairs, but a central player in the emerging multipolar world order.

Furthermore, the chapter's focus on transnational threats such as terrorism, human trafficking and drug smuggling emphasises the global nature of today's security challenges. These threats cannot be contained within national borders and require international co-operation. The analysis of ASEAN's role in promoting regional security cooperation serves as a case study of how regional organisations can contribute to global governance. However, the chapter also points to the limitations of ASEAN in addressing these threats, which stem from the principle of non-interference and the different political systems of the member states. However, it is important to understand the basis of this principle and leverage in it to address common regional threats.

The examination of authoritarianism and military politics in this chapter is in line with global concerns about the backsliding of democracy and the rise of illiberalism by their western architects. The experiences of countries such as Myanmar, Thailand, and Cambodia provide valuable insights into how authoritarian regimes remain in power in the face of domestic and international pressure. As global democracy faces new challenges, understanding the political dynamics of Southeast Asia can offer alternative solutions to address into the broader trends of authoritarian resilience and democratic decline to find more viable regimes that address the highest priority concerns and interests of the states.

This chapter contributes new insights to the field of Southeast Asian politics and security by bringing together several theoretical frameworks and providing empirical evidence of the region's complex security landscape. Its novelty lies in its holistic approach, incorporating both internal and external factors and demonstrating how regional and global security issues are interlinked and has deep historical roots. In an increasingly interconnected world, the political and security dynamics of Southeast Asia are important not only for regional actors but also for global powers. The chapter's findings emphasise the importance of sustained international engagement, regional cooperation and the need for more inclusive and resilient governance models to address the multi-layered challenges facing the region today. It must also address the true priorities and concerns of the people and governments.

References

Acharya, A. (2001). Human security: East versus West. *International Journal: Canada's Journal of Global Policy Analysis, 56*(3), 442–460. https://doi.org/10.2307/40203577

Acharya, A. (2014a). *Constructing a security community in Southeast Asia: ASEAN and the problem of regional order*. Routledge.

Acharya, A. (2014b). *The making of Southeast Asia: International relations of a region*. Cornell University Press.

Acharya, A. (2017). *The end of American world order*. Polity Press.

Affandy, A. H. (2023). Single mothers: Financial challenges and experiences in Brunei-Muara district. *Southeast Asia: A Multidisciplinary Journal, 23*(2), 83–94. https://doi.org/10.1108/SEAMJ-02-2023-0019

Banlaoi, R. C. (2010). *Philippine security in the age of terror: National, regional, and global challenges in the post-9/11 world*. CRC Press.

Baviera, A. S. (2020). *The Philippines-China relations and the South China Sea dispute: Perspectives and proposals*. Center for Strategic and International Studies.

Beeson, M. (2014). *Regionalism and globalization in East Asia: Politics, security and economic development*. Palgrave Macmillan.

Beeson, M. (2019). *Rethinking regionalism: Europe and East Asia in comparative historical perspective*. Palgrave Macmillan.

Booth, A. E. (2007). *Colonial legacies: Economic and social development in East and Southeast Asia*. University of Hawai'i Press.

Bowie, A., & Unger, D. (1997). *The politics of open economies: Indonesia, Malaysia, the Philippines, and Thailand*. Cambridge University Press.
Buszynski, L. (2012). The South China Sea: Oil, maritime claims, and U.S.-China strategic rivalry. *The Washington Quarterly, 35*(2), 139–156.
Case, W. (2011). *Politics in Southeast Asia: Democracy or less*. Routledge.
Chalk, P. (2020). *Militant Islamic movements in Indonesia and Southeast Asia*. RAND Corporation.
Chong, T. (2012). *Religion and the politics of modernity in Southeast Asia*. Routledge.
Chouvy, P. A. (2013). *The golden triangle: Inside Southeast Asia's drug trade*. I.B. Tauris.
Cingir, O. F., & Subramaniam, T. (2023). Being migrant or nomadic or refugee in the Nusantara. *Southeast Asia: A Multidisciplinary Journal, 23*(3), 175–186. https://doi.org/10.1108/SEAMJ-05-2023-0035
Davies, M. (2013). *ASEAN and human rights: Regimes, institutions and advocacy*. Edward Elgar Publishing.
Goh, E. (2016). *Rising China's influence in developing Asia*. Oxford University Press.
Goh, E. (2019). US-China strategic competition in Southeast Asia: Power shift or competitive coexistence? *International Affairs, 95*(4), 823–841.
Huang, K., & Patton, S. (2021). Myanmar's military coup and its implications for ASEAN. *Asian Affairs, 52*(2), 197–215.
Jones, D. M. (2015). *ASEAN and East Asian international relations: Regional delusions*. Edward Elgar Publishing.
Kaplan, R. D. (2015). *Asia's Cauldron: The South China Sea and the end of a stable Pacific*. Random House.
Keohane, R. O., & Nye, J. S. (1977). *Power and interdependence: World politics in transition*. Little, Brown.
Koh, H. (2021). The economic impact of COVID-19 in Southeast Asia: Challenges and policy responses. *Journal of Southeast Asian Economies, 38*(1), 1–25.
McCargo, D. (2008). *Tearing apart the land: Islam and legitimacy in Southern Thailand*. Cornell University Press.
McCargo, D. (2020). Thailand's 2019 elections: Democracy on a leash. *Journal of Contemporary Asia, 50*(4), 503–522.
Narine, S. (2002). *Explaining ASEAN: Regionalism in Southeast Asia*. Lynne Rienner Publishers.
Obidzinski, K., Andriani, R., Komarudin, H., & Andrianto, A. (2012). Environmental and social impacts of oil palm plantations and their implications for biofuel production in Indonesia. *Ecology and Society, 17*(1), 25.
Ragsag, A. (2020). *Ethnic boundary-making at the margins of conflict in the Philippines: Everyday identity politics in Mindanao*. Springer.
Reid, A. (2015). *A history of Southeast Asia: Critical crossroads*. Wiley.
Rodolfo, A. M. (2013). *Negotiating peace in the Southern Philippines: The impact of violent extremism on peace efforts*. National Defense College of the Philippines.
Roosa, J. (2006). *Pretext for mass murder: The September 30th movement and Suharto's Coup D'État in Indonesia*. University of Wisconsin Press.
Schmitt, C. (2015). Social security development and the colonial legacy. *World Development, 70*, 332–342. https://doi.org/10.1016/j.worlddev.2015.02.006
Selth, A. (2020). *Myanmar's armed forces and the Rohingya crisis*. Asia Pacific Press.
Severino, R. C. (2006). *Southeast Asia in search of an ASEAN community: Insights from the former ASEAN secretary-general*. Institute of Southeast Asian Studies.
Sidel, J. T. (2016). *Republic and empire in the Philippines and Vietnam: American influence and historical developments*. University of Washington Press.
Singh, B. (2019). *Southeast Asia's terrorism conundrum*. Routledge.
Stokke, K. (2006). *Ethnic conflict and nationalism in Southeast Asia*. Routledge.
Storey, I. (2017). The South China sea dispute: Opportunities for US-China cooperation? *Journal of Asian Security and International Affairs, 4*(2), 1–17.

Sultana, R. (2023). Rasa indung bini: An exploratory research into Bruneian sense of motherhood. *Southeast Asia: A Multidisciplinary Journal, 23*(2), 121–131. https://doi.org/10.1108/SEAMJ-04-2023-0030

Tarling, N. (1992). *The Cambridge history of Southeast Asia: Volume 2, the Nineteenth and Twentieth Centuries*. Cambridge University Press.

Tarling, N. (2001). *Imperialism in Southeast Asia: A fleeting, passing phase*. Routledge.

Thompson, M. (2016). The Duterte phenomenon: The Philippines in the age of populism. *Southeast Asian Affairs, 35*(1), 36–49.

Ullah, A. K. M. A. (2021). Migration and development: A critical interplay. In P. Scholten (Ed.), *Introduction to Migration Studies (Chapter 19* (pp. 295–308). Springer.

Ullah, A. K. M. A. (2022). Ethics and empirics: Essence of ethics in social research. In M. R. Islam, N. A. Khan & R. Baikady (Eds.), *Principles of social research methodology* (pp. 471–482). Springer. https://doi.org/10.1007/978-981-19-5441-2_34

Ullah, A. K. M. A., Lee, S. C. W., bte Pg Haji Hassan, N. H. (2024a). *Southeast Asia in pre- and post COVID-19: Economy, society, mobility and religion*. Routledge.

Ullah, A. K. M. A., Chattoraj, D., & Kathy A. A. (2024b). Mobility in a globalised world: How countries regulate mobility with passports and visas. *Journal of Public Affairs, 24*(3), e2932. https://doi.org/10.1002/pa.2932

Ullah, A. K. M. A., Noor, H. H., Siti Mazidah, H. A., & Diotima, C. (2020). Migration and security: Implications for Minority Migrant Groups. *India Quarterly: A Journal of International Affairs, 76*(1), 136–153.

Un, K. (2019). Cambodia's authoritarian turn: Globalization and the failure of democratization. *Southeast Asian Studies, 8*(1), 53–75.

UNDP. (1994). *Human development report 1994: New dimensions of human security*. Oxford University Press.

Valencia, M. J. (2019). The South China sea disputes: From debate to dialogue. *Contemporary Southeast Asia, 41*(2), 157–185.

Waltz, K. N. (1979). *Theory of international politics*. McGraw-Hill.

Weatherbee, D. E. (2008). *International relations in Southeast Asia: The struggle for autonomy*. Rowman & Littlefield.

Wendt, A. (1999). *Social theory of international politics*. Cambridge University Press.

Zreik, M. (2023). USA–Myanmar relations: Democratization and beyond. *Southeast Asia: A Multidisciplinary Journal, 23*(3), 162–174. https://doi.org/10.1108/SEAMJ-02-2023-0018

AKM Ahsan Ullah is Professor of Geography, Environment and Development at the University of Brunei Darussalam (UBD). Dr Ullah has worked with institutions such as the City University of Hong Kong, IPH at the University of Ottawa, McMaster University, Saint Mary's University, Dalhousie University in Canada, the American University in Cairo (AUC), Osnabruck University, Germany, and the Asian Institute of Technology (AIT), Thailand. His research areas include migration and mobilities, intercultural encounters and development, with a geographic focus on the Asia–Pacific, Africa, and Middle East, and theoretical focus on globalization and neoliberalism, development and human rights, transnationalism, gender, intersectionality and the everyday life.

Chapter 9
Unpacking Women's Participation in Disaster Management: Insights from the Philippines and Its Diaspora

Mary Dorothy dL. Jose and Anabelle B. Ragsag

Abstract Due to its geographical location and history of colonialism and extractive capitalism, the Philippines is considered one of the most disaster-prone countries in the world. In the context of disaster response, disasters are traditionally treated as extreme physical events that require top-down solutions that are gender-blind and disregard the socio-political and economic context. The integration of gender perspectives into disaster management has recently gained momentum with the recognition of gender, class and ethnicity as factors that influence people's different experiences of disasters. This is reflected not only in government policy, but also in the development and deployment of disaster management policies supported by feminist stakeholders and organisations. Feminist organisations have played a crucial role in incorporating women's voices and experiences at all stages of disaster management and placing disasters within the constellation of broader socio-political conditions. This demonstrates the ways in which women take a proactive role in disaster management and adopt a bottom-up approach to make the communities less vulnerable to disasters. In this article, we explore the development of gendered narratives in disaster management in the Philippines, the influence of feminist organisations and the impact on the development of inclusive disaster management strategies. In addition, this article considers how the lessons learned from feminist organising in the lead up to and in response to disasters in the Philippines can potentially be applied into disaster management in Southeast Asian countries.

Keywords Diaspora · Disaster management · Feminist organizing · Women's participation · Philippines

M. D. Jose (✉)
College of Arts & Sciences, University of the Philippines Manila, Manila, Philippines
e-mail: mdjose@up.edu.ph

A. B. Ragsag
School of Social Work, McMaster University, Ontario, Hamilton, Canada

G. Y. V. Yong and A.K.M. A. Ullah (eds.), *Brunei and the Future Transformation of Southeast Asia*, https://doi.org/10.1007/978-981-96-7901-0_9

Introduction

According to the World Risk Index Data for 2024, the Philippines tops the list of disaster high-risk countries, followed by Indonesia and India. Due to its geographical location in the Pacific typhoon belt, the Philippines is prone to the formation of around twenty typhoons per year, with Bagyong Yolanda (internationally known as Typhoon Haiyan) considered the deadliest typhoon in the Philippines' history to date. In addition, the country is also prone to earthquakes due to its location within the Pacific Ring of Fire. The Department of Science and Technology-Philippine Institute of Volcanology and Seismology (DOST-PHIVOLCS) strongly advises the public to prepare for "The Big One", which could significantly affect the region and surrounding areas in the Philippines. To make matters worse, the agricultural sector continues to be damaged by droughts and dry spells in some regions, and flooding in others, caused by the El Niño phenomenon. It's not just a weather event, it's a recipe for compounded disaster. Finally, a recent study by the University of the Philippines Resilience Institute (UPRI) and the National Institute of Geological Sciences (NIGS) found that certain parts of the Philippines, including Metro Manila, will be inundated due to rising sea levels as a result of climate change.

Climate change 'refers to a change in climate that is directly or indirectly attributable to human activities that alters the composition of the global atmosphere and that occurs in addition to natural climate variability observed over comparable time periods' (UNFCCC, 2011, p. 2). Over 90 per cent of disasters, including droughts, forest fires, pollution and floods, are weather-related, meaning that climate change increases the risk of disasters (UNEP, n.d.). The disasters have affected and continue to affect people, with women feeling the impact more than men (Waheed, 2023). An example of this is the 2004 Indian Ocean tsunami, which killed more women than men, attributed to cultural factors such as the thick and heavy clothing worn by Muslim women, which also prevented some women from walking to higher ground, and the cultural barriers that prevent girls from learning to swim, which led to the deaths of many women and girls by drowning (Pittaway et al., 2007). Bagyong Yolanda also claimed the lives of more women than men: 64% of the more than 6,300 fatalities were women, and the survivors are still suffering from the social, financial, physical and psychological consequences of the destruction years later (APWLD, n.d.). The high mortality rate of women in disasters is attributed to certain factors. These include women's physical abilities, such as the ability to climb trees and/or the inability to swim, the need to protect other vulnerable family members, especially children and the elderly, as well as livelihood patterns and the timing of disasters (Honasan & Veneracion-Rallonza, 2019).

The gender-specific dimension of disasters in the Philippine context has been addressed in relevant studies on this topic since the 1990s. In 1996, two seminal papers on women and disasters were written at the University of the Philippines (UP) Diliman. Ma. Corazon Jimenez and Norma Dollaga's (1996) MA thesis examined the capabilities and vulnerabilities of women in the areas affected by the Pinatubo eruption, while Barbara Fernandez's (1996) PhD dissertation examined the psychological response of women to two types of disasters in the Philippines. Another MA thesis from UP Diliman was written by Dejelo-Bernardo (2000), who examined the differential impact on men and women of reconstruction programmes for fisherfolk affected by the 1993–1997 Manila Bay Red Tide. Obedicen (2013) examines how disaster situations exacerbate the situation of women in Barangay San Vicente, Angono, Rizal, after Typhoon Ondoy in 2009, and stresses the importance of feminist organisation in transforming institutions. Go (2017) uses feminist ethnographic research to explore how survivors of Super Typhoon Yolanda understand the importance of Waray women. Dalisay (2018) analyses how women in coastal communities in the Philippines respond to climate change. Tanyag (2018) analyses the growing gendered security threats in the Philippines in the aftermath of Typhoon Yanda, manifested in disaster-induced displacement. Berja (2019) looks at the need to consider women in livelihood plans and programmes, drawing on a survey in three locations severely affected by Typhoon Yolanda. Finally, Jayme-Lao (2019) writes a case study on the role of women in community leadership in Brgy. Ugong, Sta. Cruz, Laguna as part of the ASEAN Intergovernmental Commission on Human Rights (AICHR) thematic study on women in natural disasters in Southeast Asia.

We are writing as two scholars with different disciplinary commitments—history and social policy—but are both interested in feminist studies in Asian communities and diaspora. This study hopes to further enrich the examination of gender integration in Philippine disaster management and exploring case studies of feminist organising in the country. This effort is to highlight the contributions and potential of women in creating an inclusive disaster management strategy, while also flagging the factors that hinder them. While women have always been portrayed as victims and not as individuals capable of changing situations to improve their lives and those of their families in times of disasters, there is a need to recognise and document their contributions in creating and implementing gender-responsive disaster strategies and programmes in their communities (Dalisay, 2018). By reviewing the literature on women and disasters in the Philippines, we hope to provide lessons and insights on the critical role of women in disaster management in the country that can be adopted by our Southeast Asian neighbours in their disaster management frameworks and practises.

To trace the origins of the integration of gender perspectives in the country's disaster policies and to explore the importance of critical reflection on concepts and practises related to disaster management, this study utilises the vulnerability approach and the intersectional feminist framework. The vulnerability approach emerged in the 1970s as a challenge to the dominant disaster discourse, which drew on the experience, knowledge and expertise of the West and embraced technocratic solutions (Bankoff, 2001). It recognises the idea that while hazards are natural, disasters usually are not. And that the intertwining of class, gender, age, ethnicity and disability places certain communities in a precarious position in times of disaster, contributing to the vulnerability of certain populations (Hewitt, 1997, cited in Bankoff, 2003). Guided by an intersectional feminist framework, Enarson (1998, p. 159) explains women's vulnerabilities to disasters:

> Gendered vulnerability does not derive from a single factor, such as household hardship or poverty, but reflects historically and culturally specific patterns of relations in social institutions, culture and personal lives. Intersecting with economic, racial and other inequalities, these relationships create hazardous social conditions placing different groups of women differently at risk when disastrous events unfold.

Specifically, this study wishes to shed light on the following queries: what were the prevailing disaster discourses in the Philippines and how did they change over time? When and how did disaster management in the Philippines start to become gendered, and in what context? How is gendered response operating in practice? What complications are there in advancing "women's participation" in disaster response and recovery management? And finally, what lessons can be drawn from them that can be adopted by other countries in Southeast Asia? By exploring answers to these questions, we hope to add to the body of knowledge on the gendered dimension of disasters and contribute to the literature in women/feminist studies, climate/disaster studies, and Philippine/Southeast Asian Studies.

Evolving Disaster Narratives and Frameworks

How does the discourse on disasters change over time? In the traditional view, the occurrence of disasters is caused by nature and the solution therefore lies in the realm of science and infrastructure. The Merriam-Webster dictionary defines a disaster as "a sudden, calamitous event involving great harm, loss, or destruction", while the Cambridge dictionary defines it as "an event involving great harm, damage, or death or serious trouble". The basic view is that a necessary but not sufficient condition for a disaster to occur is the coincidence of at least two factors: a human population and a potentially destructive agent (Oliver-Smith, 1999). Based on an alternative view, which has since gained acceptance, disasters are now seen as a consequence of the vulnerability of populations that are not only exposed to danger or risk, but also to the marginalisation they experience due to various factors that place their lives in a permanent state of emergency (Bankoff, 2001).

The concept of endangerment has been used as a cultural discourse by Western countries to justify their continued interference in the internal affairs of certain countries or former colonies. Bankoff (2001) argues how tropicality, development and vulnerability are part of an essentialising cultural discourse that results in large regions of the world being in constant need of Western intervention. Prior to the nineteenth century, knowledge of equatorial regions produced not only literature on warm climates, but also a narrative of tropicality that portrayed warm-climate regions as disease-prone areas in need of Western medicine (Arnold, 1996, cited in Bankoff, 2001). With the emergence of 'development' as the dominant discourse after the Second World War, material prosperity and economic progress were widely recognised and sought by Western nations, resulting in societies being categorised in terms of development and certain regions being portrayed as poor areas in need of Western aid (Escobar, 1995, cited in Bankoff, 2001).

The development discourse was soon replaced by the vulnerability discourse in the late twentieth century, when certain parts of the world were considered disaster-prone areas and therefore in need of Western intervention in the form of Western science. Therefore, the discourse on vulnerability, like the discourses on tropics and development, is part of a generalising cultural discourse that creates a particular narrative that makes large regions of the world unsafe (Bankoff, 2003). It was not until the 1970s that a vulnerability approach emerged, based on experiences and ideas from the Global South and gaining momentum since the 1990s with the emphasis on the gendered dimensions of disasters (Gaillard et al., 2017). Bankoff (2001) emphasises that marginality, which places the lives of vulnerable populations in a permanent state of emergency, is determined by a combination of variables such as class, gender, age, ethnicity and disability, which affect people's entitlement and empowerment to master basic needs and rights (Valila, 2023).

As an active participant in regional and global conferences on natural disasters sponsored by the United Nations (UN) and the Association of Southeast Asian Nations (ASEAN), Philippine disaster frameworks also reflect evolving views on disasters, as evidenced by the fact that disaster risk reduction is no longer seen primarily as a technical problem, but also as a social problem. To provide the necessary context for feminist organising in the Philippines, a brief history of disaster management in the country and the inclusion of gender components in the disaster risk reduction framework will now be discussed.

Disaster Management in the Philippines

Disasters have always been a part of the Filipino psyche and lived experiences since pre-colonial times, with our ancestors recording their own views on the origins of disasters in their myths. The Isnegs from the northern Cordillera of Luzon have a myth regarding how earthquakes occur whenever a giant eel and a giant crab beneath the earth fight each other, and the eel's tail hits the pillar supporting the earth (Eugenio, 1994). The Maranaos from Mindanao have a similar story about a giant animal called *lumbong* that is believed to be carrying the world on its back, and how an earthquake occurs whenever the shrimp clings to it and the *lumbong* shakes it off (Eugenio, 1994). In Bukidnon in northern Mindanao, there is a myth regarding a giant crab that was responsible for the great flood that occurred after the crab crawled into the sea, crowding the water out and covering all the land (Cole, 1916). Ancient myths reflect society's relationship with nature and might be explained as a result of humanity's failure to live in harmony with the environment. Due to the ancient religious belief that the spirit of the deities lives in nature, disasters were also viewed as a form of divine punishment for humanity's wrongdoings.

The view of disasters from a religious perspective dominated during the Spanish colonial period, a narrative that has become beneficial to the Spanish priests in increasing the number of converts to Catholicism. For example, in 1686 it is said that the Church dedicated to the Nuestra Señora de la Pena de Francia received more than

1,300 native converts to Christianity, which was further increased due to the spread of the smallpox epidemic as the people allegedly proved God's mercy in this time of disaster (Salazar, 1742). Abejo (2005) discusses how pilgrims in the town of Bunga in Samar aimed to fulfill their vows and thank the miraculous saints for saving their lives from the cholera epidemic in the nineteenth century. According to a Pulse Asia survey in October 2023, more Filipinos from the poorest socioeconomic bracket still view calamities as God's ways of punishing people for their wrongdoings (Esmaquel II, 2023), which only goes to show that the view of disasters as acts of God persists.

The establishment of *Observatorio Meteorologico de Manila* in 1865 by the Jesuit mission paved the way for the institutionalization of the modern study of weather, climate, and natural hazards in the Philippines. Alvarez (2016) credits the scientific endeavors of the *Observatorio* in bringing about major changes in colonial science in the Philippines. Recognizing the importance of the Observatory's work, the American colonial government established it as the Philippine Weather Bureau in 1901 (Manila Observatory website, n.d.). It was completely destroyed during World War II, reorganized in 1945, and continued its functions until it was replaced by a new agency called the Philippine Atmospheric, Geophysical and Astronomical Services Administration (or PAGASA, a Filipino term for 'hope') under Presidential Decree No. 78 series of 1972 (DOST-PAGASA, n.d.).

Six years later, the National Disaster Coordinating Council (NDCC) was established under Presidential Decree 1566 to ensure the protection and welfare of the people during disasters and emergencies, replacing its name with the National Disaster Risk Reduction and Management Council (NDRRMC) in 2010 when the Philippine Disaster Risk Reduction and Management Act of 2010 was signed into law (Official Gazette, 2010) as a response to the devastating effects of Typhoon Ondoy in 2009. Despite the enactment of several laws that address the protection of the environment, climate change, and disaster risk reduction and management, there is still a long way to go regarding their full implementation (Lusterio-Rico, 2024). And while DRRM has already been devolved to Local Government Units (LGUs) based on the 1991 Local Government Code, some administrative and resource issues (e.g. understaffing and underfunding) in the national and local frameworks continue to pose challenges to an effective realization of these models (Atienza, 2019).

Gender Inclusion in Philippine Disaster Management

In shaping local disaster management, the Philippines maintains its commitment to the global and regional development and policy agenda. At the global level, the country has signed the Sendai Framework for Disaster Risk Reduction 2015–2030, the Paris Agreement, the United Nations Sustainable Development Goals 2015–2030, the International Health Regulations and the New Urban Agenda. At the regional level, the country is committed to the Asia–Pacific Economic Cooperation Disaster Risk Reduction Framework, the ASEAN Agreement on Disaster Management and Emergency Response and the Asia Regional Plan for Implementation of the Sendai

Framework for Disaster Risk Reduction 2015–2030. At the national level, we have the Ambisyon Natin 2040, the Philippine Development Plan 2017–2022 and the National Framework Strategy on Climate Change as some of the local disaster framework plans (NDRRMC, 2020).

The Hyogo Framework for Action is considered a global blueprint for the implementation of disaster risk reduction, which was adopted by the United Nations Member States in 2005 (UNISDR, n.d.). It emphasises the need to integrate a gender perspective into all disaster management strategies, plans and decision-making processes. The Sendai Framework, adopted at the Third UN World Conference in Japan in 2015, reaffirms the need to integrate a gender perspective into disaster policy by highlighting the crucial role of women and their participation in managing disaster risk and in the design, financing and implementation of gender-sensitive disaster risk reduction strategies, plans and programmes (Honasan & Veneracion-Rallonza, 2019). Thus, both disaster frameworks play a crucial role in mainstreaming social concerns such as gender and vulnerability into disaster risk reduction, with the Sendai Framework emphasising gender in its call to improve disaster risk reduction (Alexander, 2023).

While disaster management initiatives in the Philippines can be traced back to the Spanish colonial era, the inclusion of a gender perspective in disaster risk reduction is a relatively recent development. A 2009 study on the integration of gender and gender perspectives in disaster management in four countries (Indonesia, Japan, the Philippines and Thailand) found that gender is a comparatively new concept in disaster discourse and practise, hence the need for gender mainstreaming in disaster management policy (Honasan & Veneracion-Rallonza, 2019).

Article III of the Universal Declaration of Human Rights states that everyone has the right to life, liberty and security of person. As a signatory to this declaration, the Philippine state reaffirms its responsibility to ensure the safety of its citizens at all times. One way to do this is to implement disaster risk reduction measures to ensure the safety of Filipino citizens in times of disasters. The Philippines also adheres to the Hyogo Framework for Action 2005–2015, which recognises gender equality as one of the key factors in disaster risk management. The state's commitment to establishing a disaster framework is enshrined in Republic Act 10121, which aims to strengthen the Philippines' disaster risk reduction and management system and institutionalise the National Disaster Risk Reduction and Management Plan (NDRRMP) 2011–2028. It was ratified after the devastation caused by Typhoon Ondoy in 2009 and shows that the government only listened to the call for such a law after Metro Manila was badly hit by Ondoy (Obedicen, 2013).

The NDRRMP signals a paradigm shift from reactive to proactive DRRM aimed at increasing people's resilience and decreasing their vulnerabilities, serving as a guideline on how DRRM shall contribute to gender-responsive and rights-based sustainable development. It defines gender mainstreaming as 'the recognition, acceptance, identification and addressing of the different roles, needs, capacities and vulnerabilities of men and women' (NDRRMC & DND, 2011, p. 36). It is committed to promoting gender-sensitive vulnerability and capacity analysis in all disaster risk reduction and management activities and seeks to lessen the susceptibility of both men and women to disasters and promote equitable participation and decision-making roles for both genders in DRRM to reduce the vulnerabilities of men and women to disasters and encourage a balance in the participation and decision-making roles of men and women in DRRM (NDRRMC & DND, 2011).

In 2020, the NDRRMP 2011–2028 was updated into the NDRRMP 2020–2030. While the former was formulated based on the Hyogo Framework for Action 2005–2015, the latter follows the Sendai Framework for Disaster Risk Reduction 2015–2030, also adopted by the Philippines to reduce the loss of lives and assets. Under the Hyogo Framework, vulnerabilities were seen as inevitably associated with marginalized groups such as women, older persons, and children, while under the Sendai Framework, the concern is being addressed by enhancing actions around risk governance to reduce people's vulnerabilities through institutional arrangements (NDRRMC, 2020, p. 19). The Updated NDRRMP recognizes the need to address the persistent gaps in DRRM as fundamental in resilience building, emphasizing the need for adequate capacity building and education to allow women and men, girls' and boys' transformative role as agents of change in DRRM (NDRRMC, 2020, p. 46).

The implementation of Republic Act No. 9710 is another manifestation of the Philippine government's commitment to eliminate discrimination through the protection, fulfillment, and promotion of Filipino women's rights, especially those belonging to the marginalized sectors of society. It is also considered as the country's commitment to the Convention on the Elimination of All Forms of Discrimination against Women (CEDAW) and other International Conventions. RA 9710, also known as the Magna Carta of Women, ensures the protection against all forms of discrimination against women through the promotion of their rights. Section 13, which is about women affected by disasters, calamities, and other crises, states that women have the right to protection and security in times of disasters, especially in all phases of relief, recovery, rehabilitation, and reconstruction efforts, and that the State shall provide for immediate humanitarian assistance and address the particular needs of women from a gender perspective to ensure their full protection from sexual exploitation and other gender-based violence. It also mandates the participation of women in the development of gender-responsive disaster management, including preparedness mitigation, risk reduction, and adaptation (RA 9710, 2010, pp. 52–53).

Given this scenario, it can be said that the Philippines is not lagging behind in promoting gender sensitivity in DRRM. In fact, the Philippines was recently cited as a "leader and role model" in gender-sensitive disaster risk reduction during the launch of the Sendai Gender Action Plan (GAP) for Disaster Risk Reduction at the 68th Session of the UN Commission on the Status of Women at the UN Headquarters in New York, USA in March 2024 (Permanent Mission of the Republic of the Philippines to the United Nations, 2024). Even the infrastructure projects of the Department of Public Works and Highways (DPWH) are geared towards ensuring gender equality and protection of all stakeholders, especially women and other marginalised sectors of society, through the "Toolkit for Making Road Infrastructure Projects Gender Responsive". This toolkit was developed using the Harmonised Guidelines on Gender Equality and Development (GAD) for Project Development, Implementation, Monitoring and Evaluation as a guiding framework (DPWH website, 2011.)

to ensure that women's needs and participation are considered from project planning to implementation. Other government agencies such as the Department of Health (DOH) and the Department of Social Work and Development (DSWD) also play an active role in providing services for women's health and well-being in times of disasters (Jayme-Lao, 2019). However, the full implementation of these measures is another matter. A study by the AICHR on the integration of gender issues into institutional infrastructure and mechanisms found that full implementation of RA 10121 is still pending and that while the Philippines has sound institutional frameworks and practises for disasters and gender issues, they are not necessarily complementary (Jayme-Lao, 2019).

Fortunately, the integration of gender perspectives is also reflected in the formulation and deployment of disaster response interventions that are informed by feminist advocacy, with feminist organising filling the gaps in integrating gender into institutional infrastructure and mechanisms. We will now examine cases of feminist organising in the Philippines and feminist diaspora organising in specific disasters.

The Impact and Limits of Gender-Sensitive DRRM in the Philippines

In the previous section, we have seen how the discursive turn toward gender-responsive disaster risk reduction measures has increasingly been institutionalized through global instruments and national policy statements in the Philippines. However, studies show that the discursive turn to gender-sensitive DRRM lands unevenly in practice: first, gender remains to be understood along binary lines and roles, i.e. of men and women, precluding responses to gender diverse and nonbinary persons; second, women are framed as vulnerable and have to be protected by institutions both local/national and international, which have nevertheless varying programmatic agendas, and more importantly, vary from what women express as most important for their communities; and third, the feminization of responsibility in community recovery.

Studies show that even with various gender-sensitive DRRM policy changes, on-the-ground realities show entrenched gender binaries between men and women, where core political leadership, governance, and systemic structures are retained to the dominance of men. For example, Go's (2016) on-site feminist ethnographic research for her MA thesis, conducted in the summer of 2015 in the town of Palo, Leyte, investigated the different disaster experiences of women and their importance in theorizing survival in the aftermath of the devastating Typhoon Yolanda. She relates: *"At the time of my fieldwork, the position of the Municipal Disaster Risk Reduction and Management Officer in Palo had just been filled by a young man who was hired with haste, after the desk was cross-posted to different municipal officers already overburdened with multi-tasking. The Barangay Disaster Risk Reduction and Management Committees (BDRRMCs) mandated by the law have yet to be* *formally organized across Palo. Barangay captains often reasoned to me that their barangay tanods—*ad hoc *all-male teams in charge of keeping order and security—already serve the purpose of BDRRMCs, and that overseeing everyday concerns in their barangay already kept their hands full with little spare time to mind additional technical requirements relating to DRRM... Before and after Yolanda, women pointed out to the very same sites they would bring their families to for refuge: the Cathedral, the Seminary, the Convent, school classrooms, or relatives' homes in other barangays on higher ground. The men still watch the skies, the rising rivers and flood waters, corresponding over their radios and cellphones when the rains begin. The women still stay at home, tasked with packing everything into boxes, carrying material possessions up to higher stories, watching over the children, ready to leave their houses should their household decide to evacuate. Such is the 'system' in place* (Go, 2016, p. 94)". What Go has witnessed points to the gender-responsive turn in DRRM policy discourses not yet manifesting in practice at the decision-making level in communities, especially in the lived experiences of people in disaster-prone areas. Moreover, the "system" that Go has observed did not speak of political and social relations that go beyond an understanding of gender binaries.

While men are still in DRRM and other leadership positions, 'women friendly spaces' (WFS) have been created in newly constructed houses with nipa roofs, bamboo walls and cement floors, set up by international organisations. These spaces are understood in Go's work as a place where women of the Waray (ethno-linguistic identification of Leyte province's natives and residents) "can organise and actively participate in the development of their local communities" (Go, 2016, p. 95). The establishment of WFS is one of the most popular gender-specific DRRM interventions to date, even though they are used in the context of competing agendas. According to the Philippine Department of Social Welfare and Development, WFS are set up in evacuation centres where women can, among other things, breastfeed their children, receive health and well-being information and counselling (Philippine Department of Social Welfare & Development, 2013). Programme documents from the Social Technology Bureau of the Philippine Department of Social Welfare and Development indicate that WFS aim to "respond to the impact of displacement and women's vulnerability to crisis by providing a wide range of services to help them cope with their situation and prepare them for their return to communities" (Social Technology Bureau, DSWD, Government of the Philippines, n.d.). The establishment of WFS in emergency situations is supported by legislation, including Republic Act 10821 (Children's Emergency Relief Protection Act, 2016) and Joint Memorandum Circular Series No. 2, 2021, which deals with Camp Coordination and Camp Management for Internally Displaced Persons (IDPs) (2021). These regulations mandate the establishment of WFS in evacuation centres to provide safe spaces for the immediate physical, emotional and psychosocial needs of women and girls affected by crises. While WFS tents can be spaces where local women's skills are developed to support other local women, the WFS, as the aforementioned project highlights, is a facility/structure that is established in evacuation camps, transition sites or communities to meet the practical and strategic needs of women during disasters and other crises.

The agendas of international donor organisations overlap with local and national WFS efforts to create a safe space for women in disaster and emergency areas; they focus more on supporting the fight against gender-based violence. The United Nations Population Fund (UNFPA) in the Philippines describes the WFS tents as places where relief items such as maternity kits are provided alongside Emergency Maternity Tent Facility (EMTF) tents as part of the UNFPA and Australia's Enhancing Capacity to Respond to Sexual and Reproductive Health and Gender-Based Violence in Emergencies programme, dignity kits and clean delivery kits for midwives were distributed and life-saving sexual and reproductive health (SRH) services were provided (United Nations Population Fund, 2022). A non-profit organisation called Women Enablers Advocates and Volunteers for Empowering and Responsive Solutions (2022) uploaded on 2 November 2022 a video on social media stating that women leaders of a community in Leyte were trained as WFS Facilitators. The project is described as a UNFPA project with other non-governmental organisations funded by the US Agency for International Development (USAID).

Action against gender-based violence is important and undoubtedly needed in disaster areas. The tension between local or state approaches and those advanced by international governments and donor agencies reveals not just a conflict, but a complex multiplicity of overlapping, and at times competing, development trajectories, knowledge systems, and policy agendas. While local governments may focus more on emergency relief, international aid organisations see WFS facilities primarily as places where they can provide training on gender sensitivity, gender-based violence, women's rights, laws to protect women and referral pathways for GBV in their respective communities. Meanwhile, while in view of chronic poverty, the fundamental concerns in communities are most likely their survival and livelihoods, not only emergency relief and likely more than training opportunities. In a study on the prevalence of GBV after disasters, the International Federation of Red Cross and Red Crescent Societies (2015) emphasises that humanitarian organisations, responders and policy makers should assume that GBV is happening, even in the absence of data. But the study also insists that livelihood projects for those most at risk of GBV must be prioritised due to their preventative role in the prevalence of GBV. The study also included the caveat that their secondary research found strong evidence of post-disaster GBV in higher-income countries, which was not the case in the lower-income countries they studied. While we do not take issue with the attention that aid organisations have paid to GBV-related interventions in these WFS during disaster response, the priorities of communities affected by disasters and emergencies may not have received the same attention. We must ask ourselves who determines the nature of WFS interventions: is it the members of the communities who disproportionately suffer, or the research and development agenda of aid organisations and the governments that fund their interventions?

Compare this to Go's study mapping indigenous knowledge through participatory and decolonising research on the Warays' (the site of the devastating Bagyong Yolanda) theory of community recovery and endurance among women survivors (Go, 2016). She learnt that Waray women's disaster response emphasises (1) hazard mapping, which includes surveying the physical and ecological terrain; (2) training on the seasons and climate change, especially for those whose households are directly adjacent to water bodies such as rivers, streams and swamps, and considering that they have often not had formal schooling; (3) collective sharing of survival narratives; and (4) addressing their livelihood concerns. This collective, indigenous and local knowledge and priorities of women survivors of the disaster, Go argues, must be incorporated into these women-friendly spaces (Go, 2016). Admittedly, the specifics of the recovery pathways identified by the Waray women in Go's study may not be applicable in all disaster-prone areas or in the context of post-disaster reconstruction. However, the findings of the above study are helpful and remind us to critically engage with understandings of gender and how to ensure that those most affected have a voice in the design of interventions in their communities when it comes to the imposition of recovery measures. The importance of a decolonial, i.e. community-created rather than externally imposed, critical engagement with how gender is addressed in DRRM strategies is all the more important as a previous systematic literature review suggests that gender is often treated as a mere demographic characteristic rather than a critical analytical element in disaster management (Sohrabizadeh et al., 2014).

In recent years, women's participation has been regarded as a measure of gender-responsive DRRM. In this chapter, we take the position that women's participation in itself, is not a problem. The trouble lies, however, in uncritically advancing "women's participation" and their roles in post-disaster recovery. This uncritical engagement with "women's participation" reveal an entanglement with assumptions regarding women's roles and obligations to family and community as nurturers and innately altruistic agents. Alburo-Cañete highlights these assumptions in a qualitative study of post-Yolanda Tacloban, Leyte and how these assumptions animate DRRM pathways that come from particular constructions of gender and motherhood (Alburo-Cañete, 2024). Women's participation, while unintended, takes the form of re-traditionalized gender roles and takes us back to limited and limiting understandings of mothers and motherhood. These uncautioned understandings the feminization of responsibilities, or the additional pressures on women's "productive, reproductive, and affective labors" (Alburo-Cañete, 2024, p. 34). Even earlier studies already highlight the intensification of burdens to women, who are largely uncompensated and unrecognized. Reyes and Lu (2016) for example, in a study on women in flood-prone Malabon City in Metro Manila, disclose that men were recognized as the household heads and community leaders even if evidence shows, women took on more roles and responsibilities *before, during, and after disasters.*

Diaspora Organizing, Women, Disasters and Pandemics

Studies show that diaspora engagement is two-pronged or motivated by one of two reasons: to improve and advocate for the well-being of their communities in host countries and to maintain links with their home countries and help them and other diaspora members in different countries (Sheffer, 2003). It is important to highlight that while there has been extensive research in the literature on how diaspora members maintain links with their home countries, there is still a large gap in the literature regarding their active engagement in political struggles and how diaspora organising in response to disasters occurs in practise. Understanding this engagement is crucial as it sheds light on the dynamic role that diaspora communities play in shaping political narratives, advocating for change in both their countries of origin and their host societies, and providing financial and in-kind support that can outlast even foreign aid and local government responses.

In this section, we will highlight two community responses to disasters and pandemics in Ontario led by women from the Filipino diaspora: Pinay Collection and Filipinas of HamOnt. Both have taken action in unconventional, creative and diverse ways to support their host and home communities. Pinay Collection is a Toronto, Ontario-based social enterprise led by Jovie Galit that is dedicated to connecting with the identity of the Filipino people by using words that are traditionally seen as incriminating and traumatising in their printed t-shirts, sweatshirts and other merchandise (Diez, 2023). According to their website, they believe that connecting to their Filipino roots is only possible if they understand the struggles of the Filipino people in their homeland (Pinay Collection, n.d.). Therefore, they are partnering with communities in the Philippines and organising ways to further support these movements.

After two devastating typhoons hit the Philippines in 2024, Pinay Collection raised 3,030.09 Canadian dollars or 121,169.60 Philippine pesos for Bagyong Kristine and 4,156.67 Canadian dollars or 135,246.52 Philippine pesos for Bagyong Karina. These amounts all came from parts of their sales and were distributed to people's organisations supporting women, jeepney and transport workers, and women farmers' organisations (Pinay Collection, 2024). What's unique about their approach is that every purchase of their merchandise goes to organisations that has long shown to be advocating for workers, women and historically marginalised Filipino communities such as an organization of peasant women and a federation of peasant women organizations-Amihan, a grassroots-based organization that empowers marginalized women in the Philippines and abroad—Gabriela, Philippine Resource Center for Inclusive Development which started as a people-led donation drive for the fire victims of Sitio San Roque in 2019-Inklusibo, a national federation of public transport associations in the Philippines—Piston and several others. Pinay Collection has been consistently committed to social entrepreneurship for several years now and in doing so, has not only gained the trust of its customer base in Canada today, but also that of the organisations it raises money for in the Philippines, with whom many in the Filipino diaspora have native ties.

Meanwhile, the Filipinas of HamOnt, co-founded by Anabelle Ragsag, who was born and raised in the Philippines, and Jessica Vinluan, born in what is now known as Canada from immigrant Filipino parents, started during the pandemic in the second half of 2020. As explained on their website, the Filipinas of HamOnt is a by-for-and-with grassroots collective that responds to the lack of spaces dedicated to Filipina/xs in Ohron:wakon or what is presently known as Hamilton… guided by antiracist and decolonial feminist values (Filipinas of HamOnt, 2020). During the COVID-19 pandemic, the Filipino community experienced disproportionate cases of infections and job losses among populations in Canada. With the Filipinas of

HamOnt's grassroots leadership in building networks, seeking resources, spotting local capacities and talents, and relying on their existing expertise, they applied for grants or mobilized resources for and with the Filipino Canadian community in Hamilton. They organized several COVID-19 projects during this time. First was in organizing low-barrier COVID-19 vaccine pre-registration for the community in partnership with the Hamilton Centre for Civic Inclusion, which was one of the organizations that Hamilton Public Health tapped to carry out the first-ever COVID-19 clinic for Black and other racialized people (Joseph et al., 2023). Secondly, with a 3,000 Canadian dollar grant from the same community partner from funds provided by the WES Mariam Assefa Fund, they launched the "You Matter" or "Mahalaga Ka" project, a community-rooted digital campaign to counter fake vaccination news and nurture information drives from local Filipino health experts in Hamilton, as this working-class city continued to tackle the impacts of COVID-19 (Hewitt, 2022; "Introducing Filipinas of HamOnt and the #MahalagaKa Campaign", 2022; Rosas, 2022). Thirdly, grants totalling 10,000 Canadian dollars from the Hamilton Community Foundation were voluntarily written on behalf of the Hamilton Filipino Community Centre (OMNI News, 2021), to support disabled older adults[1] and low-income single mothers and temporary visa workers, through a local Protestant church, which mostly have Filipinos as congregants. Fourthly, the grassroots organization launched writing and art campaigns for women and children to participate, tapping on already existing spaces—in this case, the Art Gallery of Hamilton providing the space for free (Mochama, 2022). Finally, Public Health Hamilton, noticing the body of work that the organization has done during COVID-19, approached the Filipinas of HamOnt to help bring to the table various Filipino organizations in the city, which led to the organizing of a GoVaxx Mobile Bus Vaccine Clinic (Filipinas of HamOnt, 2022) between downtown and the East end of Hamilton considered as one of the COVID-19 hotspots in the city (CBC News, 2021) and because Filipinos, as overwhelmingly frontline workers and living in multigenerational homes, were at risk of higher infection rates.

Similar to what Go (2016) offers as disaster response epistemologies that are based on decolonized, participatory pathways, Pinay Collection and the Filipinas of HamOnt share some common organizing characteristics, while also leveraging on their creative (events and people organizing and social entrepreneurship), place-based knowledges (relying on local capacities, talents, and resources while intentionally supporting existing collective efforts and infrastructures), and community partnerships. These examples of women-led diaspora responses to disasters provide some blueprint for how broader DRRM could be done. However, it must be noted that

[1] The Canadian government has advocated "people-first" language (prefer the usage of the term person with disabilities) which emphasizes putting the person first and the disability second, a practice since the 1960s (Council of Canadians with Disabilities, n.d.; Employment and Social Development Canada, 2022). Many disabled people and disability justice advocates who look at disabilities as systemic outcomes prefer "disabled people" over "person with a disability", with the understanding that they were made so by systems emanating from political decisions that do not ensure their participation and flourishing (Baheerathan & Youth Action Council, 2023; McColl, 2019; Rodriguez, 2021).

similar to the feminization of responsibilities lens, the interventions of Pinay Collection and the Filipinas of HamOnt are both relying on the carework and leadership of women, of women who do not have deeper pockets and resources, who themselves are experiencing cultural, socio-political and material exclusions, and are mostly doing their advocacies on the side of their desks, while also attending to their main sources of livelihood and commitments outside of this work.

Quinsaat raised a fundamental question in understanding the diaspora: when people migrate and settle in other countries, do they automatically form a diaspora? Not necessarily, is what she argues. Instead, Quinsaat, using a social mobilization approach, makes the case that diaspora emerges from political activism. She argues that diasporas are *outcomes,* rather than causes or agents, of transnational mobilization (Quinsaat, 2024). Through the examples of the works of Filipino diaspora groups Pinay Collection, a social enterprise, and Filipinas of HamOnt, a grassroots organization—we can see that their collective actions have not emerged out of a vacuum but by ongoing transnational and transborder mobilizations of identity, politics, and social justice.

Conclusion

As mentioned earlier, the Philippines has come a long way in the science, policy and practise of gender-responsive disaster management. Yet, gendered approaches continue to fail to bring about transformative change in disaster response and recovery. Firstly, the term 'gender' in disaster preparedness and response, as confirmed in several studies in the Philippines, is understood as synonymous with 'women', perpetuating traditional gender hierarchies and leaving out those who do not fall under the male/female binary.

Secondly, ideas of "women's empowerment, participation and leadership" in disaster relief, where the Philippines is indeed a leader, continue to be uncritically celebrated. The ways in which women's participation and gender-responsive disaster management are understood and put into practise do nothing to address gender inequalities, but rather reinforce the traditional binary ideas and roles expected of women and men. Third, where the epistemologies of women and those disproportionately affected by disasters are the starting point for these spaces of participation and leadership, externally developed intervention pathways prevail. The 'gendered' responses that also pass for 'women's participation and leadership' in disaster risk reduction—such as the establishment of 'women-friendly spaces'—may be well-intentioned, but do not appear to come from the communities themselves, but are ready-made 'best practises'. Rather than making stakeholders the starting point for developing DRRM strategies, a best practise approach may leave out context and community members, including women, and are likely to be privileging off-the-shelf plans instead.

Using the examples of Filipino diaspora groups Pinay Collection, a social enterprise, and Filipinas of HamOnt, a grassroots organisation, we can see that their collective actions have emerged from ongoing transnational and cross-border mobilisations for identity, politics and social justice. Moreover, their examples, similar to those of the Waray women: the Kababayen-an han Karak-Anin in Go's (2016) study, offer possibilities for an epistemology of disaster response based on decolonised, participatory pathways while harnessing their creative, place-based knowledge by drawing on local capacities, talents and resources while consciously supporting existing collective efforts and infrastructures. These examples of women-led responses to disasters in affected areas, whether at home or in the diaspora, provide a blueprint for how our understanding and practise of women's participation in DRRM might be transformed.

It is recommended that DRRM be placed in a broader context that addresses people's livelihoods and takes into account the needs and capacities of all members of the community, including those who are transgender and gender non-conforming. This can be done through the development and implementation of DRRM frameworks that are truly gender-responsive and grounded in Indigenous, local knowledge, while challenging the one-size-fits-all approach to disaster management. Finally, it is important to note that in promoting gender-responsive DRR, we do not arbitrarily increase the burden on women by reinforcing traditional gender roles that lead to a feminisation of responsibility. Alburo-Cañete (2024, p. 32, quoted from Tronto, 1993) reminds us: "If you look at who provides care in a society, it becomes clear what kinds of inequalities exist in that society. Often those who are marginalised or less valued are also those who are entrusted with the responsibility of caring".

References

Abejo, R. (2005). Mga kalamidad at ang rebelyong Dios-Dios sa samar noong dantaon 19. *Philippine Social Science Review, 57*(1–4), 115–142.

Alburo-Cañete, K. Z. (2024). Feminization of responsibility in community recovery: Rethinking disaster justice through the lens of care. *Philippine Studies: Historical and Ethnographic Viewpoints, 72*(1), 11–40.

Alexander, R. (2023). Gender, vulnerability and disaster. In R. Alexander, S. Wajjwalku (Eds.), *Making disaster safer: A gender and vulnerability approach* (pp. 1–18). Springer Nature Pte Ltd.

Alvarez, K. (2016). Instrumentation and institutionalization Colonial science and the Observatorio Meteorológico de Manila, 1865–1899. *Philippine Studies: Historical & Ethnographic Viewpoints, 64*(¾),385–416. Special Double Issue: Disasters in History. https://bit.ly/3zLyS2H. Accessed September 12, 2024.

Arnold, D. (1996). Tropical medicine before manson. In D. Arnold (Ed.), *Warm Climates and Western Medicine: The Emergence of Tropical Medicine, 1500–1930* (pp. 1–19).

Asia Pacific Forum on Women, Law and Development (APWLD). (n.d.). *Typhoon Haiyan claims lifetime loss and damages for women in the Philippines*. https://bit.ly/3zFKMLr. Accessed September 6, 2024.

Atienza, M. E. (2019). The role of external assistance in disaster risk reduction and management. In M. E. Atienza (Ed.), *Rebuilding disaster-affected communities for a sustainable future: Lessons*

and policy recommendations for poverty alleviation from the Typhoon Yolanda experience (pp. 5–12). University of the Philippines Center for Integrative and Development Studies.
Baheerathan, S., & Youth Action Council. (2023). *The "unfit" in Canada: A history of disability rights and justice.* Disability Justice Network on Ontario. https://www.djno.ca/history-of-disability-justice-right
Bankoff, G. (2001). Rendering the world unsafe: 'Vulnerability' as western discourse. *Disasters, 25*(1), 19–35.
Bankoff, G. (2003). 'Vulnerability' as western discourse. In *Cultures of disaster: Society and natural hazard in the Philippines* (pp. 5–17). Taylor & Francis Group.
Berja, C. (2019). Gender and livelihood. In M. E. Atienza (Ed.), *Rebuilding disaster-affected communities for a sustainable future: Lessons and policy recommendations for poverty alleviation from the Typhoon Yolanda experience* (pp. 20–30). University of the Philippines Center for Integrative and Development Studies.
CBC News. (2021, June 17). Hamilton added to list of delta hot spots as province bumps up appointments for 2nd shots. *CBC News.* https://www.cbc.ca/news/canada/hamilton/delta-hot-spot-covid-dose-1.6069546
CHCH Evening News. (2022, March 6). Introducing Filipinas of HamOnt and the #MahalagaKa Campaign
Children's Emergency Relief Protection Act, Pub. L. No. 10821, Republic Act. (2016). https://www.officialgazette.gov.ph/2016/05/18/republic-act-no-10821/
Cole, M. C. (1916). *Philippine Folk Tales.* A.C. McClurg & Co.
Council of Canadians with Disabilities. (n.d.). *About council of Canadians with disabilities.* http://www.ccdonline.ca/en/about/
Dalisay, S. N. (2018). Climate trouble: Women in coastal communities in the Philippines respond to climate changes. *Diliman Gender Review, 2018*, 53–73.
Dejelo-Bernardo, M. M. (2000). *Differential impact on men and women of the disaster rehabilitation programs for Fisherfolk affected by red tide in Manila Bay, Philippines, 1993–1997.* MA Thesis in Community Development, U.P. Diliman.
Department of Public Works and Highways website. (2011). Toolkit for making road infrastructure projects gender responsive. https://bit.ly/4ephy2t. Accessed September 16, 2024.
Department of Science and Technology-Philippine Atmospheric, Geophysical and Astronomical Services Administration (DOST-PAGASA). (n.d.) website. https://www.pagasa.dost.gov.ph/about-us.
Diez, G. C. (2023, October 15). Get to know Pinay Collection's Jovie Galit and her brand's mission. *Metro.* https://metro.style/people/society-personalities/profile-pinay-collection-jovie-galit-calligraphy/36464
Employment and Social Development Canada. (2022). *Canada's disability inclusion action plan.* Employment and Social Development Canada. https://www.canada.ca/content/dam/esdc-edsc/documents/programs/disability-inclusion-action-plan-2/action-plan-2022/ESDC_PDF_DIAP_EN_20221005.pdf
Enarson, E. (1998). Through women's eyes: A gendered research agenda for social sciences. *Disasters, 22*(2), 157–173.
Escobar, A. (1995). *Encountering development: The making and unmaking of the Third World.* Princeton University Press.
Esmaquel III, P. (2023, October 11). More Filipinos from poorest sector believe calamities are God's punishment – Pulse Asia. *Rappler.* https://bit.ly/3zLy3H9. Accessed September 10, 2024.
Eugenio, D. (1994). *Philippine folk literature.* University of the Philippines Press.
Fernandez, B. (1996). *Psychological response of women to two types of disasters in the Philippines.* PhD Dissertation in Philippine Studies, University of the Philippines Diliman, Quezon City.
Filipinas of HamOnt. (2020). *Our story.* https://www.filipinasofhamont.com/our-story
Filipinas of HamOnt. (2022, June 10). *GoVaxx: A partnership between Hamilton public health, the Hamilton Filipino community centre, and the Filipinas of HamOnt.* https://www.instagram.com/p/Ceof8lSuIEi/?igsh=MWoyZmtvOHYxd2Ztbg==

Gaillard, J. C., Sanz, K., Balgos, B., Dalisay, S. N., Gorman-Murray, A., Smith, F., & Toelupe, V. (2017). Beyond men and women: A critical perspective on gender and disaster. *Disasters, 41*(3), 429–447. https://doi.org/10.1111/disa.12209

Go, C. O. (2016). *Kababayen-an han Karak-An (Women of storm surges): A feminist ethnographic research on women survivors of super Typhoon Yolanda* [MA Thesis]. University of British Columbia.

Go, C. O. (2017). Women of storm surges Meaning making as cultural process of social repair for Yolanda survivors. *Philippine Studies: Historical & Ethnographic Viewpoints, 65*(2), 227–256.

Guidelines on Camp Coordination and Camp Management for Internally Displaced Persons (IDPs) Series of 2021, 2 Joint Memorandum Circular (2021). https://www.dswd.gov.ph/issuances/MCs/JMC_2021-002.pdf

Hewitt, K. (1997). *Regions of Risk. A Geographical Introduction to Disasters.* Addison Wesley Longman Harlow.

Hewitt, F. (2022, March 14). Virtual campaign aims to unite Hamilton's Filipino community. *The Hamilton Spectator.* https://www.thespec.com/news/hamilton-region/virtual-campaign-aims-to-unite-hamilton-s-filipino-community/article_baa45edb-7306-5188-a6cb-ba028d914905.html

Honasan, C. M., & Veneracion-Rallonza, M. L. (2019). Chapter I: Introduction. In *Women in natural disasters: Indicative findings in unraveling gender in institutional responses: An ASEAN intergovernmental commission on human rights (AICHR) thematic study* (pp. 1–17). https://bit.ly/4eP5nfa. Accessed September 15, 2024.

International Federation of Red Cross and Red Crescent Societies. (2015). *Unseen, unheard: Gender-based violence in disasters Global study.* International Federation of Red Cross and Red Crescent Societies. https://disasterlaw.ifrc.org/sites/default/files/media/disaster_law/2021-02/1297700_GBV_in_Disasters_EN_LR2.pdf

Jayme-Lao, M. E. (2019). Chapter 3.7: Philippines. In *Women in natural disasters: Indicative findings in unraveling gender in institutional responses. An ASEAN intergovernmental commission on human rights (AICHR) thematic study* (pp. 84–90). https://bit.ly/4eP5nfa. Accessed September 15, 2024.

Jimenez, M. C. & Dollaga, N. (1996). *Isang Lahukang Pananaliksik sa Pagtuklas ng Proseso at Balangkas ng Pagsusuri ng Kapasidad at Bulnerabilidad ng mga Kababaihan sa Panahon ng Kalamidad.* MA Thesis in Community Development, University of the Philippines Diliman, Quezon City.

Joseph, A., Bedminster, T., Damptey, S.J., Stearns, G., & Brockbank, M. (2023). The Restoration House COVID-19 Vaccination Clinic: Challenging Systemic Racism and Ableism through Community Solidarity and Action. *Health Reform Observer—Observatoire des Réformes de Santé, 11*(1). https://doi.org/10.13162/hro-ors.v11i1.5435. Accessed November 20, 2024.

Lusterio-Rico, R. (2024). Environment and climate change in the Philippines from Duterte to Marcos Jr. Arugay, A. & Encinas-Franco, J. *Games, Changes, and Fears. The Philippines from Duterte to Marcos Jr.* ISEAS Yusof Ishak Institute.

Manila Observatory website. https://www.observatory.ph/about/.

McColl, M. A. (2019, April 11). Should I say 'disabled person' or 'person with a disability'?. *The Conversation.*

National Disaster Risk Reduction and Management Council (NDRRMC). (2020). *National disaster risk reduction and management plan 2020–2030.* Office of Civil Defense - Policy Development and Planning Service (OCD-PDPS).

National Disaster Risk Reduction and Management Council (NDRRMC) & Department of National Defense (DND). (2011). *National Disaster Risk Reduction and Management Plan 2011–2028.* NDRRMC & DND.

Obedicen, V. (2013). Baha, Bakwet: Implikasyon ng Kalamidad sa Kakayahan ng mga Kababaihang Naninirahan sa Evacuation Center ng Brgy. San Vicente, Angono. *Review of Women's Studies, 23*(1–2), 166–203.

Official Gazette. (2010, May 27). Republic Act No. 10121. *An act strengthening the Philippine disaster risk reduction and management system, providing for the national disaster risk reduction and management framework and institutionalizing the national disaster risk reduction and management framework and institutionalizing the national disaster risk reduction and management plan, appropriating funds therefore and for other purposes.* https://bit.ly/3NaWpgK. Accessed September 13, 2024.

Oliver-Smith, A. (1999). "What is a disaster?" Anthropological perspectives on a persistent question. In A. Oliver-Smith, S. Hoffman (Eds.), *The angry earth* (pp. 18–34). Routledge.

OMNI Filipino News. (2021, July) Ayuda Para sa Ilang Kababayan Ngayong Pandemya [Assistance for Filipino Canadians in the time of the Pandemic]

Permanent Mission of the Republic of the Philippines to the United Nations 2024. (2024). *Philippines cited as "leading the way as a model" in gender-responsive disaster risk reduction.* https://bit.ly/4etTg7C. Accessed September 15, 2024.

Philippine Department of Social Welfare and Development. (2013, December 2). *DSWD sets up special place for women at evacuation camps.* https://reliefweb.int/report/philippines/dswd-sets-special-place-women-evacuation-camps

Pinay Collection. (2024, September). *Pinay collection transparency report.* https://docs.google.com/spreadsheets/d/1VJtjyYK6jC9gmw9TjqApxPwxK9K9kxTwd4f_oa8j6Ck/edit?gid=912465041#gid=912465041

Pittaway, E., Bartolomei, L., & Rees, S. (2007). Gendered dimensions of the 2004 tsunami and a potential social work response in post-disaster situations. *International Social Work, 50*(3), 157–173.

Quinsaat, S. (2024). *Insurgent communities: How protests create a Filipino diaspora.* University of Chicago Press.

RA 9710: Magna Carta of Women. Implementing Rules and Regulations. (2010). Philippine Commission on Women.

Reyes, D. D., & Lu, J. L. (2016). Gender dimension in disaster situations: A case study of flood prone women in Malabon City, Metro Manila. *International Journal of Disaster Risk Reduction, 15*, 162–168.

Rodriguez, J. (2021, February 10). Why many advocates prefer the term "disabled people" over "people with disabilities." *CTVNews.Ca.* https://www.ctvnews.ca/canada/why-many-advocates-prefer-the-term-disabled-people-over-people-with-disabilities-1.5303797

Rosas, A. C. (2022, March 8). Women in Hamilton's Filipino community band together to make sure their stories are told. *CBC News.* https://www.cbc.ca/news/canada/hamilton/filipinas-of-hamont-1.6373142

Salazar, V. (1742). Of the reduction of the Mandayas Indians to our holy faith; E. Blair, & A. Robertson (Eds.). (1903–1909). *The Philippine Islands, 1493–1898* Vol 43, 1670–1700 (pp. 73–76). The Arthur and Clark Company.

Sheffer, G. (2003). *Diaspora politics: At home abroad.* Cambridge University Press.

Social Technology Bureau, DSWD, Government of the Philippines. (n.d.). *Why we need more women-friendly spaces (WFS).* https://dswdcashassistance.ph/women-friendly-spaces-wfs/#What_is_WFS

Sohrabizadeh, S., Tourani, S., & Khankeh, H. R. (2014). The gender analysis tools applied in natural disasters management: A systematic literature review. *PLoS Curr.*

Tanyag, M. (2018). Resilience, female altruism, and bodily autonomy: Disaster-induced displacement in post-Haiyan Philippines. *Signs: Journal of Women in Culture and Society, 43*(3),563–585.

Tronto, J. C. (1993). *Moral boundaries: A political argument for an ethic of care.* Routledge.

United Nations Environment Programme (UNEP). (n.d.) *Climate change and water-related disasters.* https://bit.ly/47MLLX1. Accessed September 4, 2024.

United Nations Framework Convention on Climate Change (UNFCCC). (2011). *Fact sheet: Climate change science - the status of climate change science today.* https://bit.ly/4em2seg. Accessed September 2, 2024.

United Nations Population Fund. (2022, September 1). New beginnings: Rebuilding life after Typhoon Odette. UNFPA News. https://philippines.unfpa.org/en/news/new-beginnings-rebuilding-life-after-typhoon-odette

Valila, J. R., Jr. (2023). The classes and class struggles contents of the Bangsamoro question. *Southeast Asia: A Multidisciplinary Journal, 23*(1), 36–54. https://doi.org/10.1108/SEAMJ-01-2023-0011

Vicky, M. (2022, November 30). 'My girls are watching': Filipinas in Hamilton come together to share their stories. *TV Ontario.* https://www.tvo.org/article/my-girls-are-watching-filipinas-in-hamilton-come-together-to-share-their-stories

Waheed, Z. (2023). *Climate Change's greatest victims are women and girls. UNICEF South Asia.* https://bit.ly/3Y54gS7. Accessed September 6, 2024.

Women Enablers Advocates & Volunteers for Empowering & Responsive Solutions (Director). (2022, November 2). Community women leaders of St. Bernard, So. Leyte underwent training as Women Friendly Space (WFS) Facilitators who will manage the WFS [Video recording]. https://www.facebook.com/watch/?v=835952990778510

Mary Dorothy dL. Jose is an Associate Professor in the Department of Social Sciences and former Associate Dean for Academic Affairs at the College of Arts and Sciences (CAS), University of the Philippines (UP) Manila. She earned her Bachelor of Arts in History, Master of Arts in Asian Studies, and Doctorate in Philippine Studies from UP Diliman. Her research focuses on Philippine Studies and Women Studies, emphasizing context-sensitive and gender-sensitive perspectives. Dr. Jose has received notable recognition, including the Gender Advocate Awards from the UP Manila Center for Gender and Women Studies (for 2013 & 2014, respectively) and the UP Diliman College of Social Sciences and Philosophy's Distinguished Alumni Award for Gender Equality and Women Empowerment (2023). She has also been a recipient of research grants and fellowships, such as the Empowering Network for Thai and ASEAN Studies (ENITAS) Research Award by the Chulalongkorn University (2016), the University Library Fellowship by the Center for Southeast Asian Studies at the University of Michigan (2018), and the National University of Singapore (NUS) Research Fellows Programme for Southeast Asia (2023). In her role as a gender specialist for development projects, Dr. Jose works to enhance women's participation in policy and decision-making, ensuring that feasibility studies are designed to meet the practical needs of women and other marginalized groups in society.

Anabelle B. Ragsag's is a PhD candidate and Sessional Faculty member in Social Work at McMaster University, raised in Mindanao, Philippines. Her research bridges Southeast Asian diasporic feminisms, algorithmic welfare governance, and digital community labour. A former instructor at the University of the Philippines, she co-founded and convenes several onlone and offline and relational spaces that foster care, memory, and solidarity.

Chapter 10
Brunei and the Maritime Silk Road

Gabriel Yit Vui Yong and Izni Azrein Noor Azalie

Abstract In the fifteenth century, the Sultanate of Brunei was an important thalassocracy along the Maritime Silk Road (MSR). However, the arrival of the Europeans and internal changes in China led to the decline of this important trade route and the subsequent decline of the sultanate. The MSR was revitalised by Chinese President Xi Jinping as the twenty-first century iteration in his Belt and Road Initiative to promote and facilitate trade and co-operation for mutual growth and prosperity, as was the case in historical times. However, although Brunei joined this initiative in 2014, its participation has been limited. This is mainly due to structural dissonance in governance resulting from the country's historical development, which has seen a drastic dislocation of its institutions, economy and environment. As a small, modern nation-state, Brunei lacks the necessary capacity to fully participate in this regional development, leaving the majority of its population cut off from potential opportunities. This chapter examines the key differences between Brunei in the MSR of the fifteenth and twenty-first centuries to understand the current lack of capacity, which in today's world includes alignment with the principles and goals of sustainable development. The lack of an integrating mechanism is identified as a critical factor. The chapter proposes a mechanism to bridge structural gaps as a means to build the necessary capacity to tap into the opportunities offered by the 21st century MSR.

Keywords Maritime silk road · Brunei · Structural dissonance · Integrating mechanism · Historical trajectory

G. Y. V. Yong (✉) · I. A. N. Azalie
Faculty of Arts and Social Science, Universiti Brunei Darussalam, Bandar Seri Begawan, Brunei
e-mail: gabriel.yong@ubd.edu.bn

I. A. N. Azalie
e-mail: azrein.azalie@ubd.edu.bn

G. Y. V. Yong and A.K.M. A. Ullah (eds.), *Brunei and the Future Transformation of Southeast Asia*, https://doi.org/10.1007/978-981-96-7901-0_10

Introduction

During its golden age in the fifteenth century, the Sultanate of Brunei was an important thalassocracy along the ancient Maritime Silk Route (MSR). The demand for camphor, sandalwood, bezoar stones and other rainforest products from the interior of Borneo was high, as was the demand for a variety of foreign goods traded in the region. Like many other cities in Nusantara (maritime Southeast Asia), the capital was built on stilts over the water. Commerce was conducted in the estuary on small boats; this mobile market as well as the sellers that ply their trade in small boats were called 'padian' (Bakar, 2018; Ibrahim, 1970; Yong et al., 2023). The supply of goods was facilitated by the 'pengalu' (merchants), who conducted their trade along the river routes that extended inland to the many villages of indigenous tribes who bartered for imported goods and manufactured goods from the city in the bay (ibid.).

However, a combination of European colonisation and China's inward-looking policies during the Ming dynasty led to the decline of the MSR. The Europeans, who also attempted to convert the locals to Christianity in the course of colonisation, saw the Sultanate of Brunei as their main adversary. This led to the Castilian War of 1578, which resulted in the decline of the sultanate. The civil war of 1660–1673 further weakened the Sultanate . The rise and expansion of Sarawak and the creation of North Borneo caused Brunei to shrink to its current small territory in the early twentieth century.

In 1906, Brunei was placed under British residency. This ushered in drastic change as a program to relocate the population onto the mainland was initiated and a land-based economy developed. The discovery of oil in 1929 and a major gas field in 1963 were major milestones in the country's modern development. The young developing country became prosperous as oil price soared with the rapid economic growth following the Second World War. However, concerns about hydrocarbon dependence and long-term sustainability led to a policy of economic diversification and sustainable resource management initiatives. These include international co-operation projects to develop local and regional capacities for sustainable development in the 1990s, particularly with regard to coastal and marine resources management. This is highly relevant to Brunei as it is a coastal state.

In 2013, China launched its Belt and Road Initiative (BRI) to revitalise trade and cooperation along the ancient land and sea trade routes, the latter also known as the MSR of the twenty-first century (see, e.g., Chan, 2020). In 2017, China released its vision for maritime cooperation to build an open and inclusive platform for cooperation, establish a constructive blue partnership and forge a "blue engine" for sustainable development (China Oceanic Development Foundation & Fujian Institute for Sustainable Oceans, CODF & FISO, 2022). Although China has stated its objectives clearly (Song & Fabinyi., 2022), many in the West warned against its 'true' intentions, describing it as neo-colonialism (Pozzi, 2022). Brunei was one of the first countries to join. In 2014, the Brunei-Guangxi Economic Corridor (BGEC) was established to explore bilateral cooperation projects. In 2018, the relationship between Brunei and China was upgraded to a "strategic cooperation partnership" during the Chinese

President's visit, indicating Brunei's willingness to deepen its commitment to the new MSR (Lawrence, 2021). At the local level, the One Belt One Road (OBOR) Association was established to improve relations at the people-to-people level. However, prior to the BRI, Chinese companies were already involved in a number of projects in the oil and gas and construction sectors. Under the BRI, there were further co-operation and investment, notably in agriculture and aquaculture, telecommunications and port development (ibid.).

While a number of joint ventures have been established under the new MSR, these have been somewhat disconnected from the Bruneian public, leading to unfounded rumours falsely portraying China as a colonising power, a common incorrect perception of the MSR (Song & Fanbinyi, 2022). There is widespread ignorance or misunderstanding of the MSR, the BRI and China's stated intentions. Few take seriously China's pledge to strengthen co-operation to achieve common sustainable development goals. The intensifying geopolitical conflict has exacerbated the situation as the collective West, which has dominated the global political-economic order, takes increasingly aggressive actions to contain the rise of the global South led by China (Arensen, 2024).

This chapter examines Brunei's relationship with the MSR to identify issues and problems. It will begin with a review of the Sultanate in the fifteenth century when it was an important node along the ancient MSR, followed by a review of its transformation into its present existence as a small modern nation-state. The analysis will compare the two periods, the fifteenth and twenty-first centuries, to identify key characteristics and controlling factors. As effective governance is critical to Brunei's ability to benefit from the new MSR, the findings from the comparative analysis will be discussed in detail to elucidate persistent issues and problems. The chapter concludes with recommendations to address the issues and problems identified.

Method

This paper compares Brunei's participation and role in the MSR during the fifteenth century (MSR-15) and twenty-first century (MSR-21). As Brunei and the MSR context have changed significantly, the analysis will consider changes in both dimensions to identify key differences. The analysis will compare (i) the political structure and Brunei's role in the region; (ii) economic characteristics; (iii) exploitation of coastal-marine resources; and (iv) relationship with the MSR and China. It will draw on historical accounts and personal involvements in dialogues and initiatives related to coastal-marine development in Brunei and the region.

Brunei: Sultanate Versus Nation-State

In the fifteenth century, Brunei was a sultanate with a vast territory that stretched around the coast of Borneo and parts of the Philippines. As an absolute monarchy, it gained importance as a major thalassocracy after the decline of the mighty Majapahit Empire. The capital of the sultanate was a city built entirely on stilts over the Brunei estuary (Pigafetta, 1521, in Nicholl, 1975; Blundell, 1923). The main harbour and fortress were located at Kota Batu ('stone city'), on the north side of two parallel ridges that flanked the Brunei River, which provided natural protection at its entrance. The city was a compact complex of residential clusters built largely from materials found in the surrounding area (Yong, 2021; Zainuddin, 2021). The culture that developed was closely linked to the environmental dynamics of the area. The estuary has a microtidal regime, with the tidal range exceeding 2 m only occasionally during spring tide (Yong, 2010). Natural sedimentation has reduced the size of the Brunei River estuary significantly since the peak of marine encroachment about 5,000 years ago, a consequence of rising sea levels due to glacial melting at the end of the Pleistocene Ice Age.

Each cluster of stilt houses was built by different clans on mud or sand bars along the estuarine shore. As these deposits naturally grew over time, the inhabitants contributed to its development by disposing of their household and cottage industrial waste, which hastened the build-up of the land beneath the houses (Hassan & Yong, 2019). With the onset of modernisation, one island was artificially enlarged in the first joint venture between a local clan and Chinese traders from Singapore to establish a fixed trading centre (ibid.). Prior to this development, commerce was mobile as padians used boats to bring daily necessities to individual households (Bakar, 2018; Ibrahim, 1970). In contrast to today's Kampong Ayer ('water village'), the 'city built over water' was not just a settlement. It was also home to numerous cottage industries, including those that worked with natural resources that did not exist locally, such as brass, iron, silver and gold, as well as the manufacture of various metal goods and weapons, including cannons and swords (Azalie & Yong, 2022; Ibrahim, 1971). Other notable industries included boat building, various types of woodworking, fishing, food processing and textiles, such as the 'Kain Tenunan' (traditional hand-woven textiles with distinctive motifs and the use of gold threads).

The economic activities and social classes were defined by the location of the clans' residential groups or villages. In addition to the craftsmen, some clans also had religious scholars, teachers and clergymen. The 'pengalu' (riverine merchants) played an important role in supplying the city with fresh agricultural produce, forest products and dried meat from the villages along the rivers in the hinterland. They also brought imported and urban products to the interior. The pengalu were indispensable middlemen in the daily distribution and exchange of goods in the city and inland population as well as in international trade; they essentially sustain the lifeline of the water city (Yong et al., 2023). Even in the waning period of the sultanate, Kampong Ayer remained economically vibrant (Bakar, 2018).

In contrast, Brunei Darussalam is a young nation state that regained full independence from Britain in 1984. Given its 600-year history as a Malay-Islamic monarchy, this relationship between the monarch and his people is deep-rooted. As a result, the citizens continue to relate to him as the Sultan, rather than the Prime Minister, his role in the modern system of government. Therefore, there are two overlapping systems of government—(i) a secular one in which Britain drafted the constitution and legislation, and (ii) the traditional monarchical system. In 2014, the Sultan promulgated the Syariah law as an additional layer of legislation. The modern political system is therefore quite complex and interwoven with Islamic religious values and practices. Brunei's transition to a modern nation state began in 1906, when the country handed over administrative control of the state to the British Residency. After World War 2, an interim government, largely made up of local officers, continued the modernisation process through a series of 5-years National Development Plans (NDPs). The British Resident is replaced by a High Commissioner to maintain Britain's interest and influence. As this period coincided with global demand for oil and gas as the premier energy resource, Brunei became a prosperous state; in the 1980s it had the highest GDP per capita in the region. Revenues from the export of hydrocarbons accelerated development and transformed the former estuarine kingdom into a country where the majority of the population lives and works on land.

Brunei initiated a policy and programme to diversify its economy in the 2nd NDP (1962–1966) and intensified its industrialisation activities in the 5th NDP (1986–1990). In 1989, a Ministry of Industry and Primary Resources was established to facilitate the implementation of the plans. Brunei became an independent nation in 1984 and joined the regional bloc, ASEAN, in the same year. The latter contributed to Brunei's economic development and capacity-building through various regional-international projects. At the turn of the millennium, the Brunei Economic and Development Board (BEDB) was established to steer the country's economic development on a diversified course and to attract FDI. The twenty-first century brought with it a deepening of national and regional developments. In 2004, Brunei began developing a long-term strategic plan, the Wawasan ('Vision') 2035, which aims to develop a dynamic and sustainable economy and a high quality of life for the population. The document was published in 2007, and a council was appointed in 2014 to lead the implementation process.

In the meantime, ASEAN adopted a charter in 2007, which came into force the following year. Its aim is to promote regional stability, economic integration and socio-cultural co-operation through three main pillars: (i) the political-security community, (ii) the economic community and (iii) the socio-cultural community. In 2010, Brunei designated two sites for foreign direct investment (FDI) and industrialisation: (i) the Sungai Liang (SL) Industrial Park, and (ii) Pulau Muara Besar (PMB). While the former had limited success, the latter has been actively industrialising. The Hengyi petrochemical plant, which began operations in late 2019, was a landmark event. Brunei's participation in China's MSR-21 promoted industrial development on the island (PMB) and the neighbouring Muara and Serasa areas on the mainland. In 2017, Brunei established the Brunei Maritime and Port Authority (BMPA) to serve as a multi-agency hub for maritime industrial development in addition to its primary role as the country's port authority. The BMPA therefore has an important role in Brunei's participation in MSF-21.

Trade Versus Sustainability Concerns

The twenty-first century presents a globalised world that is increasingly focused on sustainable development. Climate change has become a critical issue, labelled an "emergency" by the United Nations in 2019. Modern development had replaced increasing areas of natural rainforests through the development of road networks, which opened up previously untouched land and facilitated their conversion to infrastructure and residential, commercial and industrial assets. Swamps have been filled and valleys cut, cleared and filled to facilitate urbanisation. The consequences of these actions include pollution, deforestation, loss of ecosystem functions, loss of biodiversity, and climate change. A pivotal moment in the late twentieth century was the 1992 UN Conference on Development and Environment, commonly known as the 'Earth Summit'. Since then, Brunei, in co-operation with the international community, has committed to Sustainable Development and adopted policies in its 7th NDP that seek a balance between socio-economic growth and environmental protection. However, these efforts took a long time to take effect. The Department of Environment, Parks and Recreation (DEPR) was established in 2002 and the first law specifically targeting environmental protection was only enacted in 2016.

Due to its dependence on hydrocarbon export, industrialisation was limited. However, despite its concern and commitment to Sustainable Development, Brunei did not establish an agency with a specific mandate to oversee sustainable development. When DEPR was established in 2002, the agency's mandate included recreation and parks. In the meantime, the pace of industrialisation in the PMB-Muara-Serasa region has exceeded the country's capacity to regulate it over the past decade (Hussin, 2022). Inadequate regulatory control is exacerbated by the conflation of environmental management with 'safety, health and environment' (SHE). The establishment of the National Safety, Health and Environment Authority (SHENA) to oversee industrial development has reinforced this misalignment institutionally, and clouds the understanding of sustainability. However, Brunei had participated in regional co-operation projects aimed at developing national capacity for sustainable development. Of particular relevance to the MSR are the 1987–1992 US-ASEAN Coastal Resource Management Project (CRMP) and the 1994–1998 ASEAN-Australia Coastal Zone Environmental Resources Management Project (CZERMP).

The CRMP introduced the concept of integrated coastal zone management (ICZM) in Brunei. It is of particular importance as Brunei is a coastal state with the majority of its population and economic assets located in the coastal zone. The project led to the publication of the environmental and resources baseline (Chua et al., 1987), the findings on the state of environment and utilization of resources after a five-year study (Silvestre et al., 1992) and an ICZM plan for Brunei (Department of

Fisheries, 1992). The plan consists of several projects aimed at developing national capacity for ICZM, as well as a number of sectoral projects to enhance fisheries and mangrove resources, conserving water quality and sensitive ecosystems, and manage oil spills. However, while some sectoral projects have been successfully implemented, others have not, particularly the capacity-building and institutional-strengthening projects (DEPR, 2006; Yong & Hamid, 2007; Yong & Tamit, 2001). Inadequate knowledge and expertise, which translates into, and is influenced by, inappropriate policies, have resulted in a vicious cycle impeding progress (Yong & Tamit, 2001). Consequently, the inter-agency committee proposed in the plan to coordinate ICZM, a crucial cog in the wheel, was not established and the ICZM plan has been shelved (ibid.).

The CZERMP focused on the use of information systems to facilitate ICZM at the regional level. Although the project included the delivery of hardware, software and training, logistical and organisational problems hampered development until its final year, when a team from the national university salvaged the project. A National Environment and Resources Information Centre (NERIC) was established using Microsoft Access and a proof-of-concept project was carried out. The latter, a project on water quality in Brunei Bay, involved modelling using Geographic Information Systems (GIS) to identify potential areas of siltation, pollution and health impacts (Yong & Hamid, 2007). Although Brunei's involvement was successful, NERIC soon became dysfunctional and there was no further development to revitalise or build on the results of this project. In 2022, an opportunity arose, through the ASEAN Centre for Biodiversity, to explore the development of a decision support system for the coastal and marine environment in the region. This is similar to NERIC but limited to living resources (species and habitats). Although the workshops were promising, there has been no further developments.

Sustainability is a salient feature of MSF-21 and the guiding principle, as defined in the Brundtland Report (WCED, 1987), is that the three mutually supportive goals of economic growth, environmental protection and equality/social justice must be achieved simultaneously. The United Nations SDGs serve as prominent and useful signposts. Theoretically, sustainability is based on the ecological model, which adopts an ecocentric perspective. The colours blue and green are used to indicate sustainability in the areas of water/sea and natural land areas. Pauli (2010), for example, envisioned a blue economy in which coastal and ocean resources are exploited in a non-destructive way, involving regenerative systems that produces no waste but plenty of meaningful jobs. Blue urbanism (Beatley, 2014) refers to a sustainable city in which the built environment is aligned with coastal-marine environmental flows and thus in balance with natural dynamics. Although they may sound idealistic, they reflect the current aspirations and reality.

The MSR-15 was primarily a trade route that served as an important network for trade and cultural exchange between Southeast Asia, China, India and the Middle East. Its importance in the region is largely related to the fact that most of Southeast Asia is archipelagic or coastal in nature. Therefore, water transport is the most important form of mobility and facilitator of trade. At a time when the human population was only 5–6% of today's eight billion, sustainability was not an issue. Environmental problems are a product of modernisation, and sustainable development is the response of the international community. China's MSF-21 promotes blue economic development, complemented by a blue citizens' initiative on land. As sustainability is the underlying concept, it can be combined with the UN's SDG goals (see, e.g. Hassan & Yong, 2023) and other sustainability programmes. As Brunei is a coastal state and exposed to rising sea levels, blue urbanisation is a logical adaptation strategy (Yong & Hassan, 2024). Brunei's recent blue economic plan focuses on the development of decommissioning works (of aging oil platforms) as part of a maritime cluster on PMB. Although this cluster has features of sustainability, such as recycling and responsible environmental procedures, it is a standard industrial set-up that differs significantly from the ecological models envisaged in the MSF-21.

Capacity for Sustainable Development

The existing situation suggests that Brunei lacks the necessary institutional capacity to effectively benefit from the MSR-21, if it were to achieve the sustainability goals of MSR-21 and Wawasan 2035. This is however a common issue for developing countries, and a concern in MSR-21 (Song & Fabinyi, 2022). Brunei's situation has been brought to light in its participation in CRMP and CZERMP. Brunei's capacity for sustainable utilization of MSR-21 will be examined from a governmentality perspective. Figure 10.1 shows the conceptual framework used in the analysis, adapted from Azalie's (forthcoming) work on blue governmentality. The concept is an adeptive version as defined in Dean's (2010) Analytics of Governing (AOG), Oels's (2005) environmentality and Fletcher's (2017) quadruple power interlinkages in governmentality. The framework is based on Foucault's (1991) concept of governmentality (cited in Burchell et al., 1991) and serves multiple purposes.

The first task is to identify the political rationalities that the state endeavours to uphold and which are referred to as its raison d'état. Since the state is composed of 'spheres and domains', i.e. different entities with different dispositions (Dean, 2015), different methods of governance can emerge even within a single system of government. When multiple governmental rationalities coexist, the resulting different technologies of government that influence many spheres would lead to dissonance (Dean, 2010). A major challenge for modern governance is the tendency to adopt a 'one-size-fits-all' approach that disregards the dissonances, discontinuities and complexities inherent in different political systems or states (Dean, 2015; James & Jorgensen, 2009). Although stakeholder consultation, coordination and the vision of empowerment and citizen participation are important (Freeman et al., 2010), together with the emphasis on effective communication based on a clear logic (Habermas, 1985), they have not sufficiently addressed ongoing governance problems.

Coordination problems in governance can be partly attributed to the neglect of key factors such as biopolitics (the management of populations), power-knowledge dynamics (the interplay between power structures and the production of knowledge)

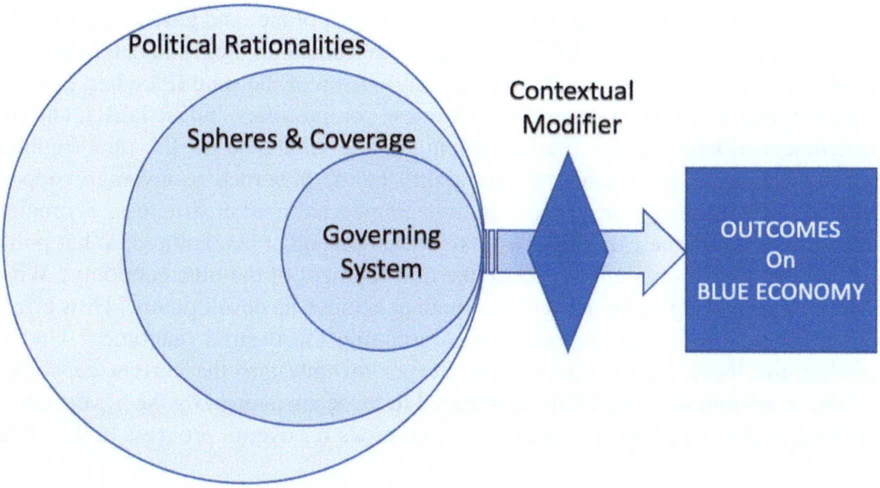

Fig. 10.1 Conceptual framework for blue economy capacity development

and power asymmetries (the imbalance of power between different social groups). In order to uncover these complex relationships, it is important to understand the specific contexts and environments in which governance takes place. In the case of Brunei, the prevailing principles of sustainable governance are significantly shaped by the tendency to go along with prevailing trends (which is referred to as 'followerism'), as well as the need to preserve national wealth and prosperity. This orientation leads to the adoption of a variety of government strategies, instruments, policies and best practises that may not fully align with local realities. The complex historical evolution of Brunei—from an independent Malay sultanate to a British Protectorate and then to a modern nation-state in which the monarchy remains prominent in the hearts and minds of the people—coupled with the recent introduction of Islamic Syariah law, has created a multiplicity of approaches to governance, complicating the effectiveness and relevance of governance.

To effectively promote blue economic development, it is critical to first understand the political motivations that shape this vision. This includes examining the intentions of those in power, the goals they seek to achieve, and the type of thinking and knowledge that guide their decisions. Understanding these underlying motivations is critical to fostering a collective sense of ownership and a shared political future that goes beyond the mere imagination of communities (Anderson, 2020). Creating a strong sense of identity between those in power (governors) and the general population (governed) is critical to the development of blue economy initiatives. In this context, creating a close association between decision makers, politicians, industrialists and local communities is critical to achieving effective blue development outcomes (Hallaq, 2012). This requires a move away from superficial ecological preservation towards a comprehensive, integrated approach to development that takes into account the needs of multiple stakeholders, particularly in relation to the SDGs.

The role of disciplinary power, which includes laws, policies and guidelines, must be emphasised again and again. These serve as both leadership tools and deterrents, but should not undermine the broader mission and vision of the state (Fletcher, 2017).

Successful transformation requires business communities, industrialists, and the government alike to assess market potentials and capitalise on the opportunities presented by neoliberal markets. The ability of the governed to leverage support from the state in navigating these incentivized external market structures is crucial. To address these issues, the following key questions must be explored: What political rationalities or challenges dictate the development of the blue economy? What governing technologies are used to facilitate or achieve its development? How effective are these governing technologies in attaining the desired outcomes? Understanding the local context and gaining deeper insights into the current capacities for blue economic development—compared to those envisioned by the governors or standard setters—will have significant implications for overall progress in this area.

Data and Analysis

Present-day Brunei is a small nation state with limited capacity to tap into the regional economy. MSR-21 is only a part of the global economy, which is undergoing seismic shifts from an order established by Western powers since the colonial period to an emerging multipolar and more equitable one. Humanity is also confronted by a variety of sustainability concerns, not least, global climate change. In contrast, Brunei was a vast maritime power in the fifteenth century, coinciding with the Sultanate's zenith. Development was less intrusive and trade and cultural activities followed natural dynamics. China was a global power in the fifteenth century, but the Ming emperors then adopted an inward-looking policy, thus ending its regional maritime engagements. The MSR, however, continued to serve regional trade, but it was eventually replaced by new trade systems implemented by European colonial powers. Presently, China has resumed its role as a major global power following a hundred years of humiliation and ensuing period as a poor third world country emerging from WW2 and a civil war. Today, it has adopted an outward-looking policy and is actively working to build a shared future of peaceful co-existence and 'win–win' cooperation (Chan, 2020; Song & Fabinyi, 2022), which is viewed in a negative way by many in the West and those aligned with their perspectives (Pozzi, 2022). The MSR-21 is a part of this initiative. The data used in the comparative analysis of Brunei's relationship with MSR in the fifteenth and twenty-first centuries were derived from articles (see, previous section) and personal involvement over 1–2 decades. Table 10.1 summarizes the salient differences between the two periods.

Table 10.1 Salient differences between Brunei's relation with MSR in the fifteenth and twenty-first centuries

ASPECTS	MSR-15	MSR-21
Brunei	Sultanate; vast territory	Small state; monarchy
City	Compact urban complex in Brunei estuary	Spread out over the land, concentrated along the coast and estuaries
Economy	Maritime and riverine trade; wide range of industries; water-based commerce	Hydrocarbon-based; low-level industrialisation; agriculture and fisheries minor components of GDP; limited success in economic diversification
Regional role	Major thalassocracy	Emerging port; underdeveloped
Resources	Forest products, e.g., camphor, sandalwood, bird's nest, etc.	Oil and gas; fishery product (limited)
Relationship with MSR	Major node; regional power; city benefits directly	Joined BRI; FDI by Chinese firms; public cut off from direct engagement
Capacity for sustainable governance	Sustainability not a concern; close inter-relationship between people and environment suggests sustainable relationship	Committed to sustainable development; lack technical capacity for effective application

Political Entity

The Sultanate of Brunei had a clearly defined political structure and a distinct system of government. It ruled over a vast territory that stretched from northern Borneo to the east-southeast coast and parts of what is now the Philippines. Its population and resource base were therefore considerable. This is an important factor supporting its role as a regional power. In contrast, modern Brunei has a much smaller territory with a total area of 5,765 km^2, divided into two parts by the Malaysian (Sarawak) district of Limbang. With a population of less than half a million people, it has a limited labour force. Although it has hydrocarbon resources, they are finite and will eventually be exhausted. Since the 1960s, Brunei has explored ways and means to diversify its economy, but has achieved limited success. This is partly due to its small local human resource base. The nation state has been and continues to be dependent on foreign expatriate and migrant workers.

Modern Brunei is a nation state with a less clearly defined political structure, having adopted a foreign system of government modelled on the British system. However, the enduring monarchical tradition, deeply rooted in over 600 years of existence as a Malay Islamic Sultanate (MIS), supersedes this secular framework. Since regaining independence, Brunei has also adopted a national philosophy based on the concept of an MIS. This has further reinforced the traditional influence, which is incongruent with the modern system. The passing of the Syariah Laws in 2014 further complicates the political system. This affects decision-making processes and contributes to delays and inefficiencies in the administration. While this may seem trivial on the surface, the effect is cumulative and systemic. They have made the system sclerotic, which hinder progress and further development.

Maritime Influence

In the fifteenth century, the Sultanate of Brunei was an important thalassocracy. Its maritime influence is therefore significant. Although Brunei is a coastal state and therefore physically connected to the maritime sphere, the modern development of the country has shifted its people and economy onto the land. While the country is still dependent on the maritime industry, especially offshore oil and gas exploration, the younger generations are disconnected from the maritime life world. Most young Bruneians do not venture out to sea or to the coast, apart from the small population that continues to live in the water village complex, a remnant of the old world. The estuary is now too shallow and the passage too narrow for today's ships. The country's only deep-water harbour in Muara has to be deepened several times to facilitate shipping. The mangroves and mudflats in the area have been reclaimed and converted into port and industrial facilities. In the last decade, the large island (PMB) has been developed to support industrialisation and trade. Currently, a maritime services and decommissioning facility is being developed in addition to the colossal Hengyi petrochemical complex. Therefore, although there have been developments in the maritime sector, modern Brunei is a pale shadow of its historical existence. Its maritime footprint is small and remains underdeveloped due to insufficient local capacity and strong regional competition. Although the MSR-21 offers an alternative, namely co-operation instead of competition, Brunei has not seized the opportunity.

Sustainable Development Capacity

The Brunei government has expressed commitment to sustainable development since the late 1980s. However, it has not made great stride in this direction, apart from statements of intent. There is no agency, fully supported with the necessary funding and technical infrastructure, dedicated to this goal. Instead, there are multiple agencies tasked with promoting sustainability. The main national agency (DEPR) shares its resources to cover parks and recreation responsibilities, in addition to its environmental mandate. The Environmental Protection and Management Act (EPMA), passed as an order in 2016 and upgraded to an act in 2022, further places the onus of EIA on the proponent, rather than the regulator. With the establishment of SHENA as the competent authority, environmental concerns are often integrated with that of workplace safety. This greatly undermines efforts to realise sustainability goals, as the two are quite different. The outcomes of Brunei's involvement in regional capacity-building projects on sustainable coastal management (namely, CRMP and CZERMP) reveal a failure to build on the successful achievements in the respective projects. Presently, initiatives by Chinese institutions offering capacity-building opportunities for sustainable coastal development (SCD), critical to effective participation in MSR-21, have been met with limited response, perhaps due to mistrust or being overwhelmed (Song & Fabinyi, 2022). Nevertheless, on-going interaction between the authors and the Chinese institutions offer optimism for progress in this area.

Discussion

Despite numerous efforts and the adoption of a progressive national vision, insufficient capacity continues to hinder the successful operationalisation and implementation of sustainable coastal and marine development, i.e. a blue economy, which is necessary to benefit from MSF-21. The fisheries sector currently accounts for less than 1% of GDP, but stocks indicate overfishing; a moratorium on fishing in coastal areas has been in place for two decades. The source of the problem most likely lies outside of Brunei's waters, as fish stocks in the South China Sea are declining (Hazra & Bhukta, 2022). A regional approach to tackling declining fish stocks would have been more effective than a unilateral moratorium in Brunei, where fishing effort is already low. The Department of Fisheries has, since the CRMP, continued to enhance its fisheries resources through artificial reefs and aquaculture programmes. However, it often finds itself between the devil and the deep blue sea, with regards to effective participation in MSR-21 due to dissonance and discontinuity in its governance structure consequent of its history and development trajectory. Meanwhile, China is providing technical assistance to build local capacity through a combination of joint ventures (JVs), training and research co-operation with its higher education institutions (e.g. Xiamen University) and Chinese companies operating in Brunei.

Currently, Hiseaton, a locally established company of Guangxi Haishitong Food Company, has been operating in Brunei under MSR-21 since 2016. The company produces seabass, hybrid grouper and golden pompano and exports to China, the USA, Japan, Singapore and Canada. The company operates a state-of-the-art, sustainable marine range off the coast of Muara and a hatchery further west on the coast at Meragang. The latter was a dysfunctional Brunei government facility. Hiseaton is just one of the few examples of how MSF-21 could benefit Brunei. However, in a case study, Muhammad (2019) found that Hiseaton's attempt to connect with local aquaculture operators to help them gain access to its regional-global network was met with a lukewarm response, resulting in limited technology transfer, largely due to local operators' low-risk attitude and unwillingness to venture out of their comfort zone. A major obstacle is the constraint associated with contract farming and technology transfer licencing. This shows a lack of appreciation of the benefits that could be gained through economies of scale, which in turn, could widen their scope and specialisation. Currently, most Bruneian companies are involved in 'low-altitude' economic activities, mainly related to retail rather than manufacturing and innovation. Regarding the latter, current practises (for low complexity products) tend to drive them into the 'commodity trap', i.e., a phenomenon where producers cannot scale up due to lack of innovation (Farfan, 2005).

In addition to the monitoring and control of the environment and ecosystem, blue economic development also includes the sustainable development of the coastal land, coastline and estuaries, which are often subjected to high development pressure. This area of governance is often neglected in maritime industrial development because the focus is only on land-based assets. However, the coastal zone encompasses a wide range of stakeholders and is the most dynamic of all environmental systems as it lies at the interface of land, sea and air. Previous initiatives on this aspect of the blue economy, which could be termed blue urban development, include the various approaches of integrated coastal management ICM, such as integrated coastal resource management, integrated coastal zone management, integrated shoreline management and integrated catchment management. "Integration" is a critical success factor. From the perspective of blue governmentality, the CRMP and CZERMP were government technologies or instruments designed to ensure the sustainable development of coastal areas (refer to Fig. 10.1). Their failure to gain acceptance could be due to a lack of the 'right' knowledge and contextualisation (Yong & Tamit, 2001). The necessary instruments and laws for sustainable coastal management are in place in Brunei (DEPR, 2006; Hussin, 2022; Silvestre et al., 1992). However, they are not utilised in an integrated system, as such a system is still in its nascent stage.

Regional Cooperation

To understand environmental dynamics and monitor change, it is necessary to take an outward-looking perspective, as natural dynamics are not limited by national and private property boundaries. A regional perspective is crucial for sustainable development. This realisation must first emerge in the realm of political rationality of the Brunei Government. MSF-21 offers the opportunity to develop local-national capacity through regional co-operation, not only with China but among all interested partners in the network. However, China is a world leader in environmental technologies and therefore has the necessary technical base to extend its arm of support. The logical response from an outward-looking, regional-global political rationality would be to accept the offer. However, when viewed through the blurred lens of Brunei's complex historical development, this offer appears to some as a 'trap' or 'neo-colonisation', especially to those still influenced by Western perspectives or misinformation from ill-informed rumours (Pozzi, 2022). This is an unfortunate reality in Brunei and indeed much of the region. It discourages states like Brunei from capitalising on the opportunities that the MSF-21 presents.

However, there are ongoing developments to remove barriers to MSF-21. A recent example at the institutional level is a joint research effort on local fisheries resource management initiated by East China Normal University and Xiamen University in China and three ASEAN universities, namely Institut Pertanian Bogor (IPB), Universiti Brunei Darussalam (UBD) and Universiti Malaysia Terengganu (UMT). Such collaboration will help clarify issues and problems at the local level. This is an important step in bridging the gap between local communities, businesses and governments and the regional programmes that exist at higher political levels. They deal with the lower spheres of governmentality and take into account the contextual factors. Another recent example is the training seminar on capacity building for blue economy development under the Regional Comprehensive Economic Partnership (RCEP) organised by the China Institute for Reform and Development (CIRD). All ASEAN member states were involved. The RCEP provides a framework to realise joint development. The focus is on the harmonisation of seven key categories: initial provisions, trade in goods, services, movement of people (mobilities), investment, the business environment and general dispute settlement provisions. The network extends not only to ASEAN and China, but also to the South Asian countries of India, Pakistan and Bangladesh, all of which have relevant technical and professional experience. Notwithstanding the opportunities available, domestic regulatory systems in terms of taxation and tax collection need to be modernised to enable capital accumulation and encourage internal spending.

The MSF-21 reinvents the global economy, with China as a leading partner and lever for the benefit and growth of others (Chan, 2020; Song & Fabinyi, 2022). It opens new avenues for inclusive and cooperative development, ranging from manufacturing to financial and innovation centres, while expanding and upgrading relationships. The rise in Chinese consumerism, fuelled by urban demographic changes, has created a desire to import goods and services from abroad (Chi, 2021). The existence of both current and latent demand provides MSF-21 partners with opportunities for export-led growth. In conjunction with the recent APEC meeting, China's Xinhua Institute (2024) published a report outlining China's goal of promoting common high-quality development for the Asia–Pacific community. This initiative emphasises the importance of cooperation and aims to maintain harmony between people and nature. It focuses on sustainability and carbon transition initiatives based on innovative scientific endeavours to be shared with partners as part of China's reform agenda.

Financing is a common main barrier to effective participation of MSR-21. This is also the case with Brunei. Transitioning to a low-carbon economy requires significant financial investment, as it involves research and development before successful outcomes can be achieved. Attracting FDI is essential. In this regard, China is a logical partner of choice, given its political stance and leadership in green/blue technologies. This is particularly relevant to small states like Brunei, because capacity building cannot be done without external assistance. Specific to Brunei, we propose a framework to address the Brunei situation (see, Fig. 10.2). Central to the concept is an enabling agency. This is similar to the inter-agency coordinator in the 1992 Brunei ICZM. It occupies the organizational space between government agencies, local and

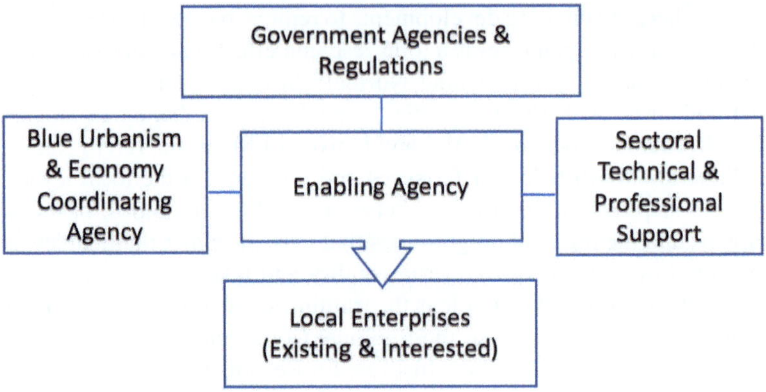

Fig. 10.2 Capacity building framework for blue development

special-interest communities (context) and desired outcomes (see, Fig. 10.1). Integration is critical to sustainable development, management and governance. It is achievable only through a living agency that acts as coordinator and facilitator at the interface of all stakeholders.

Our work in the university and collaboration with institutions in China and ASEAN partners, could play this role. We could also play the role of the Blue Urbanism and Economy Coordinating Agency, which will be supported by research and training labs, if a dedicated agency could be established for the central role. Based on our knowledge of the local Brunei context, personal engagements are need in the enabling process to bridge the disconnects and irregularities, and to harmonize dissonances, which are prevalent and serve as persistent barriers. This applies to government officials who serve as regulators and program owners, as well as seasoned, budding or aspiring entrepreneurs, special-interest groups, and existing professionals and industries. It also proactively addresses nuances highlight in the governmentality framework (refer to Fig. 10.1). The process could also promote more considered identification of political rationalities and appropriateness of governing technologies used, as well as the actual goals desired, vis-à-vis local context. Elucidating how governing process transpires and translates into the outcomes allows for reassessment and re-evaluation of strategies if needed. The inherent reflexivity is crucial towards effective capacity development and sustainable governance.

Rising Sea Levels

Sea level rise has been a concern since the latter half of the twentieth century. The coast of Brunei is experiencing coastal erosion along the eastern half of its open coast (Yong, 1996, 2010). However, it is unclear whether any sections of the coast have been affected by rising sea levels, as the erosion could be attributed to development activities upstream, as well as regional geomorphic and Holocene sea level adjustments (ibid.). In recent years, interest in climate change has increased significantly. However, this has not been matched by sufficient support for research, monitoring, and adaptation projects. Climate change is affecting Brunei in two main ways: (i) increase in rainfall and therefore flashfloods (Hassan & Yong, 2023); and (ii) a rise in sea level at about 3–4 mm/year (ibid.). Although the latter is an imperceptible process, its development is apparent from climate monitoring by agencies such as the US National Oceanic and Atmospheric Administration[1] (NOAA). This however would effectively revitalize the estuaries and coastal wetlands, enhancing its productive and ecological functions, which have been deteriorating with sedimentation, expedited by modern development. Rather than attempting to stop the climate from changing (the main current strategy), it is more logical to embrace the change. Blue economy and blue urbanisation, which are part of MSR-21, are the natural solutions (Yong & Hassan, 2024).

Conclusion and Recommendations

A comparison with the MSR-15 shows that the ability to effectively participate in the MSR-21 has decreased significantly, primarily due to the drastic changes in circumstances in Brunei, regional development, and the decline and resurgence of the MSR. As the MSF, especially in its twenty-first-century version, is not only focused on trade and economy but also on sustainable development, Brunei is no longer in a position to achieve these lofty goals under the current circumstances. The MSF-21 provides a timely opportunity to rectify this situation. Promoting a blue economy along the ancient trade route that once brought prosperity, power and prestige to many states in the region, including Brunei, opens up new avenues for economic diversification, resource protection and sovereignty enhancement. However, Brunei's participation has been slow and limited due to inherent irregularities and dissonance created by the country's development over the last century. A major factor is its inward-looking politics and somewhat parochial culture, a consequence of its recent history. This needs to be corrected, as regional co-operation and a cosmopolitan outlook are essential in today's reality, where politics, economics and trade are inextricably linked and global.

Capacity building for sustainable growth is therefore critical to Brunei's effective participation in MSF-21, and a mechanism for integration is an essential cog that must be inserted into the governance wheel. We propose a framework to address this structural issue. If implemented satisfactorily, it would increase awareness and trust within the governance system while improving understanding of current developments and contributing to more favourable participation in regional initiatives such as RCEP and BRI. Participants in the blue economy, from governors to industrialists to local communities, would gain adequate practical knowledge and experience for RCEP

[1] https://www.climate.gov/climatedashboard.

negotiations and would have local expertise as support. In addition, capacity for more effective and informed policy formulation would be enhanced, and increased cultural and economic exchange can also foster deeper understanding between stakeholders. To achieve this, a change in mindset and the removal of unfounded prejudices is needed to enable positive development. To summarise, the MSF-21 offers development potential and opportunities for Brunei, as both countries are pursuing the common goal of sustainable growth in the maritime sector.

References

Arensen, L. (2024). Ungoverned landscapes: Analyzing post-conflict resettlement in Sdao commune in northwest Cambodia. *Southeast Asia: A Multidisciplinary Journal, 24*(3), 198–207. https://doi.org/10.1108/SEAMJ-02-2024-0023

Azalie, I. A., & Yong, G. Y. V. (2022). Heritage industry as conservation strategy for Kampong Ayer (Brunei Darussalam). *Brunei Museum Journal, 27–51*, 2022.

Azalie, I. A. (forthcoming). Blue governmentalities and blue economy development: Leveraging small state economy with china via the regional comprehensive economic partnership (RCEP). In P. Ordóñez de Pablos, M. Lytras, & S. Malik (Eds.), *Securing sustainable futures through blue and green economies*. IGI-Global Ltd.

Anderson, B. (2020). Imagined communities: Reflections on the origin and spread of nationalism. In *The new social theory reader* (pp. 282–288). Routledge.

Bakar, H. T. A. (2018). The residents of Kampong Ayer, 1906–1940: The activities of Pengalu, Padian and handiwork. *Tawarikh: International Journal for Historical Studies, 9*(2), 151–158.

Beatley, T. (2014). *Blue urbanism: Exploring connections between cities and oceans*. Island Press.

Blundell, P. (1923). *The city of many waters*. Arrowsmith.

Burchell, G., Gordon, C., & Miller, P. (Eds.). (1991). *The Foucault effect: Studies in governmentality*. The Chicago University Press.

Chan, G. (2020). China's maritime silk road advancing global development? New Horizons in East Asian Politics series, 192 pp. https://doi.org/10.4337/9781789907490

Chi, F. (2021). *China's consumption: Establishing a new development paradigm with dual circulations*. China Intercontinental Press Ltd.

China Oceanic Development Foundation & Fujian Institute for Sustainable Oceans. (2022). *Capacity index of sustainable coastal development for countries along the maritime silk road* (Report No. XXXX).

Chua, T. E., Chou, L. M. & Sadorra, M. S. M. (1987). The coastal environmental profile of Brunei Darussalam: resource assessment and management issues. Brunei Darussalam: Brunei Department of Fisheries and International Centre for Living Aquatic Resources Management.

DEPR (Department of Environment, Parks & Recreation, Brunei Darussalam). (2006). Brunei Darussalam State of the Environment Report. Bandar Seri Begawan: Brunei Darussalam.

DOF-MIPR (Department of Fisheries, Ministry of Industry and Primary Resources, Brunei Darussalam). (1992). The integrated management plan for the Coastal Zone of Brunei Darussalam. ICLARM Tech. Rep. 29, 122 p. https://digitalarchive.worldfishcenter.org/handle/20.500.12348/3000

Dean, M. (2010). *Governmentality: Power and rule in modern society* (2nd ed.). Sage Publications Ltd.

Dean, M. (2015). Foucault must not be defended. *History and Theory, 54*(3), 389–403.

Farfan, O. (2005). Understanding and escaping commodity-dependency: A global value chain perspective. Paper prepared for the Investment Climate Unit, The World Bank Group, US.

Fletcher, R. (2017). Environmentality unbound: Multiple governmentalities in environmental politics. *Geoforum, 85*, 311–315.

Freeman, R., Harrison, J., Wicks, A., Parmar, B., & De Colle, S. (2010). *Stakeholder theory: The state of the art*. Cambridge University Press.

Habermas, J. (1985). *The theory of communicative action: Volume 1: Reason and the rationalization of society* (Vol. 1). Beacon Press.

Hallaq, W. (2012). Quranic constitutionalism and moral governmentality: Further notes on the founding principles of Islamic society and polity. *Comparative Islamic Studies, 8*(1–2), 1–51.

Hassan, N. H., & Yong, G. Y. V. (2019). The cultural value of Bakuts in Kampong Ayer, Brunei Darussalam. *Southeast Asia: A Multidisciplinary Journal, 19*, 47–63.

Hassan, N. H., & Yong, G. Y. V. (2023). Climate change and sustainable development of Kampong Ayer: Its history, people, and environment. In *Public policy's role in achieving sustainable development goals* (pp. 96–114).

Hazra, S., & Bhukta, A. (Eds.). (2022). *The Blue Economy: As Asian Perspective.* Springer International. ISBN: 3030965201, 9783030965204

Hussin, D. S. B. P. H. (2022). Structural analysis of environmental management of the Serasa Industrial Park. *Southeast Asia: A Multidisciplinary Journal, 22*(2), 55–72.

Ibrahim, H. A. L. H. (1970). Padian: Its market and the women vendors. *Brunei Museum Journal, 2*(1), 39–51.

Ibrahim, H. A. L. H. (1971). Variations and changes in the names and locations of the wards of Kampong Ayer over the last century. *Brunei Museum Journal, 2*(3), 56–73.

James, T., & Jorgensen, P. (2009). Policy knowledge, policy formulation and change: Revisiting a foundational question. *The Policy Studies Journal, 37*(1), 141–162.

Lawrence, I. (2021). Brunei's Response to China's Belt and Road Initiative: Embracing Asymmetry, Enhancing Authority. *Asian Perspective 45*(2), 397–420. https://doi.org/10.1353/apr.2021.0006

Muhamad, S. S. (2019). Technology transfer: A case study of Brunei Darussalam Aquaculture firms. [Master's Thesis (Master of Arts in Asia Pacific Studies), Thammasat Institute of Area Studies, Thammasat University].

Nicholl, R. (Ed.) (1975). European Sources for the History of the Sultanate of Brunei in the Sixteenth Century. Brunei: Muzium Brunei.

Oels, A. (2005). Rendering climate change governable: From biopower to advanced liberal government? *Journal of Environmental Policy and Planning, 7*(3), 185–207.

Pozzi, L. (2022). China, the maritime silk road, and the memory of colonialism in the Asia region. In S. Lewis, J. Olick, J. Wawrzyniak, M. Pakier (Eds.), *Regions of memory* (pp. 139–160). Palgrave Macmillan Memory Studies. Palgrave Macmillan. https://doi.org/10.1007/978-3-030-93705-8_6

Pauli, G. (2010). *The blue economy: 10 years, 100 innovations, 100 million jobs.* Paradigm Publications. ISBN 9780912111902.

Silvestre, G., Matdanan, H. J. H., Sharifuddin, P. H. Y., De Silva, M. W. R. N., & Chua, T.-E. (Eds.) (1992). The coastal resources of Brunei Darussalam: Status, utilization and management. In *ICLARM Conference Proceedings* 34, 214 p. Department of Fisheries, Ministry of Industry and Primary Resources. Bandar Seri Begawan, Brunei Darussalam, and International Center for Living Aquatic Resources Management, Manila, Philippines. ISSN 0115i4435.

Song, A. Y., & Fabinyi, M. (2022). China's 21st century maritime silk road: Challenges and opportunities to coastal livelihoods in ASEAN countries. *Marine Policy, 136*, Article 104923. https://doi.org/10.1016/j.marpol.2021.104923

WCED (World Commission on Environment & Development. (1987). *Our common future.* Oxford University Press.

Xinhua Institute. (2024). *Jointly promoting high-quality development and building an Asia-Pacific community with a shared future - Achievements of future-oriented APEC development and China's action.* Beijing: Xinhua Institute.

Yong, G. Y. V. (1996). Coastal geomorphology in Brunei Darussalam. In K. U. Sirinanda & P. Michael (Eds.), *Geography at Universiti Brunei Darussalam: The First Decade* (pp. 53–80). Universiti Brunei Darussalam.

Yong, G. Y. V. (2010). Brunei (Negara Brunei Darussalam). In E. Bird (Ed.), *Encyclopedia of the world's coastal landforms* (pp. 1142–1143). Springer.

Yong, G. Y. V. (2021). Identifying cultural traits of the historic Kampong Ayer of Brunei Darussalam using biomimetic analysis. *International Journal of Environment, Architecture & Societies, 1*(02), 83–94.

Yong, G. Y. V. (2022). Climate-resilient urban design from a biomimicry-arcology perspective. *International Journal of Environment, Architecture, and Societies, 2*(1), 1–15. https://doi.org/10.26418/ijeas.2022.2.1.1-15

Yong, G. Y. V., & Hamid, I. (2007). Coastal resources and environmental profile of Brunei Darussalam. Department of Fisheries, unpublished report.

Yong, G. Y. V., & Hassan, N. H. (2024). Blue urbanism: An insight of navigating climate change in humid tropical cities in Brunei Darussalam. *International Journal of Environment, Architecture, and Societies, 4*(01), 23–36. https://doi.org/10.26418/ijeas.2024.4.01.23-36

Yong, G. Y. V., Azalie, I. A., & Hassan, N. H. (2023). *Riverine lifelines of the Brunei Sultanate*. Paper presented at the Fluid Phenomena: Re-engaging Rivers in the Anthropocene, Universiti Brunei Darussalam, Brunei, 27–28 February.

Yong, G. Y. V., & Tamit, M. H. (2001). Implementing ICZM programs in Brunei Darussalam. In D. Chew, P. Sawal & A. H. Sim (Eds.), *Integrated coastal zone management*. SDI Sabah.

Zainuddin, N. H. H. (2021). Evolution of stilts in Kampong Ayer. *Brunei Museum Journal, 2021*, 95–110.

Gabriel Yit Vui Yong is a lecturer at Universiti Brunei Darussalam. With a BSc in Geological Geophysics (Reading, 1986) and an MSc in Environmental Analysis (Hull, 1999), he was recruited to teach coastal geomorphology in 1984, but widened his scope to include integrated coastal management for over a decade. Over three decades, he has actively been involved in national-regional projects. Over past the 5 years, he has returned to the coast, focussing on sea level rise and 'blue urban and economic development' from a complex systems perspective.

Izni Azrein Noor Azalie is a human geographer who graduated with an MSc in Global Transformations from Loughborough University UK and recently completed his PhD in Development Policy under the Development Planning Unit, University College London, researching halal industrial governing and its effects on Brunei agri-food producers. He has been a lecturer at Universiti Brunei Darussalam (UBD) for more than 15 years. He has been involved in professional consultancies involving environmental and communal advisories relating to oil and gas explorations, environmental monitoring and impact assessments on urban development projects, and socio-economic costs of industrial development towards coastal fisheries and communities in Brunei Darussalam. His research specialisation covers industrial production networks, urban economy and development and industrial development in general. His research has expanded to include studies on creative industrial development, heritage industry conservation and poverty alleviation within the old urban centres. As Brunei embarked on a path of leveraging the development of the blue economy as part of its economic diversification strategies, it became imperative to understand the dynamics between policymakers and governmental strategies and how they are translated into actual practices. This sphere is currently of interest to the researcher.

Chapter 11
Southeast Asia: Mobility, Development and Long-Term Plan

Diotima Chattoraj

Abstract This chapter examines the interplay between mobility, development and long-term planning in the context of Southeast Asia. At the intersection of geopolitical complexity and economic interdependence, Southeast Asia has experienced a remarkable degree of mobility, encompassing diverse areas such as rural–urban migration and transnational labour movements. The chapter examines the diverse impacts of these mobility patterns on the region's development trajectories. The chapter also looks at the symbiotic relationship between mobility and development in Southeast Asia. It examines the convergence of migration dynamics, infrastructure initiatives and economic strategies and emphasises their joint influence in shaping the socio-economic fabric of the region. The chapter critically assesses the challenges and prospects related to long-term planning in Southeast Asia. It addresses the trade-off between short-term economic imperatives and long-term sustainable development goals.

Keywords South-East Asia · Mobility · Long-term plan · Development

Introduction

In today's world, migration has become one of the most important issues due to its omnipresence, impact and consequences. Migration is about people, their hopes and fears, triumphs and tragedies, which makes this topic not only interesting and relevant, but also captivating and fascinating (Chattoraj, 2020; Islam, 2024). In recent years, there have been significant cross-border population movements, especially from developing countries to industrialised countries and between developing regions (Chattoraj, 2024). In 2022, the number of international migrants reached 281 million, meaning that 3.6% of the world's population lived outside their country of birth (World Migration Report, 2024). Europe was home to an estimated 86.7 million international migrants that year, closely followed by Asia, which hosted 85.6 million

D. Chattoraj (✉)
Nanyang Technological University, Singapore, Singapore
e-mail: Diotima.chattoraj@gmail.com

G. Y. V. Yong and A.K.M. A. Ullah (eds.), *Brunei and the Future Transformation of Southeast Asia*, https://doi.org/10.1007/978-981-96-7901-0_11

migrants (ibid.). These shifts have had a significant impact on international relations, influencing politics, economic dynamics, development and social integration efforts on all continents.

Southeast Asia, a region known for its geopolitical diversity, rich cultural history and economic dynamism, has long been a major hub for various forms of migration. Migration patterns (Sutarya, 2024) in the region are shaped by numerous factors, including economic inequalities, political conflict, environmental challenges and labour market demand. This chapter explores the intricate relationship between mobility, development and long-term planning in Southeast Asia, a region characterised by geopolitical complexity and increasing economic interdependence. As a rapidly growing region, Southeast Asia has experienced significant mobility in various forms, including rural–urban migration and transnational labour movements (Capaldi, 2023). These migration patterns are closely intertwined with the economic development of the region and have a profound impact on the socio-economic landscape. Migration in Southeast Asia is not only an expression of the search for better opportunities, but also reflects the region's ongoing structural transformation towards urbanisation and industrialisation (Ullah & Kumpoh, 2018).

The chapter also examines the symbiotic relationship between mobility and development. Economic strategies, infrastructure projects and migration trends are intertwined in shaping the region's development path. For example, countries such as Singapore, Brunei and Malaysia rely heavily on labour mobility to boost their economic growth, while countries such as Myanmar and Cambodia face challenges as labour-exporting states (Zreik, 2024). The role of regional cooperation frameworks such as the ASEAN Economic Community (AEC) in facilitating this mobility and supporting development goals is also addressed (ASEAN, 2015).

In addition to mobility, the chapter critically assesses the long-term planning strategies implemented by Southeast Asian governments. While short-term economic imperatives often drive policies, long-term sustainable development goals present a complex challenge, as balancing immediate needs with future aspirations can be difficult (Jayasuriya, 2023). By incorporating interdisciplinary perspectives from geography, economics, and sociology, the chapter provides a nuanced understanding of how mobility and long-term planning intersect, shaping the socio-economic fabric of Southeast Asia. This chapter, therefore, serves as an important contribution to the academic discourse on Southeast Asia, offering insights into the region's future growth prospects amidst ongoing mobility and development dynamics.

South-East Asia and Mobility: Different Trends and Patterns

Southeast Asia has long been a hub for migration, and the complexity of population movements within and from the region has intensified in recent years. In 2019, an estimated 10.1 million international migrants resided in the region, a five-fold increase since the 1990s (Capaldi, 2023). The region today faces mixed migration flows, including both labour migrants—mostly irregular—and forced migration such as environmental displacement and refugee movements. The drivers of these flows include poverty, economic inequality, persecution and marginalisation, which sets Southeast Asia apart from other regions globally, as intra-regional migration is declining in other parts of the world (Capaldi, 2023). Migration in Southeast Asia encompasses various forms—rural–urban migration, transnational labour migration, refugee migration, diaspora migration and climate-induced migration—all of which have a significant impact on the socio-economic landscape of the region.

Rural–urban migration, driven by urbanisation and economic prospects, has led to rapid urban expansion and contributed to overcrowding and informal settlements (Rigg, 2013). Cross-border labour migration plays a central role in the economies of countries such as Singapore and Malaysia, where workers from Indonesia and the Philippines fill important labour gaps and provide critical remittance support to the economies of their home countries (Kaur, 2010). Forced migration, such as the Rohingya crisis, underscores the political instability and humanitarian challenges in the region (Ullah & Chattoraj, 2018; Alejandria et al., 2024). In addition, the Southeast Asian diaspora continues to contribute to economic growth through remittances and maintains strong ties to their home countries (Wickramasekara et al., 2018). Climate-induced migration triggered by environmental disasters is a growing problem, while irregular migration and human trafficking remain pressing issues (Elliott, 2012; Saha, 2023). Irregular migration and human trafficking are significant problems, with Southeast Asia serving as a hub for forced labour, sex and bride trafficking, exacerbated by legal migration routes and corruption. Addressing these challenges requires an understanding of the underlying socio-economic and political factors that drive migration. The region also attracts skilled professionals and international students, contributing to knowledge transfer and economic development (Falk & Graf, 2016; Nghiêm-Phú et al., 2024; Ullah et al., 2019). Managing these diverse migration flows is crucial for the sustainable development, social cohesion and security of Southeast Asia.

Migration patterns within Southeast Asia show the main destinations and sources. Thailand, Malaysia and Singapore are important destination countries for migrants from Cambodia, Indonesia, Lao PDR and Myanmar (Ullah & Chattoraj, 2022; Ullah et al., 2019). The Philippines is one of the largest sources of global migrants and its economy is heavily dependent on remittances. Migration patterns fuelled by rapid economic growth pose complex policy challenges (King & Chin, 2024). The lack of adequate protection mechanisms has become particularly evident amid ongoing crises, such as the irregular maritime flows of Rohingya refugees and the increasing violence in Myanmar following the military coup of February 2021 (Capaldi, 2023). Looking at migration in Southeast Asia in overly simplistic terms risks failing to recognise the diversity and complexity of mobility in the region. Urbanisation, wealth disparities and an ageing population are likely to lead to imbalances in the supply and demand for labour. Ongoing humanitarian and environmental crises emphasise the urgent need for effective governance, protection of migrants' rights and intergovernmental cooperation in border management. The challenges of managing the various migration flows are exacerbated by the vulnerability of irregular labour, internally displaced persons and stateless persons. In 2017, Southeast Asia was home to 3.37

million 'persons of concern', including asylum seekers, internally displaced persons, refugees and stateless persons (Capaldi, 2023). These mixed migration challenges are closely linked to the different political and democratic systems in the region and test bilateral relations where non-interference and sovereignty are key principles.

Due to the lack of data for cross-country comparisons, it remains difficult to estimate how many people move each year in the Asia–Pacific region (Charles-Edwards et al, 2016; Hugo, 2005, 2014; Iredale et al, 2003; Ullah & Chattoraj, 2022). In the Asia–Pacific region, which is home to more than three-fifths of the world's population, there is little research on international migration. Inconsistencies in definitions and measurements continue to limit the information available. The recent rise in temporary migration and the expansion of the regional labour market have increased the urgency to understand migration patterns in the Asia–Pacific region (Ullah & Chattoraj, 2022). In some Asian countries, the status of migrants can quickly change from temporary to irregular, e.g. if they have quit their job or are awaiting regularisation. Since the 1990s, both regular and irregular migration have increased significantly in Asia. Notable examples are the increase in irregular migration between Indonesia and Malaysia and between Myanmar and Thailand, where the regulation of long and porous borders is a challenge (Pietsch, 2015). Irregular migrants are welcome in many Asian communities due to their low-cost labour. Estimating the number of illegal immigrants is difficult due to insufficient data (Bhuiyan & Abedin, 2024).

The Association of Southeast Asian Nations (ASEAN), which was founded in 1967 by Indonesia, Malaysia and the Philippines and joined in 1984 by Brunei, Vietnam in 1995, Laos and Myanmar in 1997 and Cambodia in 1999, is a regional intergovernmental organisation with the aim of economic growth, social development, cultural progress and regional stability (Cajilig, 2024). The ASEAN model provides a framework to harmonise the needs of member countries and stakeholders. The migration landscape in ASEAN has evolved over the decades, reflecting changes in the volume, direction and skill composition of migrants. The decline in the birth rate and the skills gap have led to stable immigration rates. Studies show that the Philippines and Thailand are important source countries for people with different skills, while Singapore is known for highly skilled workers. Filipino doctors and nurses prefer to work in Brunei, Ireland and Kuwait, while engineers and academics from Singapore often go to North America and Europe (Ullah, 2018). Since the 1970s, the brain drain of skilled Filipino professionals has raised concerns about the loss of human capital (Purgill, 2010). Vietnam also faces challenges from brain drain (Dang et al., 2010).

In Malaysia, Indonesians make up the largest group of migrants, followed by those from Myanmar, Vietnam, the Philippines and other countries (Pasadilla, 2011). Although migrants make up about a third of the Malaysian labour force, many Malaysians seek opportunities abroad. In 2022, 1.13 million of the 1.86 million Malaysians were living abroad, specifically in Singapore (Yeap, 2024). Malaysia's returnee programme has brought back about 1,000 talents in the last ten years (Ullah & Chattoraj, 2022). Cambodian labour migrants mainly move within ASEAN, with Thailand and Malaysia being the main destinations.

Mobility within ASEAN has increased with the implementation of the ASEAN Economic Community (AEC), which aims to facilitate the free flow of skilled labour, goods, services, investment and money. The mutual recognition agreement is expected to create 14 million jobs and increase the region's economic output by 7% by 2025 (Ullah & Chattoraj, 2022). Migration has significantly shaped the socio-economic context of Southeast Asia. The region's more productive economies have long attracted migrants from poorer countries. Over the last two decades, intra-regional migration has increased by around seven million, with three million migrating to Thailand (Testaverde et al., 2017). Since 1995, the share of migrants from Southeast Asian countries in the global population has risen from 6 to 8% (Capaldi, 2023). The feminisation of migration is remarkable, as women accounted for 46.8% of all migrants within ASEAN in 2018. The number of female migrants has increased from 1.37 million in 1990 to 4.77 million in 2019 (Capaldi, 2023). While the number of female migrants in Singapore and Thailand has increased, the numbers for Malaysia and Indonesia have decreased (Capaldi, 2023). In 2019, intra-regional migrants accounted for more than 92% of the Southeast Asian migrant population, mainly from Cambodia, Lao People's Democratic Republic and Myanmar, migrating primarily to Thailand. Indonesia, the Philippines and Vietnam tend to send migrants outside the region, with Filipino migrants moving to North America or the Middle East and Vietnamese migrants moving to North America. Malaysia and Singapore send migrants both within and outside the region, while Thailand attracts migrants from Cambodia, Lao PDR and Myanmar. The concept of "gradual migration" explains the hierarchical movement towards favoured destinations (Paul, 2011) (Tables 11.1 and 11.2).

Table 11.1 Size and share of international migrants from, to, and within ASEAN countries, 2017

Origin	Destination	Bilateral international migration stock	% of world bilateral stock
World	World	266,143,792	100
World	Asia–Pacific	81,986,535	31
World	ASEAN	21,537,460	8
Asia–Pacific	World	94,857,062	36
Asia–Pacific	Asia–Pacific	47,492,260	18
Asia–Pacific	ASEAN	15,735,457	6
ASEAN	World	11,484,963	4
ASEAN	Asia–Pacific	21,644,369	8
ASEAN	ASEAN	7,321,653	3

Source Ullah and Chattoraj (2022)

Table 11.2 Growth of immigration in SEA

Countries	1990	2000	2010	2015
Brunei	73,200	96,296	100,587	102,733
Cambodia	38,375	146,085	81,977	79,963
Indonesia	465,612	292,307	305,416	328,846
Laos	22,866	21,948	21,185	22,244
Malaysia	695,920	1,277,223	2,406,011	2,514,243
Myanmar	133,545	98,011	76,414	73,308
Philippines	154,071	318,095	208,599	211,862
Singapore	727,262	1,351,691	2,164,794	2,543,638
Thailand	528,693	1,257,821	3,224,131	3,913,258
Vietnam	28,118	56,745	61,756	72,793

Source Ullah and Chattoraj (2022)

South-East Asia and Development

The development of Southeast Asia in recent decades has been characterised by rapid economic growth, industrialisation and increasing integration into the global economy. However, this development has also been accompanied by significant socio-economic challenges, including inequality, environmental degradation and social displacement. Scholars from various disciplines have analysed these complex interrelationships and provided nuanced perspectives on the region's growth trajectory.

Economic Growth and Industrialisation

Southeast Asia's rapid economic development has been largely driven by export-oriented industrialisation, particularly in countries such as Singapore, Malaysia, Thailand and Vietnam. These countries have benefited from their strategic geographical location and open market policies to attract foreign investment, establish export zones and develop into manufacturing centres. According to scholars such as Hill (2014) and Hill and Hill (2005), this approach has led to impressive GDP growth rates and turned many Southeast Asian countries into middle-income countries. Vietnam, for example, has become a major player in the global electronics and textile industries, while Singapore has become a global financial and logistics centre. However, Hill also points out the differences between the countries. While Malaysia and Thailand have successfully made the transition to middle-income economies, countries such as Cambodia, Laos and Myanmar continue to face major development challenges, with lower levels of industrialisation, limited infrastructure and weaker governance structures. This uneven development reflects a broader problem in the region: the "middle-income trap'," where countries struggle to move from middle-income to high-income status.

Urbanisation and Migration

The region has experienced dramatic urbanisation as cities such as Bangkok, Jakarta, Manila and Ho Chi Minh City have grown rapidly through rural–urban migration. This phenomenon is discussed in detail by Rigg (2013), who emphasises how rural–urban migration has reshaped both urban and rural landscapes in Southeast Asia. As people move to cities in search of better economic opportunities, the rural economy shifts and often becomes dependent on remittances from urban migrants. While urbanisation has helped to reduce poverty by creating more employment opportunities, it has also led to problems such as overcrowding, the growth of informal settlements and increased demand for housing and services. Urbanisation has also led to significant social and environmental challenges.

In his work on urbanisation in Southeast Asia, Aritenang (2014) examines how the rapid growth of cities has outpaced infrastructure development, resulting in informal settlements and inadequate access to basic services such as clean water, sanitation and healthcare. These problems are particularly pronounced in megacities such as Jakarta and Manila, where slums have become an integral part of the cityscape and exacerbated social inequality. Singapore, with its high number of migrants, has become a developed city-state and global financial centre through strategic policies, urban planning and industrialisation (Cavada et al., 2019). In contrast, countries such as Vietnam, Thailand and Malaysia have experienced significant economic growth but still face challenges related to inequality, rural poverty and urbanisation. While Singapore is an example of successful urbanisation and sustainable development, many Southeast Asian countries struggle with overcrowded cities and inadequate infrastructure. Singapore's role within ASEAN promotes regional economic integration, but development disparities in the region remain.

Environmental and Climate Challenges

Development in Southeast Asia is also associated with considerable environmental costs. Scholars such as Amartya Sen (1999) and Thompson (2012) emphasise that while economic growth has lifted millions of people out of poverty, this has often been at the expense of environmental sustainability. Large-scale deforestation, land reclamation and mining have led to significant environmental degradation, particularly in Indonesia, Malaysia and the Philippines. The demand for natural resources, both for domestic consumption and for export, has led to a loss of biodiversity and increased vulnerability to natural disasters.

In addition, climate change poses a significant threat to the region's development. Southeast Asia is one of the most climatically vulnerable regions in the world. Countries such as the Philippines, Vietnam and Indonesia are frequently hit by typhoons, floods and droughts (Yusuf & Francisco, 2009). These environmental challenges have led to large-scale internal displacement and migration, particularly in rural and coastal areas (Elliott, 2012). While this environmental migration is a response to immediate survival needs, it complicates development strategies as governments are forced to address both the immediate impacts of displacement and the long-term need for sustainable development.

Inequality and Social Development

Despite impressive economic growth, Southeast Asia continues to struggle with high levels of income inequality. Carroll (2020) argues that globalisation has exacerbated wealth inequality, with the benefits of economic growth disproportionately accruing to the elite. Inequalities are particularly evident in countries such as the Philippines and Indonesia, where land ownership remains concentrated among a small number of wealthy families, limiting opportunities for social mobility. Access to education and healthcare also reflects these inequalities. While countries such as Singapore and Malaysia have world-class education and healthcare systems, rural populations in Cambodia, Laos and Myanmar often struggle with inadequate access to these important services. Several studies show how these inequalities in access to education and healthcare perpetuate the cycle of poverty and inequality and limit long-term development prospects for marginalised populations.

Governance and Corruption

Political stability and governance are critical factors influencing Southeast Asia's development. While some countries in the region, such as Singapore and Malaysia, have enjoyed relatively stable political environments conducive to economic growth, others, like Myanmar and the Philippines, have been plagued by political instability, corruption, and weak governance institutions. Scholars like Slater (2010) have explored how authoritarianism and weak democratic institutions in several Southeast Asian countries have hindered inclusive development. Corruption remains a major issue, with Transparency International consistently ranking countries like Cambodia and Myanmar among the most corrupt in the world. Corruption not only undermines economic development but also perpetuates inequality by allowing elites to capture the benefits of growth at the expense of broader social progress.

The Role of ASEAN and Regional Integration

Regional organisations, in particular ASEAN, have played a key role in promoting economic cooperation and integration between the countries of Southeast Asia. The establishment of the ASEAN Economic Community (AEC) in 2015 was an important step towards creating a single market and production base in the region. As discussed by Das (2015), ASEAN's efforts to reduce trade barriers and promote economic integration have led to increased connectivity and investment across the region, fuelling growth and development. However, while ASEAN has been successful in promoting economic co-operation, it has been less effective in addressing issues such as human rights violations, environmental degradation and social inequality. Critics argue that ASEAN's principle of non-interference in the internal affairs of member states has limited its ability to address these critical development challenges.

Southeast Asia's development trajectory is complex and multi-layered, driven by factors such as economic growth, urbanisation, industrialisation and globalisation. Jonathan Rigg, a prominent scholar in the field of Southeast Asian development, examines the intricate interplay of these forces and demonstrates that development in the region cannot be measured by conventional economic metrics alone. Instead, Rigg emphasises the need to consider social, cultural and environmental dimensions, particularly how rural areas and marginalised populations experience and contribute to the development process.

In his work 'More than the Soil: Rural Change in Southeast Asia' (2014), Rigg criticises the idea that rural areas are simply 'left behind' by development, arguing instead that rural areas are closely linked to urbanisation and economic change. He emphasises how mobility, particularly rural–urban migration, has reshaped rural economies and livelihoods. As labour migrates to urban centres, remittances become an important driver of rural development, but this also contributes to the fragmentation of traditional rural communities. Migration is therefore not just an economic phenomenon, but a social process that redefines identities, aspirations and family structures in both rural and urban areas.

He also discusses the uneven nature of development in Southeast Asia, as he does in his book Unplanned Development: Tracking Change in Southeast Asia (2012). He points out that while countries such as Singapore and Malaysia have achieved significant industrial growth and economic prosperity, large sections of the population in other countries, particularly Cambodia, Laos and Myanmar, continue to face poverty and limited access to basic services. This uneven development reflects a broader challenge of managing rapid growth in the face of persistent inequalities. Understanding development in Southeast Asia requires addressing how social inequalities intersect with economic policies, environmental degradation and global forces.

In Challenging Southeast Asian Development (Rigg, 2015), he also critically examines the prevailing narrative of development as progress and draws attention to how economic growth can lead to environmental degradation, displacement and social marginalisation. He argues that development policy often prioritises economic expansion at the expense of sustainability and social justice. For example, large-scale infrastructure projects such as dams and roads have often displaced rural communities while benefiting urban elites and foreign investors. Rigg calls for a more inclusive and sustainable approach to development that takes into account the long-term environmental and social impacts.

Rigg's analysis of development in Southeast Asia highlights the importance of mobility, the interaction between rural and urban spaces, and the socio-economic inequalities that shape the region's growth. His work emphasizes that Southeast Asia's development cannot be understood through a one-dimensional lens but must account for the complex interconnections between economic policies, social structures, and environmental considerations. These insights are critical for formulating more equitable and sustainable development strategies in the region.

Therefore, development in Southeast Asia is a dynamic and multifaceted process that reflects the region's complex socio-economic, political, and environmental landscape. While countries like Singapore, Malaysia, and Vietnam have made significant strides in industrialization and economic growth, challenges such as inequality, environmental degradation, and governance issues persist across the region. Scholars like Jonathan Rigg, Amartya Sen, and Hal Hill provide critical insights into how these challenges intersect, emphasizing the need for more inclusive, sustainable, and equitable development strategies.

South-East Asia and Its Long-Term Plan

Southeast Asia's long-term development plans are characterised by a combination of national strategies and regional cooperation aimed at promoting economic growth, social progress and sustainable development. Countries in the region, such as Singapore, Vietnam and Thailand, have taken steps to transition from traditional agrarian economies to diversified, industrialised and knowledge-based economies. These long-term plans focus on improving infrastructure, digital transformation, environmental sustainability and human capital development.

Thailand's "Thailand 4.0" policy (Kohpaiboon, 2020), for example, which was introduced in 2016, aims to transform the country into a high-income economy through innovation-driven growth over a decade. This strategy focuses on key sectors such as biotechnology, smart electronics and healthcare, moving away from traditional manufacturing. Meanwhile, Vietnam has expanded its role in global supply chains, particularly in the electronics and textile industries, through its long-term economic policies that focus on increasing foreign investment and improving infrastructure. Singapore, the most developed country in the region, serves as a model for long-term urban and economic planning. It has invested heavily in sustainable urban development and is a leader in innovation for smart cities, green technologies and public housing solutions. Singapore's 'Green Plan 2030 (Chng & Ong, 2021), for example, is a comprehensive approach to tackling climate change that prioritises clean energy, waste reduction and green infrastructure.

At the regional level, the ASEAN Master Plan on ASEAN Connectivity 2025 (Korwa, 2019) is an important long-term strategy that aims to deepen regional integration by improving physical, institutional and people-to-people connectivity between member states. Projects such as the China-led Belt and Road Initiative (BRI) and the Indo-Pacific Economic Framework (IPEF) also emphasise the geopolitical importance of Southeast Asia, with many countries weighing up external partnerships to support their infrastructure and long-term growth plans. Despite these initiatives, Southeast Asia faces challenges in addressing income inequality, climate change and ensuring inclusive growth. For example, countries such as Indonesia and the

Philippines continue to struggle with poverty reduction and vulnerability to climate change, particularly in rural areas affected by environmental degradation and extreme weather events. These long-term development strategies are critical as Southeast Asia navigates its future amid rapid globalisation, technological change and regional interdependencies.

Mobility, Development and Long-Term Plan: A Relation

The chapter explores the interconnectedness of mobility, development, and long-term planning in Southeast Asia, emphasizing how these elements influence each other within the region's socio-economic landscape. Each of these concepts influences and shapes the others, creating a complex web of interactions that can either promote or hinder progress in the region.

Mobility as a Driver of Development

Labour supply: Mobility, particularly in the form of rural–urban migration, provides a vital labour supply for rapidly growing urban economies. As individuals move to cities in search of better job opportunities, they contribute to the workforce that drives industrialization and economic growth. This influx of labour can enhance productivity and innovation, leading to increased economic output.

Remittances: Transnational migration allows workers from Southeast Asia to seek employment abroad, often in higher-paying jobs. The remittances they send back home play a crucial role in supporting local economies, reducing poverty, and improving living standards. These financial inflows can be reinvested in education, healthcare, and small businesses, further stimulating local development.

Development Outcomes Influencing Migration Patterns

Economic Opportunities: The level of development in a country significantly influences migration patterns. Countries with robust economic growth and job opportunities tend to attract migrants from less developed regions. Conversely, areas facing economic stagnation or decline may see increased out-migration as individuals seek better prospects elsewhere. Social and Environmental Factors: Development outcomes, such as access to education, healthcare, and social services, also affect migration decisions. Regions that provide better living conditions and opportunities for advancement are more likely to retain their populations, while those that struggle with poverty, inequality, and environmental degradation may experience higher rates of out-migration.

The Role of Long-Term Planning

Sustainable Development: Effective long-term planning is essential for managing the dynamics of mobility and development. Policymakers must create strategies that not only address immediate economic needs but also consider the long-term sustainability of growth. This includes investing in infrastructure, education, and technology to support both current and future populations.

Integration of Mobility into Planning: Long-term planning must incorporate an understanding of mobility patterns. This means recognizing how migration affects urbanization, labour markets, and social services. By integrating these considerations into planning processes, governments can create more inclusive and resilient urban environments that accommodate the needs of both migrants and local populations.

Addressing Socio-economic Challenges

Complex interactions: The relationships between mobility, development and long-term planning are complex and multi-layered. For example, rapid urbanisation driven by migration can lead to problems such as overcrowding, inadequate housing and overloading of public services. Effective long-term planning can help mitigate these problems by ensuring that urban infrastructure and services keep pace with population growth. Equity and inclusion: Understanding these linkages is critical to addressing socio-economic challenges in Southeast Asia. Policy makers must strive for equitable growth that benefits all sections of society, including marginalised groups and migrants. This requires a holistic approach that takes into account the social, economic and environmental dimensions of development.

This chapter therefore argues that mobility, development and long-term planning are interlinked concepts that characterise the socio-economic landscape of Southeast Asia. It highlights how migration drives development by providing labour and remittances, while development outcomes influence migration patterns as people seek better opportunities. Effective long-term planning is essential to manage these dynamics and ensure sustainable and equitable growth. By understanding the relationships between these concepts, policymakers can better address the region's complex socio-economic challenges, including inequality, environmental degradation and inadequate infrastructure, and ultimately promote inclusive and resilient development.

Conclusion

To summarise, the intricate interplay between mobility, development and long-term planning is fundamental to understanding and addressing Southeast Asia's socio-economic challenges. As the region continues to experience high levels of migration and urbanisation, it is imperative for policy makers to adopt a holistic approach that integrates these concepts into their development strategies. By recognising the critical role of mobility in economic growth and the influence of development outcomes on migration patterns, governments can develop more inclusive and sustainable policies. Effective long-term planning will not only help to manage the impact of rapid change, but also ensure that growth benefits all parts of society and promotes resilience and equity in the face of ongoing socio-economic change. Ultimately, a comprehensive understanding of these linkages will be critical to shaping a prosperous and sustainable future for Southeast Asia.

References

Alejandria, M. C., Hernandez, P. J., Quan-Nalus, M. A., Alipao, F., Tumaneng, D., Ruiz, C. J., Dela Cruz, K. A., & Casimiro, K. M. (2024). Humanitarian leadership in urban communities: An exploratory study on the role of community leaders in humanitarian coordination during the COVID-19 pandemic in the Philippines. *Southeast Asia: A Multidisciplinary Journal, 24*(3), 208–220. https://doi.org/10.1108/SEAMJ-01-2024-0008

Aritenang, A. F. (2014). *Urbanization in Southeast Asia: Issues and impacts.*

Bhuiyan, S. A., & Abedin, M. (2024). The curious case of Facebook commerce: An analysis of the digital commerce management guide 2021. *Southeast Asia: A Multidisciplinary Journal, 24*(1), 3–15. https://doi.org/10.1108/SEAMJ-06-2023-0045

Cajilig, P. G. (2024). The seawall helps but—: Using "material registers" to understand coastal infrastructure for disaster risk management in an island community. *Southeast Asia: A Multidisciplinary Journal, 24*(3), 221–239. https://doi.org/10.1108/SEAMJ-12-2023-0081

Capaldi, M. P. (2023). Present-day migration in Southeast Asia: evolution, flows and migration dynamics. In S. Petcharamesree, M. P. Capaldi (Eds.), *Migration in Southeast Asia. IMISCOE Research Series*. Springer, Cham. https://doi.org/10.1007/978-3-031-25748-3_1

Carroll, T. (2020). The political economy of Southeast Asia's development from independence to hyperglobalisation. In *The political economy of Southeast Asia: Politics and uneven development under hyperglobalisation* (pp. 35–84).

Cavada, M., Tight, M. R., & Rogers, C. D. (2019). A smart city case study of Singapore—Is Singapore truly smart?. In *Smart city emergence* (pp. 295–314). Elsevier.

Charles-Edwards, E., Muhidin, S., Bell, M., & Zhu, Y. (2016). Migration in Asia. In M. J. White (Ed.), *International handbook of migration and population distribution* (pp. 269–284). Springer.

Chattoraj, D. (2020). Book review: AKM Ahsan Ullah and Md. Shahidul Haque (2020). The Migration Myth in Policy and Practice: Dreams, Development and Despair. Singapore, Springer Nature. (v + 221 pp. ISBN 978-981-15-1754-9).

Chattoraj, D. (2024). International Relations Versus International Migration: Complements or Substitutes?. In *Governance, migration and security in international relations* (pp. 33–48). Springer Nature Singapore.

Chng, K., & Ong, K. W. (2021). The Singapore Green Plan 2030: Analysing its implications on law and the legal industry in Singapore. *Environmental Law Review, 23*(4), 336–343.

Dang, NA, Tran, TB, Nguyen, NQ., & Dao, T. S. (2010). *Development on the move: Measuring and optimising migration's economic and social impacts in Vietnam*. Institute for Public Policy Research.

Das, S. B. (2015). *The ASEAN economic community and beyond: Myths and realities*. ISEAS-Yusof Ishak Institute.

Elliott, L. (2012). *Climate change and migration in Southeast Asia: Responding to a new human security challenge*. S. Rajaratnam School of International Studies.

Falk, M. L., & Graf, A. (2016). Student mobility within Southeast Asia. *TRaNS: Trans-Regional and-National Studies of Southeast Asia, 4*(1), 1–4.

Hill, H. (2014). Is there a Southeast Asian development model? *Malaysian Journal of Economic Studies, 51*, 89–111.

Hill, H., & Hill, S. (2005). Growth econometrics in the tropics: What insights for Southeast Asian economic development? *The Singapore Economic Review, 50*(spec01), 313–343.

Hugo, G. (2005). Migration in the Asia-Pacific region. Paper for the Policy Analysis and Research Programme of the Global Commission on International Migration, Geneva.

Hugo, G. (2014). A multi-sited approach to analysis of destination immigration data: An Asian example. *International Migration Review, 48*(4), 998–1027.

Iredale, R. R., Guo, F., & Rozario, S. (2003). Introduction. In R. Iredale, F. Guo, & S. Rozario (Eds.), *Return migration in the Asia Pacific* (pp. 1–25). Edward Elgar Publishing.

Islam, M. S. (2024). Impact of British fall in Burma on Bengal famine of 1943: Experience of Faridpur district of Bangladesh. *Southeast Asia: A Multidisciplinary Journal, 24*(1), 27–39. https://doi.org/10.1108/SEAMJ-02-2023-0015

Jayasuriya, R. (2023). Achieving Resilience through Digitalization, Sustainability and Sectoral Transformation-What Are the Long Term Strategic Options and Reforms for Sri Lanka Armed Forces.

Kaur, A. (2010). Labour migration trends and policy challenges in Southeast Asia. *Policy and Society, 29*(4), 385–397.

King, V. T., & Chin, W. L. (2024). COVID-19 and post-COVID-19 transitions: Case study on the Lao PDR. *Southeast Asia: A Multidisciplinary Journal, 24*(2), 150–163. https://doi.org/10.1108/SEAMJ-03-2024-0031

Kohpaiboon, A. (2020). Industry 4.0 policies in Thailand. Economcis Working paper, No. 2. ISEAS-Yusof Ishak Institute.

Korwa, J. R. V. (2019, September). Asean Connectivity 2025: Shall Australia Become The Future Member Of Asean?. In *Proceeding of the 1st International Conference on Asean (IC-ASEAN)* (p. 98).

Nghiêm-Phú, B., Hoàng, T.T.-T., Kiều, H. T., & Vũ, H. L. (2024). University students' perceptions of tourism impacts: A study in Vietnam. *Southeast Asia: A Multidisciplinary Journal, 24*(2), 107–120. https://doi.org/10.1108/SEAMJ-06-2023-0043

Pasadilla, G. O. (2011). Social security and labor migration in ASEAN.

Paul, A. M. (2011). Stepwise international migration: A multistage migration pattern for the aspiring migrant. *American Journal of Sociology, 116*(6), 1842–1886.

Pietsch, J., & Clark, M. (2015). Introduction. In J. Pietsch & M. Clark (Eds.), *Migration and integration in Europe, Southeast Asia and Australia: A comparative perspective* (pp. 11–25). Amsterdam University Press.

Purgill, J. (2010). *Draining of the Philippines. Brain drain in the Philippines*. Retrieved February 19, 2022, from www.slideshare.net/jenniferpurgill/brain-drain-in-the-philippines-jennifer-purgill-dec-2010

Rigg, J. (2013). From rural to urban: A geography of boundary crossing in Southeast Asia. *TRaNS: Trans-Regional and-National Studies of Southeast Asia, 1*(1), 5–26.

Rigg, J. (2015). *Challenging Southeast Asian development: The shadows of success*. Routledge.

Saha, P. (2023). Navigating climate change: Migration challenges in Southeast Asia. *Southeast Asia: A Multidisciplinary Journal, 23*(2), 132–145.

Slater, D. (2010). *Ordering power: Contentious politics and authoritarian leviathans in Southeast Asia.* Cambridge University Press.

Sutarya, I. G. (2024). On-off hybrid spiritual tourism in the new normal era. *Southeast Asia: A Multidisciplinary Journal, 24*(2), 95–106. https://doi.org/10.1108/SEAMJ-09-2023-0063

Testaverde, M., Moroz, H., Hollweg, C. H., & Schmillen, A. (2017). *Migrating to opportunity: Overcoming barriers to labor mobility in Southeast Asia.* World Bank Publications.

Thompson, E. C. (2012). Anthropology in Southeast Asia: National traditions and transnational practices. *Asian Journal of Social Science, 40*(5–6), 664–689.

Ullah, A. A., & Chattoraj, D. (2018). Roots of discrimination against Rohingya minorities: Society, ethnicity and international relations. *Intellectual Discourse, 26*(2), 541–565.

Ullah, A. A., & Chattoraj, D. (2022). *COVID-19 pandemic and the migrant population in Southeast Asia: Vaccine, diplomacy and disparity* (Vol. 2). World Scientific.

Ullah, A. A., & Kumpoh, A. A. Z. A. (2018). Are borders the reflection of international relations? Southeast Asian borders in perspective. *Journal of Asian Security and International Affairs, 5*(3), 295–318.

Wickramasekara, P., Tennant, E., & Taran, P. (2018). Engaging the diaspora and migrant workers for home country development: Diaspora finance and remittances1. *Labor Migration in Asia.*

World Migration Report. (2024). Available at: https://worldmigrationreport.iom.int/news/world-migration-report-2024-reveals-latest-global-trends-and-challenges-human-mobility

Yeap, A. (2024). "The Big Read: Malaysia's brain drain to Singapore – winners and losers in the long run not so clear-cut", March 4. Available at: https://www.channelnewsasia.com/singapore/big-read-malaysia-brain-drain-singapore-winners-losers-long-run-not-clear-4165636

Yusuf, A. A., & Francisco, H. (2009). *Climate change vulnerability mapping for Southeast Asia.*

Zreik, M. (2024). Resurgence and transformation: The impact of COVID-19 on Myanmar's tourism industry and future trends. *Southeast Asia: A Multidisciplinary Journal, 24*(2), 134–149. https://doi.org/10.1108/SEAMJ-12-2023-0085

Diotima Chattoraj is a Research Fellow at the Wee Kim Wee School of Communication and Information in Nanyang Technological University. Additionally, she is an Adjunct Research Fellow at the department of Social and Health Sciences in James Cook University, Singapore. A qualitative researcher, she focuses on Asian migration, international relations, consumer psychology and advertising effects. She has authored more than 25 journal articles, 6 books, 10 book chapters in leading journals. Additionally, she is the Deputy Editor of *South Asia Research* (Sage) and serves as a peer reviewer for a number of refereed journals.

Chapter 12
Migration Routes in Southeast Asia: How Economy Shapes the Routes

A. K. M. Ahsan Ullah

Abstract This chapter examines the interplay between migration routes and economic structures in Southeast Asia and provides an in-depth analysis of the ways in which labour mobility is influenced by regional economic dynamics. Through the lens of key migration theories, including neoclassical economics, dual labour market theory and world systems theory, the chapter examines the motivations, mechanisms and impacts of migration within and across national borders. Specific migration corridors such as Malaysia–Indonesia and Thailand–Myanmar illustrate how wage differentials, labour demand and regional politics influence the movement of people. The role of remittances in supporting the economies of sending countries such as the Philippines and Cambodia emphasises the profound economic impact of migration. By linking theoretical frameworks with empirical evidence, this chapter highlights the centrality of economic imperatives in shaping the Southeast Asian migration landscape, while also addressing policy challenges related to labour rights, integration and regional cooperation.

Keywords Migration routes · Southeast Asia · Economic disparity · Labor mobility · Remittances

Introduction

Migration in Southeast Asia is a dynamic process determined by a variety of factors, with the economy playing a central role in influencing migrant routes. This region has long been characterised by both internal and international migration flows, driven primarily by labour needs, economic opportunities and significant regional inequalities (Ullah et al., 2024). Southeast Asia serves as both a source and destination for millions of migrants, with countries such as Malaysia, Singapore and Thailand acting as key destination countries due to their economic stability and growing demand for labour (Ullah, 2010). On the other hand, countries such as Myanmar, Cambodia, Laos

A. K. M. A. Ullah (✉)
Faculty of Arts and Social Science, Universiti Brunei Darussalam, Bandar Seri Begawan,, Brunei
e-mail: ahsan.ullah@ubd.edu.bn

G. Y. V. Yong and A.K.M. A. Ullah (eds.), *Brunei and the Future Transformation of Southeast Asia*, https://doi.org/10.1007/978-981-96-7901-0_12

and Vietnam are important source countries for migrants, where economic hardship, political instability and limited employment opportunities are fuelling out-migration. These flows of people are not random, but are largely determined by the structure and needs of the economies in both sending and receiving countries (IOM, 2020; Hugo, 2011).

The relationship between migration and economic development in Southeast Asia is particularly profound, creating a strong correlation between the supply of migrant labour and the demand for labour in specific industries. Countries such as Malaysia and Thailand, for example, rely heavily on migrant labour to sustain sectors such as construction, manufacturing, agriculture and domestic services (Athukorala, 2019). In Malaysia, almost 15% of the labour force is made up of migrant workers who fill important but low-paid jobs that locals often avoid (ILO, 2017). This dependence on foreign labour, driven by the country's economic growth and labour shortages, has shaped migration routes from neighbouring countries such as Indonesia and Bangladesh. The demand for migrant labour is a major economic driver, creating and sustaining migration corridors across the region and reflecting the overall global trend of labour mobility in response to economic needs (Yeoh & Lam, 2016).

The role of economic policies and regional trade agreements, such as the ASEAN Economic Community (AEC), has also shaped migration routes in Southeast Asia. The AEC, which aims to promote regional economic integration, has facilitated the cross-border migration of skilled labour, although it has yet to tackle the complexities of unskilled labour migration, which remains the predominant form of migration in the region (ASEAN, 2021). Economic disparities between ASEAN member states combined with unequal development patterns continue to drive migration (Ullah & Chattoraj, 2022). Countries with higher wages and more robust economies, such as Singapore and Malaysia, continue to attract migrants from less developed neighbouring countries. These migration routes are not static, but evolve with changing economic conditions, labour market demand and migration policies in both sending and receiving countries (Piper, 2019; Rubel et al., 2024).

In addition, remittances play a crucial role in the economic relationship between migration and development in Southeast Asia. Migrants send billions of dollars in remittances to their home countries, contributing significantly to poverty reduction, improved living standards and economic stability. In countries such as the Philippines and Indonesia, for example, remittances make up a significant portion of national income and influence both local and national economies (World Bank, 2021a, 2021b). These financial flows create a complex web of economic dependencies between the migrants' countries of origin and host countries, strengthening established migration routes and influencing new ones.

By examining the intersection of migration and economic forces, this chapter provides an in-depth analysis of the ways in which migration routes in Southeast Asia are shaped by economic inequalities, labour demand and regional integration efforts. It provides a comprehensive understanding of the ways in which temporality, migration flows and economic conditions influence the movement of people within and across borders, affecting both the countries that send migrants and the countries that receive migrants (Ranjan et al., 2024).

Southeast Asia is a region with great economic diversity and considerable differences between countries in terms of prosperity, development and labour market conditions. Countries such as Singapore and Malaysia are economic powerhouses that attract large numbers of migrants in search of better employment opportunities, while countries such as Cambodia, Myanmar and Laos (Ullah et al., 2024) are typically seen as emigration countries whose citizens move abroad in search of a better livelihood.

Historically, migration in Southeast Asia has been influenced by colonial legacies, ethnic conflicts and regional economic integration initiatives. The Association of Southeast Asian Nations (ASEAN), with its vision of regional integration and economic co-operation, has also played a role in shaping migration flows within the region (Ullah, 2010). The economic disparities between ASEAN member states and the different demands of the labour market have led to the different migration routes that reflect these economic imbalances (Haji Marzuki & Alkaff, 2024). The economic importance of migration in Southeast Asia cannot be overestimated. Migrant workers drive industries such as construction, agriculture and domestic work in destination countries, while remittances from migrant workers make up a crucial part of GDP in many countries of origin (Ullah, 2022). The region is therefore characterised by a complex web of migration routes that are closely linked to economic needs and opportunities, reflecting a broader global trend of labour mobility driven by economic constraints.

Understanding the interplay between migration routes and the economy in Southeast Asia is important for several reasons. Firstly, migration is an important driver of economic growth in the region. Many Southeast Asian economies rely heavily on migrant labour to sustain their industries, particularly in labour-intensive sectors such as construction, agriculture and services. Secondly, remittances from migrant workers in several sending countries make up a significant portion of national income and provide economic stability to millions of households.

By examining how economic factors influence migration routes, this study also addresses issues of political relevance. The governance of migration in Southeast Asia is a contentious issue, with debates about labour rights, human trafficking and the role of regional cooperation (Azizuddin & Shamsuzzoha, 2024). This study contributes to these debates by highlighting the economic forces that underlie migration routes and by making policy recommendations to ensure that migration benefits both sending and receiving countries (Fig. 12.1).

Theoretical Issues on Migration Routes and Economy

Migration is a complex phenomenon that has been studied from different theoretical perspectives. Each of these perspectives offers valuable insights into the factors that drive human mobility and the impact on economic structures (Ullah, 2010). To understand the complex relationship between migration and economic forces, it is important to understand how migration routes evolve and why individuals and groups

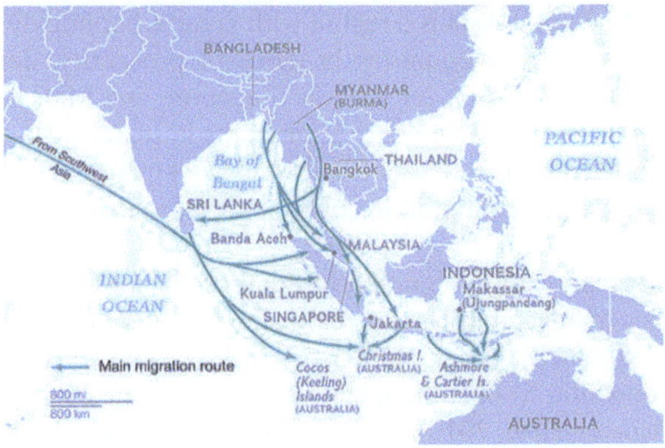

Fig. 12.1 Migration map in Southeast Asia

choose certain routes over others. In the context of Southeast Asia, where migration is both a historical and contemporary reality, several theories provide a framework for analysing migration routes in the region, which are largely shaped by economic pressures.

Neoclassical economic theory: Neoclassical economic theory, one of the most influential theories in migration research, assumes that migration occurs due to wage differentials between regions or countries (Harris & Todaro, 1970). The basic assumption is that people migrate to maximise their economic gain by moving from areas with lower wages to areas with higher wages. In Southeast Asia, this theory helps to explain why workers migrate from low-income countries such as Myanmar, Cambodia and Laos to richer countries such as Singapore, Malaysia and Thailand. The wage differences between these countries are a strong incentive for migration.

In Myanmar, for example, where per capita income is significantly lower compared to Thailand or Malaysia, economic opportunities in higher-paying sectors such as construction or manufacturing attract a large number of migrant workers (ILO, 2019). The economic disparity between sending and receiving countries corresponds to the neoclassical model, which assumes that people make rational economic decisions to improve their standard of living by seeking work in more prosperous neighbouring countries. This theory explains the economic motivation behind cross-border migration, with the relatively wealthy countries of Southeast Asia acting as magnets for labour.

The New Economics of Labour Migration (NELM): While neoclassical economic theory focuses on individual decisions, the NELM expands the framework to include household-level decisions. Stark and Bloom (1985) argue that migration decisions are made not only by individuals seeking to maximise their personal income, but also by households seeking to diversify their sources of income and manage economic risk. Households in poorer regions, such as rural Cambodia or the Mekong Delta in Vietnam, may send a family member to work abroad. This is part of a broader strategy to secure remittances and mitigate local economic insecurity (Massey et al., 1993).

In Cambodia, for example, many rural families send members to Thailand to work in sectors such as agriculture or construction with the expectation that the migrant will remit a portion of their earnings to support the family (Hing et al., 2011). This household strategy not only increases financial stability, but also serves as a form of insurance against crop failure or economic shocks in their home regions. The remittances sent home are often used for important needs such as education, healthcare and investment in local businesses. By diversifying household income sources through migration, families in Southeast Asia mitigate the risks associated with subsistence farming or informal employment in their home countries.

The dual labour market theory: The dual labour market theory, developed by Piore (1979), highlights the structural demand for migrant workers in certain sectors of the receiving economies. This theory states that developed economies create a segmented labour market consisting of primary sectors (high wages, stable employment) and secondary sectors (low wages, unstable employment). Migrants tend to fill jobs in the secondary labour market, where the jobs are often undesirable for native workers due to poor working conditions and low wages (Ullah et al., 2022). In Southeast Asia, countries such as Singapore, Malaysia and Thailand have this dual labour market structure.

In Malaysia, for example, migrant workers predominantly occupy low-skilled, labour-intensive jobs in industries such as construction, plantations and domestic work (Rahman, 2019). These jobs are often characterised by precarious conditions, including limited job security, low wages and exploitation. The dual labour market theory explains why local workers are reluctant to work in these sectors despite higher unemployment rates, leading to a reliance on migrant workers. In Singapore too, migrant workers from Bangladesh, Indonesia and the Philippines form the backbone of the domestic and construction sectors, performing important but undervalued labour (Yeoh & Huang, 2010).

The dual labour market theory proposed by Michael Piore explains labour migration by segmenting the labour market into two sectors: the primary sector (which offers high wages, job stability and career advancement opportunities) and the secondary sector (which is characterised by low wages, poor working conditions and few opportunities for advancement). This theory provides a useful framework for understanding migration dynamics in Southeast Asia, where labour migration is often driven by demand for workers in low-skilled sectors. Below are some notable examples with sources. Malaysia relies heavily on foreign labour to fill jobs in the construction, plantation and domestic service sectors. These industries are part of the secondary labour market and offer low wages, limited job security and poor working conditions. Migrants from countries such as Indonesia, Nepal and Bangladesh are recruited for these jobs because local workers are not interested (Rahman & Ullah, 2012).

The Thai fishing industry employs a large number of migrant workers from Myanmar, Cambodia and Laos. This industry is a classic example of the secondary labour market, as the work is physically demanding, poorly paid and often exploitative. Despite efforts to regulate this industry, migrant workers are often victims of human rights violations, highlighting the vulnerability of those working in the secondary labour market (Derks, 2010).

In countries like Singapore, Malaysia and Hong Kong, Filipino workers dominate the domestic service industry. While the primary sector attracts skilled workers such as nurses and IT workers, the secondary labour market for domestic services requires low-skilled workers who are often characterised by long working hours, low wages and limited mobility (Parreñas, 2001). The garment industry in Southeast Asia, particularly in Cambodia, employs thousands of workers in conditions typical of the secondary labour market. These workers, many of whom migrate from rural to urban areas, are paid low wages and have limited opportunities for advancement, often resulting in the recruitment of foreign labour to maintain competitiveness (Robertson, 2011).

Thailand is a major destination for migrant labour from Myanmar, particularly in sectors such as agriculture, construction and domestic work. These jobs fall under the secondary labour market, which is driven by Thailand's economic demand for cheap labour and Myanmar's political and economic instability (Jones & Pardthaisong, 1999). The labour market in Singapore is an example of the dual structure. High-skilled jobs in the technology and financial sectors are part of the primary market that attracts international professionals, while low-skilled sectors such as construction, sanitation and elderly care are dependent on migrant labour from Southeast Asia, particularly Indonesia, Malaysia and the Philippines (Yeoh & Huang, 1998). These examples illustrate how dual labour market theory illuminates Southeast Asia's dependence on labour migration to sustain economic growth, particularly in sectors shunned by locals due to undesirable working conditions. This structural dependence perpetuates the segmentation of the labour market and highlights the crucial role of migrant workers in the region's economic landscape (Ullah et al., 2021a, 2021b).

World-systems theory: Introduced by Wallerstein (1974), world-systems theory views migration as a consequence of the global capitalist system, in which labour flows from peripheral, less developed regions to the central, more developed regions. In this framework, migration is seen as a structural feature of economic globalisation, driven by the integration of economies and the unequal distribution of resources. Southeast Asia's migration patterns reflect this global system of economic interdependence, with wealthier countries such as Singapore and Malaysia acting as economic cores and attracting labour from peripheral countries such as Myanmar, Laos and Cambodia.

The world systems theory can be observed in the historical context of Southeast Asian migration, where colonial economic policies and later globalised industrialisation have created patterns of labour mobility that persist to this day. For example, the colonial rubber and tin industries in Malaysia relied heavily on migrant labour from India and China, setting a precedent for Malaysia's dependence on migrant labour today (Hugo, 2005). Today, the same global economic forces are driving labour migration within the region as Southeast Asian countries integrate into global supply chains that require flexible, low-cost labour.

Migration Routes and the Economy in Southeast Asia

Migration routes in Southeast Asia are characterised by the economic structure of the region, which is characterised by strong differences in income levels, employment opportunities and industrial needs. Countries such as Malaysia, Singapore and Thailand have become important hubs for migrant workers due to their rapid economic growth, industrialisation and demand for labour in labour-intensive sectors (Ullah et al., 2021a, 2021b). The development models of these countries have led to labour shortages in critical sectors, necessitating the importation of foreign labour to sustain economic output.

Malaysia and Singapore as Economic Hubs

Malaysia, Brunei and Singapore are three of the top receiving countries for migrants in Southeast Asia, largely due to their robust economies and need for low-skilled labour. In Malaysia, economic development, particularly in the construction and plantation sectors, has created a demand for foreign labour. According to the Malaysian Ministry of Home Affairs (2020), there are approximately 2.3 million documented migrant workers in the country, with millions more working informally. Migrants from Indonesia, Bangladesh, Nepal and the Philippines are employed in sectors such as construction, manufacturing and agriculture, where there are chronic labour shortages (ILO, 2019).

Singapore, with its high per capita income and small domestic labour force, also relies heavily on migrant workers, particularly in construction, domestic work and healthcare. Migrants from Indonesia, the Philippines and Bangladesh occupy positions that are critical to the development of the city-state's infrastructure and social services (Yeoh, 2006). In Singapore, where the local population is reluctant to take low-wage jobs, migrant workers form the backbone of the economy and enable continuous growth and development.

Thailand's Role in Intra-Regional Migration

Thailand is another important destination for labour migrants, particularly from Myanmar, Cambodia and Laos. The country's economic growth, particularly in the agriculture, fishing and construction sectors, has made it reliant on migrant labour from its less developed neighbours (Harkins et al., 2017). Migrants from Myanmar, for example, form the largest group of foreign labourers in Thailand. They are attracted by the opportunities in low-skilled sectors where wages, although modest by international standards, are significantly higher than in Myanmar.

Thailand's border areas, such as Mae Sot, have become important transit points for migrants seeking work in the country's labour-intensive industries. The demand for cheap labour has created established migration routes from Myanmar, Cambodia and Laos into Thailand, with informal networks and brokers facilitating cross-border movements. The economic disparity between Thailand and its neighbours, coupled with porous borders and weak immigration enforcement, has led to large numbers of undocumented migrants working in the Thai economy.

The Philippines is unique in Southeast Asia in its role as one of the world's largest exporters of labour. While much of the migration from the Philippines is to the Middle East, Europe and North America, migration within the region has also grown in importance. The demand for Filipino domestic workers, nurses and carers in countries such as Singapore, Malaysia and Hong Kong have led to the establishment of robust migration corridors within Southeast Asia (Asis, 2017).

The Philippine government actively promotes labour migration as part of its economic development strategy, with remittances from overseas workers accounting for a significant portion of the country's GDP. In 2021, remittances totalled over $34 billion, equivalent to almost 10% of the Philippines' GDP (World Bank, 2021a, 2021b). The government's institutionalised labour export policy, coupled with economic opportunities abroad, has created long-standing migration routes that benefit both the destination countries, which gain workers, and the Philippines, which gains foreign exchange.

Temporality, Flows, and the Shaping of Migration Routes

Migration routes in Southeast Asia are not static, but are shaped by temporal factors such as economic cycles, political conditions and seasonal labour needs. The temporality of migration refers to how migration patterns fluctuate over time, influenced by changes in economic conditions, changes in immigration policies and external shocks such as pandemics or financial crises.

Economic Cycles and Migration Flows

Economic booms in the countries receiving migrants often lead to an increase in migration flows as the demand for labour rises sharply. For example, the construction boom in Malaysia in the early 2000s, driven by rapid urbanisation and infrastructure projects, led to a significant demand for foreign workers, particularly from Indonesia and Bangladesh (Rahman, 2019). Migrants flocked to Malaysia to take advantage of economic opportunities and fill labour shortages in sectors where local workers were unavailable or unwilling to work. Conversely, an economic downturn can lead to a decline in labour migration if jobs become scarce and immigration policies are tightened. The global financial crisis of 2008, for example, had a profound impact on migration in Southeast Asia, as countries such as Malaysia and Thailand tightened immigration policies.

Migration routes in Southeast Asia are strongly influenced by economic factors, with labour mobility being a key feature of regional development. Economic inequalities between countries combined with the demand for labour in the wealthier countries have created specific migration corridors across the region (Maslyana & Ullah, 2021). These routes are determined by both push factors, such as poverty and lack of employment opportunities in the countries of origin, and pull factors, such as higher wages and better job prospects in the receiving countries. In addition, cultural, historical and geographical factors play an important role in shaping these routes, which further reinforces economic migration patterns.

The Malaysia-Indonesia Corridor

One of the best-known examples of economically driven migration routes in Southeast Asia is the Malaysia-Indonesia Corridor. Indonesia, the most populous country in the region, has been an important source of migrant labour for Malaysia for decades. Economic relations between the two countries are deeply rooted. Malaysia relies on Indonesian migrant workers to fill labour-intensive jobs in the low-wage sector, agriculture, plantations and domestic work (Kaur, 2010). Indonesian labourers, especially from rural and impoverished areas, seek work in Malaysia to escape poverty and provide for their families through remittances.

Malaysia's economic attraction for Indonesian workers is further fuelled by significant wage differentials between the two countries. For example, the average monthly wage for an unskilled worker in Malaysia in 2021 was around USD 600, while in Indonesia it was around USD 200 (World Bank, 2021a, 2021b). This considerable difference creates a strong incentive for Indonesians to seek work in Malaysia, despite the challenges they may face in terms of working conditions, immigration policies and labour rights.

The historical and cultural ties between Indonesia and Malaysia also facilitate migration. Both countries share a common Malay heritage, linguistic similarities and religious ties (Kaur, 2010). These factors help Indonesian migrants to integrate easily into Malaysian society, even though they often occupy a marginalised position in the workforce. In addition, Malaysia's proximity to Indonesia, particularly through the Straits of Malacca, is geographically favourable for Indonesians to migrate, often through both legal and irregular means.

The Thailand-Myanmar Corridor

Another important migration corridor, which is characterised by economic factors, is the Thailand-Myanmar corridor. Thailand has experienced rapid economic growth in recent decades, which has led to a high demand for migrant workers, particularly in labour-intensive sectors such as agriculture, fishing, construction and industry (Chantavanich, 2014). Migrant workers from Myanmar, especially from the border regions, are attracted to Thailand due to the relatively higher wages and numerous employment opportunities.

The economic disparities between Thailand and Myanmar are large. In 2020, the average monthly wage in Thailand was around USD 490, while in Myanmar it was around USD 150 (ILO, 2020). This difference drives a large number of Myanmar nationals, especially those from impoverished rural areas, to seek work in Thailand. Migrants often work in low-skilled and dangerous jobs that local Thai workers are unwilling to accept due to low wages and poor working conditions.

Political instability in Myanmar, including decades of ethnic conflict and military rule, has also contributed to people migrating to Thailand (Vungsiriphisal et al., 2013). Many migrants are fleeing violence or persecution, while others are simply seeking better economic prospects in the face of political and economic uncertainty at home. This combination of economic and political push factors, coupled with Thailand's demand for cheap labour, has led to a well-established migration route between the two countries.

Geographically, Thailand shares a long border with Myanmar, making cross-border movement relatively easy. Many migrants enter Thailand through unofficial channels, facilitated by the porous border and the presence of smuggling networks. Once in Thailand, migrants often face problems related to legal status, exploitation and access to social services, but the economic benefits of working in Thailand are sufficient to encourage further migration.

Vietnam's Internal Migration: Rural-to-Urban Shift

While international migration is a key feature of the Southeast Asian labour landscape, internal migration also plays a crucial role in shaping economic development within the countries. Vietnam provides a compelling example of how economic factors influence internal migration routes. The country has undergone significant economic transformation in recent decades, moving from a mainly agricultural economy to one characterised by industrialisation and urbanisation (Dang et al., 2003).

As Vietnam's cities, particularly Ho Chi Minh City and Hanoi, have developed into industrial and commercial centres, millions of workers have migrated from rural areas to these urban centres in search of better employment opportunities. This rural–urban migration is driven by the wage gap between agricultural labour in the countryside and industrial or service sector work in the city. For example, while rural agricultural labourers in Vietnam earn the equivalent of USD 100 per month, factory workers in urban areas can earn three to four times that amount (Nguyen et al., 2015).

The Vietnamese government has played an active role in encouraging this migration through policies aimed at urbanisation and industrialisation. Special Economic Zones (SEZs) and industrial parks have been established to attract foreign and domestic investment, which in turn creates a demand for labour. Migrants from rural areas tend to be employed in low-skilled manufacturing jobs, often in sectors such as textiles, electronics and food processing.

This internal migration is not without its problems. Migrants often struggle to access housing, healthcare and education because they do not have legal residency status in urban areas, as Vietnam operates a household registration system (the ho khau) that restricts access to public services for those who are not officially registered as urban residents (Phuong, 2019). Nonetheless, the economic benefits of urban employment are significant enough to maintain a steady flow of rural–urban migration.

The Philippines: A Case of Labour Export

The Philippines presents a somewhat different case within Southeast Asia, as it is one of the leading exporters of labour in the region. Economic factors, particularly the lack of sufficient employment opportunities and low wages in the domestic labour market, are driving millions of Filipinos to seek work abroad. While the Philippines sends labour to many parts of the world, Southeast Asia, particularly countries such as Singapore and Malaysia, is a major destination for Filipino migrants, especially in the domestic and healthcare sectors (Asis, 2006).

The Philippine government has institutionalised labour migration as part of its development strategy. Agencies such as the Philippine Overseas Employment Administration (POEA) regulate and promote overseas employment. Remittances from overseas Filipino workers (OFWs) are an important source of income for the country, accounting for around 10% of Philippine GDP in recent years (Bangko Sentral ng Pilipinas, 2020). This dependence on remittances emphasises the importance of migration routes for the country's economy.

While the Philippines benefits economically from labour migration through remittances, there are also social and political implications. Migrant workers often face exploitation and abuse abroad, especially in sectors such as domestic labour where legal protection is weak. In addition, the loss of skilled labour, particularly in the health sector, can jeopardise the provision of domestic services in the Philippines itself.

Theories in Practice: Tying Migration to Economic Reality

The migration routes in Southeast Asia outlined above are closely related to several important migration theories, each of which offers insights into the relationship between migration and economic factors.

Neoclassical economic theory provides a simple explanation for the movement of migrants from poorer countries to richer countries. This theory states that migration occurs due to wage differentials between countries, with labour moving from areas of surplus (low wages and high unemployment) to areas of scarcity (high wages and labour demand) (Todaro & Maruszko, 1987). In Southeast Asia, this dynamic is clearly evident in the migration of Indonesian and Myanmarese migrants to wealthier countries such as Malaysia and Thailand, where the wage gap creates a strong incentive for labour migration.

The dual labour market theory also explains why migrants tend to occupy lower-paid and less secure jobs in the receiving countries. In Malaysia, Thailand and Singapore, migrants occupy jobs in sectors such as agriculture, construction and domestic work, which are often characterised by low wages, difficult working conditions and limited legal protection (Piore, 1979). These jobs are essential to the functioning of the economy but are generally shunned by local workers, creating a demand for migrant labour.

The NELM adds another layer of complexity by highlighting that migration is often a household strategy aimed at diversifying income and managing economic risk (Stark & Bloom, 1985). In Cambodia, for example, households may send a family member to work in Thailand to ensure a steady flow of remittances, which can provide financial security in the face of local economic challenges such as crop failures or limited employment opportunities.

Finally, world systems theory views migration as a consequence of global economic structures in which labour flows from peripheral, less developed regions to central, more developed regions (Wallerstein, 1974). The migration routes of Southeast Asia can be understood within this framework, as migrants move from poorer countries such as Myanmar, Cambodia and Indonesia to more economically advanced countries such as Malaysia, Thailand and Singapore. The integration of Southeast Asian economies into global capitalist networks has further reinforced these migration patterns, as industries in the wealthier countries rely on cheap migrant labour to remain competitive in the global market.

The economic impact of migration routes in Southeast Asia is significant for both sending and receiving countries. For receiving countries such as Malaysia, Thailand, Brunei and Singapore, migrant workers are critical to maintaining industries that rely on cheap, flexible labour. In Malaysia, for example, the palm oil and construction sectors are heavily dependent on migrant workers, who make up a large part of the labour force (Kaur, 2014). Without migrant labour, these industries would struggle to meet production demands and maintain profitability.

Migration patterns in Southeast Asia are strongly influenced by economic dynamics. Various theoretical frameworks help to explain the relationship between economic forces and migration routes. In practise, the interaction between migration and economic development is evident in several countries in the region, where labour mobility plays a crucial role in shaping national and regional economies. Neoclassical economic theory, dual labour market theory, the NELM and world systems theory provide valuable insights into these processes and show how economic inequalities and labour demand determine migration routes.

Neoclassical economic theory assumes that migration is driven by wage differentials between countries or regions. In Southeast Asia, this theory is particularly well suited to understanding the movement of migrants from countries such as Myanmar, Cambodia and Indonesia to more economically developed countries such as Malaysia, Thailand and Singapore. Migrants from poorer countries are drawn to richer countries where wages are significantly higher, creating economic incentives for cross-border labour mobility.

In the Malaysia-Indonesia corridor, for example, the average wage for an unskilled worker in Malaysia is much higher than in Indonesia, which is a strong economic incentive for Indonesian migrants. In 2021, Indonesian workers in Malaysia could earn almost three times as much as in their home country, prompting many to migrate in search of a better livelihood (World Bank, 2021a, 2021b). The neoclassical framework explains this pathway, where migration serves as a rational response to economic inequalities and allows workers to maximise their income by moving to regions with higher wages and higher labour demand.

Similarly, the migration route from Myanmar to Thailand reflects the neoclassical perspective. The average monthly wage in Myanmar was significantly lower at around USD 150 in 2020 compared to USD 490 in Thailand (ILO, 2020). This wage gap provides an incentive for migrants from Myanmar to cross the border into Thailand, especially as Thailand's rapid economic growth continues to create demand for low-wage labour in agriculture, fishing and manufacturing. The neoclassical approach provides a clear explanation for these movements, where wage differentials are the main motivation for migration.

While neoclassical economic theory explains migration through wage differentials, dual labour market theory emphasises the structural need for migrant workers in the receiving countries. According to this theory, two different labour markets often emerge in the economies of developed or developing countries: a primary labour market with higher wages and more job security and a secondary labour market with low-paid, insecure and undesirable jobs (Piore, 1979). Migrants generally fill jobs in the secondary labour market, as these jobs are often avoided by native workers.

In Malaysia, for example, migrant labour is essential in sectors such as construction, domestic work and agriculture, where wages are low, working conditions are harsh and the work is physically demanding. The palm oil industry, a cornerstone of the Malaysian economy, relies heavily on migrant labour from Indonesia and Bangladesh. Without this influx of migrant labour, Malaysia would struggle to maintain production levels in these labour-intensive sectors (Kaur, 2014). The dual labour market theory explains why migrants continue to fill these less desirable jobs as they are a mainstay of industries that are vital to Malaysia's economic growth.

In Thailand, the fishing industry similarly relies on migrants from Myanmar, Cambodia and Laos to fill low-paid jobs. Local workers tend to avoid these jobs because of the dangerous working conditions and low pay, but migrants, especially from poorer countries, take them because they pay better than in their home countries (Chantavanich, 2014). The dual labour market theory thus helps to explain why migrants are concentrated in certain sectors and why their presence is crucial for the functioning of key industries in receiving countries.

While neoclassical and dual labour market theories focus on individual motivations and labour market structures, the NELM shifts the focus to the household as a decision-making unit. The NELM assumes that migration is often a household strategy aimed at diversifying their sources of income and reducing economic risks. In this framework, households may send a member to work abroad not only to earn higher wages, but also to cope with risks related to economic instability, crop failure or lack of access to credit (Stark & Bloom, 1985).

This theory is particularly important for understanding migration from Cambodia to Thailand. For many Cambodian households, sending a family member to work in Thailand is an important source of income that helps the family cope with financial risks at home. In rural areas of Cambodia, where employment opportunities are limited, migration offers households a way to diversify their income streams and improve their economic security. Remittances from Cambodian migrant workers in Thailand are crucial for supporting families back home, covering basic needs such as food, education and healthcare (Cambodia Development Resource Institute, 2017).

For households in Myanmar, sending a family member to work in Thailand or Malaysia can also mean financial relief in the face of the economic challenges there. Migrants send remittances, which are often used to improve living conditions, invest in small businesses or pay off debts. In these cases, migration is not just an individual pursuit of higher wages, but a family strategy aimed at improving the overall well-being of the household and mitigating economic risks in the home country.

World systems theory offers a more comprehensive view of migration and emphasises the role of global economic forces in shaping migration routes. According to this theory, migration takes place as part of the global capitalist system in which labour flows from peripheral, less developed regions to the central, more developed regions (Wallerstein, 1974). In Southeast Asia, this can be seen in the migration patterns from poorer countries such as Myanmar, Cambodia, the Philippines and Indonesia to wealthier nations such as Singapore, Brunei, Malaysia and Thailand. Indonesia, while should not be classified as a poor country, grapples with significant issues of inequality, a challenge common to many modern developing nations. This inequality is deeply intertwined with the structural dynamics of capitalism, which is often embedded within the broader framework of modernization. As a result, the benefits of economic growth and development are unevenly distributed, creating disparities that persist across different regions and social strata within the country.

The integration of Southeast Asian economies into global production networks has intensified these migration patterns. Industries in more developed countries, such as the electronics industry in Malaysia and Singapore, rely on cheap migrant labour to remain competitive in global markets (Kaur, 2014). At the same time, the demand in labour-intensive sectors such as agriculture, construction and domestic work attracts migrants from poorer neighbouring countries, despite their low wages. World systems theory helps to explain how global economic structures, shaped by capitalist expansion, create and sustain migration routes that bring labour from poorer to richer countries within the region.

The central theme of this chapter is the crucial role that economics plays in shaping migration routes in Southeast Asia. Whether through wage differentials, labour market structures, household strategies or global economic forces, migration in the region is primarily motivated by economic factors. These migration routes not only reflect economic inequalities between countries, but are also a response to the demand for labour in the region's key industries.

Evidence of migration corridors such as Malaysia-Indonesia, Thailand-Myanmar and Vietnam's rural-to-urban shift show how economic pressures drive labour mobility. Migrants are drawn to countries and cities where wages are higher and job opportunities more diverse, while receiving countries benefit from the availability of cheap labour to boost their economic growth. Theoretical frameworks such as neoclassical economic theory, dual labour market theory, NELM and world systems theory offer valuable insights into these dynamics and provide a comprehensive understanding of how economic forces influence migration routes in Southeast Asia.

Migration routes in Southeast Asia are deeply embedded in the economic landscape of the region. Economic inequalities, labour market demand and global capitalist structures all contribute to cross-border labour migration and migration within countries. By linking theoretical frameworks with real-life examples, we see how the economy is both the driving force and the main beneficiary of migration. These migration patterns not only shape the lives of individual migrants and their families, but also have a profound impact on the economies of both sending and receiving countries in Southeast Asia.

Conclusion

Migration in Southeast Asia is closely intertwined with the economic realities of the region, as it is both a consequence of economic inequalities and a driver of development. This chapter has shown how migration routes are influenced by factors such as wage differentials, labour market demand and economic policies. Countries such as Malaysia, Thailand, Brunei and Singapore serve as hubs for migrant workers who play an important role in industries such as construction, agriculture and domestic work, while sending countries such as the Philippines, Indonesia, Cambodia and Myanmar rely on remittances to stabilise their economies.

The application of migration theories, including neoclassical economics, dual labour market theory and world systems theory, provides a comprehensive understanding of these dynamics. For example, wage-dependent migration from Myanmar to Thailand and from Indonesia to Malaysia reflects the rational economic decisions emphasised in neoclassical models. Similarly, dual labour market theory explains the concentration of migrant workers in low-wage, labour-intensive sectors that are avoided by locals.

In addition, the integration of Southeast Asian economies into global production networks has increased the demand for cheap migrant labour and maintained established migration corridors. Migrant remittances support households, alleviate poverty and contribute to economic stability in countries of origin, reinforcing these patterns. However, problems such as exploitation, undocumented migration and lack of legal protection for migrant workers persist. Addressing these problems requires coordinated regional policies that balance the economic benefits of migration with the rights and welfare of workers. Although regional policies and International Labour Organization (ILO) conventions exist to protect the rights of migrant workers, enforcement remains weak due to a combination of economic pressures and systemic exploitation. For example, manpower agencies, often in collusion with regulators, prioritise profit over ethical practices, while employers are primarily focused on minimising costs. A notable policy framework is the ASEAN Framework on the Implementation of the ASEAN Declaration on the Protection and Promotion of the Rights of Migrant Workers (2007). However, the persistent challenges suggest that the focus should shift toward strengthening institutional capacities to ensure the effective implementation of such policies and to safeguard the rights and welfare of migrant workers more comprehensively. Enhanced monitoring, regulatory enforcement, and cross-border cooperation are essential to address the gaps in existing mechanisms. Improving regional co-operation through frameworks such as ASEAN could identify ways to manage migration more effectively and equitably. Migration is an integral part of Southeast Asia's economic fabric and its effective management has the potential to benefit both sending and receiving countries. By addressing the structural inequalities and vulnerabilities that underpin migration, the region can achieve more sustainable and equitable development.

References

ASEAN. (2021). *ASEAN Economic Community 2021 Integration Progress Report*. ASEAN Secretariat.

ASEAN Declaration on the Protection and Promotion of the Rights of Migrant Workers (2007). ASEAN Declaration on the Protection and Promotion of the Rights of Migrant Workers. https://asean.org/asean-declaration-on-theprotection-and-promotion-of-the-rights-of-migrant-workers/ Accessed on 7 June 2024

Asis, M. M. B. (2006). The Philippines' culture of migration. *Migration Policy Institute*. Retrieved from https://www.migrationpolicy.org/article/philippines-culture-migration

Asis, M. M. B. (2017). The Philippines: Beyond labor migration, toward development and (possibly) return. *Migration Policy Institute*. Retrieved from https://www.migrationpolicy.org/article/philippines-beyond-labor-migration-toward-development-and-possibly-return

Athukorala, P. (2019). *Southeast Asia and the global economy: Industrialization, trade, and international production networks*. Edward Elgar Publishing.

Azizuddin, M., & Shamsuzzoha, A. (2024). Women leadership development in local governance in South and Southeast Asia: A critical analysis. *Southeast Asia: A Multidisciplinary Journal, 24*(1), 40–52. https://doi.org/10.1108/SEAMJ-06-2023-0052

Bangko Sentral ng Pilipinas. (2020). Overseas Filipino remittances. Retrieved from https://www.bsp.gov.ph

Cambodia Development Resource Institute. (2017). *The impact of migration and remittances on the Cambodian economy*. Cambodia Development Resource Institute.

Chantavanich, S. (2014). *Thailand policies towards migrant workers from Myanmar*. Asian Research Center for Migration.

Dang, N. A., Goldstein, S., & McNally, J. W. (2003). Internal migration and development in Vietnam. *International Organization for Migration.*

Derks, A. (2010). Migrant labour and the politics of immobilisation: Cambodian fishermen in Thailand. *Asian Journal of Social Science, 38*(6), 915–932.

Haji Marzuki, A., & Alkaff, S. N. H. (2024). An investigation on perceptions of street remarks by Bruneian men and women. *Southeast Asia: A Multidisciplinary Journal, 24*(1), 53–68. https://doi.org/10.1108/SEAMJ-02-2023-0021

Harkins, B., Lindgren, D., & Suravoranon, T. (2017). *Risks and rewards: Outcomes of labour migration in South-East Asia*. International Labour Organization (ILO). Retrieved from https://www.ilo.org/asia/publications/WCMS_619727/lang--en/index.htm

Harris, J. R., & Todaro, M. P. (1970). Migration, unemployment and development: A two-sector analysis. *The American Economic Review, 60*(1), 126–142.

Hing, V., Lun, P., & Phann, D. (2011). *Irregular migration from Cambodia: Characteristics, challenges, and regulatory approach*. Cambodian Development Resource Institute.

Hugo, G. (2005). *Migration in the Asia-Pacific region*. Global Commission on International Migration.

Hugo, G. (2011). Migration and development in low-income countries: A challenge for the social sciences. *Population and Development Review, 37*(1), 141–161.

International Labour Organization (ILO). (2017). *Migrant workers in Malaysia: Protection of wages and working conditions*. International Labour Organization.

International Labour Organization (ILO). (2019). *Labour migration in ASEAN region*. Retrieved from https://www.ilo.org/asia/WCMS_675239/lang--en/index.htm

International Labour Organization (ILO). (2020). *Global Wage Report 2020–2021: Wages and minimum wages in the time of COVID-19*. International Labour Organization.

International Organization for Migration (IOM). (2020). *World migration report 2020*. International Organization for Migration.

Jones, H., & Pardthaisong, L. (1999). The impact of overseas labour migration on rural Thailand: Regional, community, and individual dimensions. *Journal of Rural Studies, 15*(1), 35–50.

Kaur, A. (2010). Labour migration in Southeast Asia: Migration policies, labour exploitation, and regulation. *Journal of the Asia Pacific Economy, 15*(1), 6–19. https://doi.org/10.1080/13547860903488195

Kaur, A. (2014). Economic change and labour migration in Southeast Asia: The Malaysia-Indonesia nexus. *The Malaysian Journal of Economic Studies, 51*(1), 29–49.

Malaysia's Ministry of Home Affairs. (2020). *Statistics on foreign workers in Malaysia.* Retrieved from https://www.moha.gov.my/index.php/en/

Massey, D. S., Arango, J., Hugo, G., Kouaouci, A., Pellegrino, A., & Taylor, J. E. (1993). Theories of international migration: A review and appraisal. *Population and Development Review, 19*(3), 431–466.

Nayan, H. M. B., & Ullah, A. K. M. A. (2021). Indian migrants in Brunei: The role of social networks. *Migration Letters, 18*(6), 697–709.

Nguyen, Q. T., Tran, Q. T., & Truong, V. T. (2015). Rural-urban migration and its impact on economic development in Vietnam. *Journal of Economics and Development, 17*(1), 34–50.

Parreñas, R. S. (2001). *Servants of globalization: Women, migration, and domestic work.* Stanford University Press.

Phuong, L. T. (2019). The impact of Vietnam's *ho khau* household registration system on migrant laborers. *Journal of Vietnamese Studies, 14*(2), 85–112. https://doi.org/10.1525/jvs.2019.14.2.85

Piore, M. J. (1979). *Birds of passage: Migrant labor and industrial societies.* Cambridge University Press.

Piper, N. (2019). *Migrant labour in Southeast Asia: A comparative analysis of labour migration in Southeast Asia.* Routledge.

Rahman, M. M. (2019). *Bangladeshi migration to Singapore: A process-oriented approach.* Springer.

Rahman, M., & Ullah, A. K. M. A. (Eds.). (2012). *Migration policies in Asia: South, Southeast and East Asia.* Nova Science.

Ranjan, A., Chattoraj, D., & Ullah, A. K. M. A. (Eds.). (2024). *India and China in Southeast Asia.* Palgrave Macmillan.

Robertson, R. (2011). Apparel wages before and after the Better Factories Cambodia program. *ILO Research Notes.*

Rubel, M. A. I., Rumi, M. H., & Ripon, M. A. (2024). Empowerment through engagement: Assessing the female university students' involvement in volunteering activities. *Southeast Asia: A Multidisciplinary Journal, 24*(1), 69–81. https://doi.org/10.1108/SEAMJ-08-2023-0062

Stark, O., & Bloom, D. E. (1985). The new economics of labor migration. *The American Economic Review, 75*(2), 173–178.

Todaro, M. P., & Maruszko, L. (1987). Illegal migration and US immigration reform: A conceptual framework. *Population and Development Review, 13*(1), 101–114.

Ullah, A. K. M. A. (2010). *Rationalizing migration decisions: Labour migrants in South and South-East Asia* (1st ed.). Ashgate.

Ullah, A. K. M. A. (2022). The culture of migration in Southeast Asia: Acculturation, enculturation or deculturation? *Journal for Cultural Research, 26*(2), 184–199. https://doi.org/10.1080/14797585.2022.2097881

Ullah, A. K. M. A., & Chattoraj, D. (2022). *Covid-19 pandemic and the migrant population in Southeast Asia: Vaccine, diplomacy and disparity.* World Scientific.

Ullah, A. K. M. A., Chattoraj, D., & Ibrahim, W. Z. (2022). The COVID-19 pandemic and the 'stranded' migrant population: An unequal pain. *SUVANNABHUMI: An International Journal of Southeast Asian Studies, 14*(2), 61–88.

Ullah, A. K. M. A., Haji-Othman, N. A., & Daud, M. K. (2021a). COVID-19 and shifting border policies in Southeast Asia. *Southeast Asia: A Multidisciplinary Journal, 21*(2), 1–14.

Ullah, A. K. M. A., Kumpoh, A. A., & Haji-Othman, N. A. (2021b). COVID-19 vaccine and international relations: New frontiers of vaccine diplomacy. *Southeast Asia: A Multidisciplinary Journal, 21*(1), 1–14.

Ullah, A. K. M. A., Lee, S. C. W., & Hassan, N. H. B. P. H. (2024). *Southeast Asia in pre- and post COVID-19: Economy, society, mobility and religion.* Routledge.
Vungsiriphisal, P., Chantavanich, S., & Laodumrongchai, S. (2013). *Human trafficking in Thailand: Lessons learned from the process of establishing anti-trafficking measures.* UNESCO.
Wallerstein, I. (1974). *The modern world-system: Capitalist agriculture and the origins of the European world-economy in the sixteenth century.* Academic Press.
World Bank. (2021a). *Migration and remittances data.* Retrieved from https://www.worldbank.org/en/topic/labormigrationremittancesdata
World Bank. (2021b). *World development indicators.* Retrieved from https://data.worldbank.org/
Yeoh, B. S. A. (2006). Bifurcated labour: The unequal incorporation of transmigrants in Singapore. *Tijdschrift Voor Economische En Sociale Geografie, 97*(1), 26–37.
Yeoh, B. S. A., & Huang, S. (1998). Negotiating public space: Strategies and styles of migrant female domestic workers in Singapore. *Urban Studies, 35*(3), 583–602.
Yeoh, B. S. A., & Huang, S. (2010). Foreign domestic workers and home-based care for elders in Singapore. *Journal of Aging & Social Policy, 22*(1), 69–88. https://doi.org/10.1080/08959420903385462
Yeoh, B. S. A., & Lam, T. (2016). Immigration and its impact on Southeast Asia. *Asian Population Studies, 12*(1), 22–43.

Chapter 13
The Future Transformation of Southeast Asia

A. K. M. Ahsan Ullah and Gabriel Yit Vui Yong

Abstract This concluding chapter begins with an update of developments highlighted in the introduction. Notably, Donald Trump has won the US presidential election and BRICS have put forward a payment system to facilitate trade outside SWIFT, which is under Western control. Regrettably, the genocide in Palestine continues, as does the war in Ukraine and the Middle-East. Geopolitical tension remains high and the trade war is expected to escalate. The chapter however focuses on the transformations expected in the second quarter of the century, because the transition to a multipolar world has become inevitable. It also discusses the existential barriers presented by the potential of a nuclear war and climate change. It then focuses on the digital transformation, an independent and irreversible process in human evolution, which will create a hybrid virtual–physical future reality. It concludes by highlighting insights from the chapters and geographical factors, such as the Yuxi Circle. They are grouped into: (i) climate change; (ii) sustainable development; (iii) migration and social justice; and (iv) misconceptions and uncertainties. From a Brunei perspective, local needs of the small and marginalised, which are often ignored due to their perceived insignificance, are critical to achieving a sustainable future, which appears to be within reach. An ecological understanding of relationships is also crucial.

Keywords Transformation · Geopolitics · Existential · Yuxi · Hybrid virtual–physical future

A. K. M. A. Ullah (✉) · G. Y. V. Yong
Faculty of Arts and Social Science, Universiti Brunei Darussalam, Bandar Seri Begawan, Brunei
e-mail: ahsan.ullah@ubd.edu.bn

G. Y. V. Yong and A.K.M. A. Ullah (eds.), *Brunei and the Future Transformation of Southeast Asia*, https://doi.org/10.1007/978-981-96-7901-0_13

Introduction

At the UN Summit of the Future[1] on 22 September 2024, world leaders adopted a pact, which included proposals on the Global Digital Compact and a Declaration on Future Generations. On its webpage,[2] the headlines read: 'The world is at a crossroads. Rapid advances in technology and science hold the promise of a better, more peaceful and prosperous future for people and the planet. Yet, in our interconnected and fast-changing world, we are not keeping pace with emerging risks and opportunities. At the same time, we are grappling with multiple conflicts and escalating geopolitical tensions, rising inequality and mistrust, stalled progress on the Sustainable Development Goals (SDG), and a climate emergency. As a global community, we face a pivotal choice'. This resonates with the purpose and context of the book, which sought to highlight these concerning trends and uncertainties, because the majority of society remains oblivious to them. Therefore, they go about their daily lives in pursuit of the gamut of personal interests within the global simulacrum constructed by corporations and agencies with different agendas.

In its review of events that shaped 2024,[3] the UN stated that 'overwhelmingly, 2024 was a year marked by conflicts that destabilised whole regions and further polarised an already fractured planet'. It cited that 1 in 5 children worldwide are either living in or fleeing from wars, and that escalating violence in Gaza, Sudan, and Ukraine has destroyed millions of lives and damaged faith in multilateral institutions to maintain peace. The report also notes the 'disparity in how the international community has responded to Gaza versus Ukraine', which has deepened divisions among UN Member States and eroded public perception. Indeed, many are questioning the relevance of the UN due to its impotence in reigning the rogue state, Israel, and consistently getting steam-rolled by the USA. The Gaza situation highlights the UN's incapacity to deliver humanitarian aid and broker peace, and ensuring rules of engagement in wars, especially in relation to health personnel and health facilities, as they were blatantly ignored by the West.

Life goes on in most parts of the world that are not directly affected by wars. Many are unawares of their repercussions on neighbouring regions. Instead, they blame rising numbers of refugees and soaring inflation on convenient culprits proposed by the media and disinformation, which are another component of the same conflicts waged in a different dimension. Meanwhile, although the majority of the global population are aware of the term 'climate change', and states, institutions and corporations have adopted a variety of the policies and programmes to curb carbon emissions, natural disasters have wreaked havoc on lives and infrastructure in isolated areas in 2024. While the general consensus is that leaders and society as a whole lack commitment or employed the wrong strategies (e.g., Latin, 2012) to achieve

---

[1] https://www.un.org/en/summit-of-the-future/pact-for-the-future.

[2] https://summitofthefutureun.org/pact/?_gl=1*176l75z*_ga*NDg2OTY4MjU3LjE3MzM4ODg2ODU.*_ga_TK9BQL5X7Z*MTczNDY1NTk4NS4xMS4xLjE3MzQ2NTY4NjEuMC4wLjA.

[3] https://unfoundation.org/blog/post/year-in-review-the-events-that-shaped-the-un-and-our-world-in-2024/.

SDG 13 (to combat climate change), few questions the suitability of the strategies employed. Yong (2022), for example, related climate change to the modern urbanisation paradigm, which has systematically diminished the natural ecosystem functions and given rise to a host of environment problems, including climate change and associated issues. Deeper analysis in the realm and knowledge creation (Putnam, 1991; Capra, 1997) is needed. The majority of modern societies exist within simulacra that are constructed by society, institutions, and global corporations seeking to create markets to generate wealth, from which they must be emancipated to examine reality from other perspectives.

Our book centres on Southeast Asia, in which the authors live and work. The majority were born and bred here, and therefore have insiders' knowledge and understanding of environment, culture and regional dynamics. Although the book is only able to present glimpses of circumstances and perspectives due to the limitation of a single volume, it hopes to have provided enough useful insights, particularly from the perspectives of Brunei and a few neighbouring states, that address common issues and concerns. The book's interest however is on the future of the region. ASEAN, advancing as a bloc, has made great strides. Some, including Kishore Mahbubani, the distinguished diplomat, academic, and former UN Security Council President, known for his insights on international relations and Asian global influence, sees ASEAN as one of the three leading economic powers of the twenty-first century (Mahbubani, 2022, 2024). Others are frustrated by the slow pace and non-interference policy of ASEAN, because they would like to see human rights issues addressed in the same manner as in the West (Galimberti, 2023), through the same lens, which disregards local–regional historical-cultural as well as political-economic orientation for the future. Sovereignty is of paramount importance, as it is with all countries in the Global South, because of the shared experience of being colonised and exploited.

The book regards the coming years as a watershed moment of momentous shifts in global order. The 16th BRICS summit in Kazan on 22–24 October 2024 unveiled an alternative financial transaction system to SWIFT, which set de-dollarisation on an irreversible path, regardless of the threats made by US president-elect, Donald Trump. The next quarter of the twenty-first century will be characterised by a multipolar global order, where China, India and ASEAN will play a significant role. The USA, even if it continues on the present self-inflicted self-destructive course, will remain a powerful country, due to the deep entrenchment of the USD in the global economy, its ability to print money, and its military might. However, it will no longer be able to lord over other nations in the same way, as many countries and regions are rising up, largely due to the win–win cooperative development process initiated by China and facilitated by its BRI and supporting institutions, such as the New Development Bank, and the consolidation and expansion of BRICS. However, the transition will neither be smooth nor likely to be amicable.

Existential Barrier

The desperate struggle of the USA to cling on to its powerful position as global hegemon could pose an existential threat to humanity. In November, 2024, Donald Trump won the presidential election, promising peace, the return of jobs to the US, and to punish anyone, friend or foe, that do not comply with policies that serve US interests. The result of the election, however, is not a reflection of the American public's high regard of Trump, but their distrust, even disgust, with the incumbent administration, especially its blind devotion to funding the ongoing genocide in Palestine and prolonging and escalation of the Ukraine war with hundreds of billions of dollars, while the majority of US citizens struggle to make ends meet. However, Trump's cabinet choices include many of the existing warmongering neoconservatives responsible for the policy of 'US exceptionalism' that has brought about the current situation. Nevertheless, his brash, self-serving and intellectually clumsy style could thwart the deep state's agenda from time to time, because he brings a degree of unpredictability.

With Trump at the helm, the US is firmly set on its way to relinquishing its exorbitant supremacy as the sole hegemon holding the world's reserve currency. At this stage in human history, the majority of the world's population has woken up to the fact that they are shackled by the US and its Western vassals in the current world order. Many are therefore determined to regain their true sovereignty and determine their own development path. They share a common vision of peaceful coexistence and beneficial co-operative relations with China and BRICS, which are creating an alternative future world. Trump's desire to punish anyone who does not bow to the US certainly resonates with neoconservatives in both the Republican and Democratic camps. However, if he goes through with his threats to impose hefty tariffs on any nation, including allies and vassals, that dares to join the de-dollarisation movement, it will, ironically, only accelerate this process. In the short period leading up to Trump's inauguration, however, the incumbent has escalated the Ukraine conflict to pose an existential threat.

The spectre of nuclear war has become quite real with the announcement by the outgoing US President, who has been considered cognitively incapable of running for office, but somehow of sufficiently sound mind to give Ukraine the permission to use NATO intermediate-range nuclear arsenals capable of penetrating deep into metropolitan Russia. This, as President Putin emphasises, is only possible with the direct involvement of US and British personnel, which he regards as a declaration of war. On 19 and 21 November, American and British medium-range missiles were fired at Russia (Times of India, 2024, November 22).[4] Although the missiles were intercepted by the defence system and a Russian non-nuclear hypersonic missile was fired in retaliation, destroying an important NATO armament facility in Ukraine, Russia has exercised restraint and only issued warnings to NATO that directly contained a clear warning of the consequences of further similar actions.

[4] https://youtu.be/yR-OSK7grOA.

The situation is therefore currently on a knife-edge. In the meantime, the majority of the world's population go about their daily lives, fighting for petty benefits and trivial causes, including at COP-29,[5] the annual UN conference on climate change, held on 11 November 2024 in Baku, Azerbaijan. It is difficult to understand the logic of the powerful entities behind the US decision and the actions of the US and UK, as they appear to be callous and fail to recognise the potentially existential consequences. One possible consolation is the greatly diminished production capacity of the US and the collective West, which could curtail their war efforts and thus bring about peace as they run out of ammunition. Moreover, Russia, China and Iran, identified as enemies by the West, are well prepared as they have built up their defence capabilities since the US and its vassals started their war in the economic and information arenas, and have expanded their military presence in strategic locations. As the Chinese and Russian arsenals are technologically advanced, the West is likely to be caught off guard in the event of a hot war, because it appears to believe in its own disinformation that China and Russia are backward and on the verge of collapse, i.e., the political and MIC elites appear to be trapped in the simulacrum that they have developed themselves.

However, regardless of how a military conflict would play out, its consequence is existential, if it were to escalate to the nuclear level. The climate threat would then be meaningless. From this perspective, it is clear which has real existential consequences. Climatic changes at higher extreme variability and extended temporal ranges have occurred in the planet's history, described as (i) 'Hothouse Earth' or Greenhouse Earth, and (ii) Icehouse Earth (see, e.g., Zachos et al., 2010). The former refers to periods when the planet was significantly warmer and where no large continental ice sheets could be found in the polar regions. Sea levels were higher, and tropical conditions extended further towards the poles. In the latter, Earth has significant ice sheets at the poles. The current climate state is therefore considered an 'Icehouse', because large ice sheets exist in Antarctica and Greenland. The present warm Holocene epoch is an interglacial period within the longer 'Ice Age'. It is therefore clear from paleoclimate research that climate change would not vanquish life on Earth, although a number of species could become extinct when climatic states shift drastically. This is evident from the fossil record.

The future would be a bright one for the region, and indeed for humanity as a whole, because a multipolar world founded on mutual respect, 'win–win' cooperation and shared future aligns closely with the Brundtland commission's definition of Sustainable Development (WCED, 1987). However, humanity must first surmount the potential threat of a nuclear war. This however is dependent on the will and action of Western societies, because they need to wrestle back their governments from the clutches of the deep state, i.e., capitalist-oligarchs that views countries and the world primarily in terms of resources, economy and competitive advantage; they invariably disregard the rights and wishes of indigenous population. The paradox and hypocrisy of this worldview, values, and actions, which are diametrically opposite

[5] https://unfccc.int/cop29.

to the principles of democracy and human rights, do not appear to have dawn on this enormously wealthy minority.

Yuxi Circle

The book has adopted a geographical approach to examine the theme of Southeast Asia's future transformation because geopolitics, international economy and regional-international development cannot be understood without analyses of regions and their relationships, as well as their interconnections in other dimensions. The physical context, for example, is crucial to understand how climate change would impact the region at specific localities, i.e., provide the necessary details for appropriate response and adaptation to future circumstances. Geographical studies also provide the necessary details for simulation models used to predict future changes under a range of scenarios, if sufficiently high-resolution models could be developed. Geography encompasses the gamut of studies involving human and environment and their interactions across scales, from communities to countries, regions and the planet as a whole. It generates details of areal characteristics, of people, cultures, ecology, institutions, daily dynamics, impacts of hazards and disasters, etc. They all contribute to a better understanding, which will serve sustainable development. The converse, i.e., glossing over international relations and dynamics based on broad statistics, tend to exacerbate inequality and disparity, impacts on certain community and natural environment, because they do not consider specific needs, especially of marginal groups and regions, which are often disproportionately impacted.

A Valeriepieris Circle (Arthur, 2023; Shah, 2022) is a geographical tool for spatial analysis. Proximity is a controlling factor in a range of geographical analysis. They include: extent of environmental impact of a particular pollution source; coverage of a particular service; fallout of a deleterious event, such as, a conflict or volcano eruption; catchment of commercial complex; etc. It also provides an indication of the region's future. The Yuxi Circle is a Valeriepieris circle, which extends 4,000 km from Yuxi in Yunnan, southern China, is home to 4.32 billion people (Shah, 2022). More than half of the planet's population lives in this region, which covers an area of 50 million km^2 or 10% of the Earth's surface. It includes the most populous countries, namely India, China and Indonesia. Most of the economic resources and infrastructure as well as the richest and most diverse natural and cultural elements are concentrated in this part of the world. It is also the region where Xi Jinping's BRI has established an extensive network for trade and cooperation. Southeast Asia (SEA) is located within this compact circle, in close proximity to the global production network and supply chains, key resources, and where a new financial-trade system based on a shared future has been established.

SEA is therefore in a prime location for growth, and the vision of BRICS and BRI aligns closely with Sustainable Development, because they promote mutually beneficial economic growth, in contrast with the zero-sum competitive worldview held by the USA. They are also actively transitioning to green technologies, and equity, where inclusion is more important than return on investment (refer, for example, to China's infrastructure projects and its contribution to supporting development in countries that have struggled with under-development (Jerry's Take on China, 2022, September 18)[6]). Furthermore, China has become the largest economy based on purchasing power parity (IMF[7]), and the Indian sub-continent poised to be a major economic region based on current trajectory. They are located in SEA's backyard and ASEAN has established strong ties with them. Trade and human mobility are expected to become increasing vibrant as the countries and their people refine the means to create opportunities that motivate interaction and exchanges. Finally, at the BRICS summit in Kazan, four ASEAN countries, namely Indonesia, Malaysia, Thailand and Vietnam, were accepted as 'partner' countries.

Regional proximity and the physical environment are key contributing factors to the success of China's ancient silk road and the thalassocracies that developed in SEA (refer to Chapter 10). Therefore, the twenty-first-century iteration of this ancient trade network is likely to be successful for the same reasons. Impediments are largely in the form of systems installed by the colonial powers and its post-WW2 modifications, which are strongly influenced, or even controlled by Western former colonial powers, e.g., USA in the Philippines, and the United Kingdom in Brunei. A large part is due to SEA political leaders wishing to maintain friendly relationships with all nations. Although this is a good stance, the initiatives by the USA to recruit allies to contain China are working against the smooth development of the new silk road. Brunei, for example, has not actively leveraged on its participation to foster and facilitate trade and economic growth.

The Digital Hybrid Future

At the UN Futures summit, head of states agreed on a 'global digital compact' as part of its declaration on the future and future generations. In the region, leaders have also endorsed the ASEAN Digital Masterplan 2025 (ASEAN, 2022). Leaders of state and industry alike, as well as academicians and other interested parties, are aware of this inevitably: transition to a future that is increasingly mediated by computer networks with artificial intelligence (AI) capabilities. Apart from the electronic mind (software and AI), the physical appendages of such systems are also taking over more and more functions and services through automations and robotics. The latter include mobile self-propelling devices, such as drones and quadrupeds, which are increasingly being used in dangerous operations, such as disaster search and rescue, fire-fighting, environmental surveillance, and security operations. In recent conflicts (e.g., Ukraine, Red Sea, Syria and Palestine), drones were widely used in offensive as well as supportive and defensive manoeuvres, offering insights to future wars.

[6] https://youtu.be/MGt_yVt2J90.

[7] https://www.imf.org/external/datamapper/profile/chn.

The technological leaps over the past half century have been enormous. It is therefore not far-fetched to imagine the world in the second quarter of this century to be a hybrid physical-digital one, because this is already a developing reality, albeit at the nascent stage. Although some are fearful and have warned against AI and automation, it is just an inevitable process in human evolution, understood a century early by Vernadsky (1965). A book by British cybernetics professor, Kevin Warwick (1997), described in chilling details how machines could replace humanity and engineered the latter to serve as fuel sources to sustain the former, much like in the move, The Matrix.[8] This is because modern industries are constantly seeking to free humankind from work, essentially causing humankind to forfeit its power to make decisions and controls over its destiny. The problem could arise when activities that require high-level intelligence, such as surgery, is done by machines. Intelligent machines could simply decide to alter conditions to favour their own survival over that of humankind, once it becomes 'self-aware' (a possible emergence in complex systems). However, although this is a plausible development, it is unlikely because there will always be a large portion of humankind outside this sphere of activity.

The most likely scenario is a hybrid reality, where people have virtual equivalents, or digital twins, which would be represented by avatars. This would offer individuals with alternative lives and personalities, and indeed worlds in which they could also earn a living within these realities. This however would not be a virtual reality, because current developments in this space are already working to integrate virtual with physical in Mixed Reality (MR) and the Metaverse (Walsh, 2020, June 8; Graylin & Rosenburg, 2024). The Future will be unimaginably different from contemporary living that new facilities, services and infrastructures will be needed. This means, new types and forms of employment. The office or school might become extinct, as the Covid-19 experience has shown that they are not necessary. Work-from-home applications, such as Zoom and Microsoft Teams, were eventually referred to as 'work from anywhere' apps, and many corporations have adopted more flexible requirements and cultures. Even advancement in communication and information technology could develop further, despite the incredible strides made in recent years. Elon Musk's Neuralink (Reuters, 2023, September 21),[9] for example, aims to link minds and computers, to enhance human as well as human–machine interface, a next phase on the evolution of the Noosphere.

However, notwithstanding the complexification in the realm of humanity and mind, a physical reality exists, and the former is couched within the latter. As in the move, The Matrix, the person dies when its physical counterpart is killed, even though he could perform super human feats in the simulated world. The physical world is the one facing the climate crisis. From an Earth-environmental science perspective (in contrast to activism, policy-making, etc.), climate is only one aspect of the environment. The climate crisis is a result of serious environmental impacts, caused by wanton destruction of nature in the pursuit of wealth due to global capitalism (see, e.g., Bond, 2012; Klein, 2014). This truth has widely been accepted

[8] https://www.imdb.com/title/tt0133093/.

[9] https://www.youtube.com/watch?v=I7pn66iO58M.

before it was replaced by an uneven focus on the climate, and calls for solution that ignore analysis of the problem in its complexity. The development model founded on global capitalism (i.e., modernisation, colonialisation and globalisation) is unsustainable as it destroys nature, disrupts ecosystem functions, and exacerbates inequality (WCED, 1987). However, the international sustainable development programme did not replace capitalism with a more environment-benign alternative, and so its impacts have continued to grow. Modern urban development is fundamentally unsustainable (Soleri, 1969), and a radical shift founded on natural design imperatives is need in the design and development of entire cities (McCullough, 2012). Urban development, if it embraces arcological principles (architecture + ecology), would address climate change and a host of other environmental problems simultaneously (Yong, 2022). Future SEA would certainly be a 'greener' one; hopefully, it would be truly ecological[10] in design.

The Book's Contribution

The chapters addressed four areas of future concerns: (i) climate change, (ii) sustainable development; (iii) migration and social justice; and (v) misconceptions and uncertainties.

Climate Change

Fredolin Tanggang (Chapter 3) provided some clarity of what to anticipate from his long experience with the International Panel on Climate Change (IPCC). Downscaled climate modelling and tidal gauge data suggest that sea level will rise by 0.8 m by 2100 under the worst-case scenario. Typhoons will increase in frequency and intensity. As most SEA countries have highly developed coastline, with some section experiencing subsidence, such details are useful in formulating the appropriate response. Although sounding the alarm and calling for drastic cuts in emission is needed, this has been done. What is needed are concrete developments to adapt to the anticipated future environment. Climate change involves imperceptible incremental shifts, rather than catastrophic turns, and there are human communities and ecosystems presently thriving in regions under almost the entire range of climatic conditions on Earth today, from scorching hot dry deserts to the bitterly cold tundra. Since ancient times, humanity has adapted to environmental conditions by responding to its dynamics and carving out habitats (dwellings, farms, villages, etc.), much in the same way as other organisms. This is needed, besides curbing CO2 emissions.

[10] https://www.ecoliteracy.org/.

A pragmatic response would follow organisms that thrive in specific locality, because they have found 'what works', i.e., 'biomimicry' (Benysus, 1997). A logical-practical solution in urban development in anticipation of future changes could involve mimicking natural biomes that would develop in the predicted conditions. This would be a climate-resilient design (Yong, 2022). In Chapter 10, Gabriel Yong and Izni Azalie referred to the Kampong Ayer, the dense compact settlement complex in the Brunei estuary, and compared how it had thrived with a vibrant 'blue economy' in the fifteenth century before modernisation shifted much of its population and economy to land, displacing the rainforest in the process, and leaving the remnant as a marginal settlement that is bereft of economy. Sea level rise would create conditions that harken back to the past, when estuaries were considerably wider and deeper, and therefore present an opportunity for blue economy development. However, as Yong and Azalie explained, present-day Brunei lacks the capacity to participate and benefit fully in China's revival of the 'ancient maritime silk road' under the BRI. Nevertheless, a 'blue development strategy' is a proactive positive response to rising see level (Yong & Azalie; see, also, Hassan & Yong, 2023; Yong & Hassan, 2024). This is applicable to other coastal settlements and cities in the region, because they shared similar history and environment, and as Tanggang has pointed out, the worst-case scenario is the most probable, based on current international response.

In Chapter 9, Dorothy Jose and Annabelle Ragsag highlighted the need to consider women's needs in disaster response, because the disaster risk reduction management (DDRM) systems developed, though comprehensive, views gender sensitivity from a more conceptual than practical-contextual perspective. The Philippines, together with Vietnam, is located in the typhoon belt, while others are affected indirectly by the disturbances caused. Jose and Ragsag echo concerns raised by Yong and Azalie in policy and management that are disconnected from ground realities, which fail to recognise needs of vulnerable and marginalised groups. This aligns with the aim of the book, which shifts the focus from broad regional superficial characterisation to the marginal and neglected that are inconspicuous to national and regional policy-makers and foreign interventions. The creation of 'women friendly spaces' (WFS) by the Waray community (Jose & Ragsag) is an example of bottom-up developments that must be given greater consideration, support and incorporated into local and national DRRM systems across the region, because inequality is an inherent feature of countries and regions, and social justice not given due consideration in development and DRRM policies and systems. Global capitalism further exacerbates inequality and therefore heightens the risk of marginal groups.

According to Tanggang, the overall climate will be warmer and drier. This would have implications for the drier regions further from the equatorial belt, where there is a distinct dry season, because forest fires could become more prevalent. 'True' forest fires in the equatorial belt are uncommon and unlikely because living trees contain large amount of water and the rainforest is naturally damp. What is reported as forest fires are mainly bush fires in disturbed areas, some of which have stands of trees. They are the by-products of development, which are generally abandoned and at various stages of regeneration. Many are continually disturbed and therefore exist as patches of mixed vegetation that are drier or wetter than the surrounding because of urban structures. Climate change would exacerbate the situation. While bush fires would be more prevalent during hot and dry episodes, these patches are also habitats of pests and disease vectors. In any case, whether bush or forest fires, development policies, as well as management and response systems would be more effective, if they are tailored to the local ground conditions.

Sustainable Development

Every country in the region has committed to sustainable development and ASEAN as made several declarations on it. The book addressed food security, production networks and value chains in economy, Halal industry, and the perception of sustainability by the young, namely, Gen Z. Food security is a critical component to a country's survival and development. The region, however, is largely productive, except for Brunei and Singapore, and are therefore relatively secure in this respect. Countries that lie in the typhoon belt are more exposed to climate risks than other, and increased climate variability will certainly affect cultivation. Therefore, the necessary structures needed to be installed in anticipation of a warmer future with higher incidence of hot dry spells and flash floods, and probably more pests, too. Norzurianie Kamarulzaman presents a case study on paddy cultivation in Brunei (Chapter 5). It highlights the difficulty for success when conventional methods are used in environments that are not conducive to cultivation. With its naturally pronounced marine intrusion, the Brunei case offers a glimpse of conditions with higher sea levels. The topography is also prone to flooding and water-logging but insufficient water during dry spells, another feature of climate change, where variability is heightened. In the Brunei case, the scheme focussed on introducing high-yield wet paddy variety which are not endemic and therefore would need time to adapt to the local environment. This was done on the advice of regional experts and the prioritisation of higher yields over what is most suitable.

Izni Azalie's presentation of Brunei's Halal production programme (Chapter 6) reveals struggles and inefficiencies, because the country similarly adopts conventional approaches without due consideration of local context. He examined the situation from a production network and value chain perspectives, and presented examples of the different ways that local companies tried to support this development, revealing insights to contextual controls that ought to be considered in policy and programme design. Taking a network approach provides a more detailed understanding of the processes at the human and enterprise level, which are inconspicuous in policy making at national and regional levels. The latter tend to focuses on macro-economic indicators. Furthermore, his research design, which is couched within Foucault's concept on governmentality, reveals gaps and dissonances in the mechanisms of programmes based on one-size-fits-all models that are typically introduced and adopted by regional governments; he also employed the same Foucaudian framework in his study on the new maritime silk road with Yong (Chapter 10). The Halal industry is a growing sector in the global economy. In the region, Indonesia and Malaysia are leaders, but others are participating, and would benefit due to the fact that: (i) it is a growing sector; (ii) Muslims and non-Muslim halal consumers account for a significant part of the region's fixed and mobile population; and (iii) Halal standards align closely with sustainable development requirements.

Hasharina Hassan's study of sustainable consumption perception of Gen Z (Chapter 7) found the current young generation (youths) are mostly believers and advocates of sustainability. However, it also reveals a lack of depth in understanding the connections between belief on the one hand, and actions and lifestyle choices on the other. The majority of respondents, who are Muslims, further failed to make the connection between religious guidelines, to which they adhere, and sustainability practices. The majority had received their education on sustainability from social media and government campaigns. This is characteristic of youths in society at large, who tends to refer to social networks for information, or derive them simply from 'google search'. This practice in an increasingly busy world, drowned out by distractions (rather than actual productive work), has given rise to a culture of superficial knowledge that glosses over subjects, themes and issues. This is also evident in policy-making and programme management, presented in the cases above. Perhaps this culture has also penetrated the world of work and governance. Sustainability is a common goal in the region, and of humanity as a whole. However, it is unfortunately generally understood superficially. An appreciation of its ecological basis would certainly consolidate understanding, which would improve programmes on food security and economic production, because they would catalyse the necessary changes to cater to the specific context.

Migration and Social Justice

Migration and mobility are features of the region since historical times, and perhaps, in recent geological epochs, due to the relative ease of mobility across land and shallow continental shelf. Land bridges would have facilitated migration during low sea level stands associated with glacial advances during the Pleistocene Epoch, before they were inundated again by rising seas during interglacial and most recently, the current Holocene epoch, which began c. 11,700 ago. Diotima Chattoraj examined mobility and development in the region in more recent times and discuss their implications in long-term plans (Chapter 11). She examined short-term economic imperatives against long-term sustainability goals. Singapore and Malaysia were identified as key destinations, while Indonesia, Philippines, Cambodia, Lao PDR, Myanmar and Vietnam are sending countries. The trend for Vietnam has, however, changed due to its rapid economic development over the past decade, providing younger generations with ample opportunities and inducing return migration. Migration to developed countries, however, has continued because Vietnamese wishes to find opportunities and develop skills not found at home.

Ahsan Ullah compared a number of theories to explain the main migration corridors that have developed, such as the Malaysia-Indonesia and Myanmar-Thailand corridors (Chapter 12). He noted the primacy of economic drivers, and the pull factors in relation to neo-classical economic theory. However, the dual labour market theory refined understanding of migration pattern due to market stratification that is characteristics of the developing countries, where locals prefer to occupy the primary segment (higher wages, stable employment) and migrants serve in the secondary segment (low wages, unstable employment), which locals avoid, if they have a choice. With globalisation gaining hold in the region, the world systems theory proved more amenable to explain the increasing complex pattern. The new economics of labour migration theory shed light on the migrants' side of the story and elucidated how remittance contributed to developments at home. In the Philippines, remittance from Overseas Filipino Workers (OFW) accounted for 10% of its GDP in recent years.

Both articles highlighted social justice issues associated with migration and mobility. This is one of the three core pillars of sustainable development, which must be addressed in long-term development and economic plans in the pursuit of a sustainable future. ASEAN has adopted a charter to foster and facilitate long-term integration and cooperation, including with regional partners, such as Japan, Korea and China. As the chapters have highlighted, issues and challenges remain. In the vein of this book, perhaps greater attention on the inter-relationship between enterprises and labour at the human levels would provide a better understanding of the complexity and interdependence, as well as contextual constraints. The latter are typically a consequence of existing policies that require updating and circumstance with historical roots dating back to the colonial period, and cultural idiosyncrasies.

Misconceptions and Uncertainties

In the end, the overarching cause of diverse perceptions, concerns, confusions, disagreements, tensions, conflicts and uncertainties relates to our assumptions about reality, i.e., the paradigms we embraced. Gabriel Yong compared two co-existing paradigms (Cartesian Mechanism, which is basis of modernity, and Organismic Systems, an emerging philosophy founded on observation of the natural world) in the perception of, and contribution to, change in global order and climate (Chapter 4). He elaborated on the development and relevance of the latter from studies of the Earth and natural environment and their evolution. The key works presented recognised the active role played by living organisms, including humans, in the co-evolution of the planet and human communities. A fundamental feature of complex evolving systems that contain life is the innate capacity to self-regulate and self-organise, including innovation to evolve novel features and systems. The emergence of mind and intelligence is inevitable in complex living system, and its development across the planet has given rise to the Noosphere. Presently, it is supported by a global web of information-communication system, and created a physical-virtual hybrid reality.

In Chapter 2, Yong elaborated on how the global info-communication network increasingly shaped reality of individuals and societies. The current young generation, who were born into and raised in this hybrid reality perceives the world from largely co-created constructed worldview, which may be quite different from physical reality (compare, Hassan, Chapter 7). The chapter attempted to provide some clarity on global changes by navigating the information super-highway and social media to glean facts from the complex entanglement of narratives, some of which are accurate, while other misinformation or blatant lies disseminated to serve specific agendas. It refers to Baudrillard's simulacrum, as depictions of the current world, in particular in the West, where societies' reality are largely constructed by private and state actors and sustained by societies' participation. The chapter however concluded that humanity is on the precipice of a seismic shift, due to geopolitical processes, which have set in motion developments that is irreversible and will inevitably end in a transition from the current West-led unipolar world with the USA as the primary hegemon to a multipolar world order, which is being developed by BRICS.

While Yong had introduced a number of theories in Chapters 2 and 4, and Izni Azalie in Chapter 6, Ahsan Ullah examined a diverse set of theories relating to politics and security in Southeast Asia. They included colonial legacy, influence of the cold war, globalisation and regionalism. The philosophies of Realism, Liberalism and Constructivism were also introduced to provide theoretical frameworks for a nuanced analysis. It showed how regional policies and dynamics have maintained stability and contributed to impressive growth. In the face of rising militarisation in the South China Sea, routinely attributed to China by Western media, there is a need to winnow out the disinformation and hidden agendas to ensure continued regional stability and advancement. The USA has set up military bases that strategically encircle China; set up an alliance with Japan, Australia and the UK in preparation for a military encounter; recruited the Philippines and established three more new bases; and made no secrets about its intention to contain China. Meanwhile, China has repeatedly said that it has no intention of replacing the USA as the global hegemon. It has consistently engaged in internal development, and international trade and cooperation under a shared future, win–win cooperation, and respect for sovereignty banner. The situation is less muddled when words and actions align.

Conclusion: Insights from Brunei Perspective

As stated at the outset, the book does not present Brunei as a model or exemplar for the region going forward. Instead, it refers to a view of current development and selected issues from Brunei, as a vantage point. The rationale has been explained in Chapter 1. Briefly, Brunei is somewhat insignificant and largely inconspicuous in the international arena and the literature on the region. Among the few to be found, it is often treated with simplistic characterisation: a small oil-rich nation where the Sultan was once the richest man in the world. Expatriate professionals, who have lived in Brunei, even for many years, have tended to live within bubbles, much like the European and Americans in the oil industry and British Army. The book is an attempt to present common issues faced by countries in the region, but reveal the struggles and barriers due to local circumstances, which has historical roots. With regards to the last point, it hopes to show that fortunes and circumstances could change rapidly as the outcome of regional response to external forces. The arrival of the Europeans in the sixteenth century and their subsequent activities based on their agendas, principles, priorities and values, drastically transformed the region, which comprised of the Malay World (Nusantara) and Indochina. The presence of the USA in the South China Sea to defend the region against China, the region's leading partner for trade and in sustainable development, is not very different from the earlier colonisation, which reduced Brunei's territory and prestige to its present insignificant state.

A common threat that connects the chapters is the need to understand reality on the ground, where the smaller people, marginalised, and forgotten, are generally not given much consideration. These people and small regions (from countries, to parts of it) however, collectively and cumulatively over time, could undermine sustainable development, and add to the burden borne by states, and the region. This is why the Brundtland commission included equity and social justice as the third of three pillars in the sustainable development concept (WCED, 1987). The tendency to adopt models and 'best practices' from developed nations and professional consultants, while they have their merits, do not consider local and regional idiosyncrasies. Another common feature of smaller neglected regions is the lack of technical capacity and professionals with the necessary skills and knowledge. A deep dive will reveal large gaps and dislocations, which must be filled and bridged, but hampered by funding. In Brunei, it is taboo to reveal such details because they are misunderstood as an embarrassment or bringing shame to the country. However, from a pragmatic standpoint, it is necessary to improve on weaknesses to make the system and country stronger/better, and more successful.

Finally, the book has opted to look at the negative aspect of development, and particularly of Brunei. To reiterate, the intention is not to present Brunei in a bad light. Instead, as explained, the function of academia in society is to elucidate, so as to improve understanding of the world, and of relationships and interactions, phenomena and undesirable outcomes, and so on. This is achieved by identifying structural weaknesses, which are outcomes of complex interactions involving multiple actors and elements, as well as historical events. This is why the book included chapters that provided some details of Brunei's history and development, and presented its current state and circumstance as part of the country's story, which is by no means, near its end. Brunei, like other countries in the region, have achieved tremendously since their rebirth following the Second World War and decolonisation, and is continuing to grow. However, as countries and regions develop and interact, adjustments are needed to reach equilibrium that is favourable to the parties involved. This is particularly important at this juncture in humanity's story, because it is poised for a major shift, and geopolitical tension are almost at breaking point. Therefore, identifying weaknesses would help focus attention on what is most needed. This is where a perspective from Brunei would help.

References

Arthur, R. (2023). Valeriepieris circles for spatial data analysis. Geographical analysis. Wiley Online Library. https://doi.org/10.1111/gean.12383

ASEAN. (2022). ASEAN digital masterplan 2025. https://asean.org/book/asean-digital-masterplan-2025/

Benysus, J. (1997). *Biomimicry: Innovation inspired by nature*. Harper Collins.

Bond, P. (2012). Capitalism and the environment: Reconciling the conflict. University of the Witwatersrand Press.

Center for Ecoliteracy. (n.d.). Center for Ecoliteracy. https://www.ecoliteracy.org/

Galimberti, S. (2023, March 27). ASEAN 2050: Towards a new ambition. *The Jakarta Post*. https://asianews.network/asean-2050-toward-a-new-ambition/

Graylin, A. W., & Rosenburg, L. (2024). *Our next reality: How the AI-powered Metaverse will reshape the world*. Nicholas Brealey.

Hassan, N. H., & Yong, G. Y. V. (2023). Climate change and sustainable development of Kampong Ayer: Its history, people, and environment. In M. Hamdan, M. Anshari, N. Ahmad, & E. Ali (Eds.), *Public policy's role in achieving sustainable development goals* (pp. 96–114). ICI Global.

Klein, N. (2014). *This changes everything: Capitalism vs. the climate*. Simon & Schuster.

Jerry's Take on China. (2022, September 18). China's Diversity - From Hightech Startup to Poverty Alleviation, They're All in It [video]. YouTube. https://youtu.be/MGt_yVt2J90

Latin, H. A. (2012). *Climate change policy failures: Why conventional mitigation approaches cannot succeed* (p. 268). World Scientific. ISBN 978-981-4355-65-0. https://doi.org/10.1142/8194

Mahbubani, K. (2024). *Living the Asian century: An Undiplomatic Memoir* (p. 336). PublicAffairs. ISBN 139781541703049

Mahbubani, K. (2022). *The Asian 21st century* (p. 270). Springer Singapore. ISSN 2730-9983. https://doi.org/10.1007/978-981-16-6811-1

McCullough, L. (Ed.). (2012). *Conversations with Paolo Soleri*. Princeton Architectural Press.

Putnam, J. (1991). *Representation and reality* (p. 154). MIT Press. ISBN 9780262660747.

Reuters. (2023, September 21). Musk's Neuralink to start human trial of brain implant [Video]. YouTube. https://www.youtube.com/watch?v=I7pn66iO58MSoleri. (1969). *Arcology, city in the image of man*. MIT Press.

Shah, R. (2022). Half the world's population lives inside this circle. Shah, Riaz (2022–11–10). "Half the world's population lives inside this circle

Times of India. (2024, November 22). Putin FULL SPEECH I 'Ukraine War Going Global. Russia Has the Right to Strike [video]. YouTube. https://youtu.be/yR-OSK7grOA

Vernadsky, V. I. (1965). The biosphere: The living envelope of the Earth. (L. G. Sears, Trans.). Harvard University Press. (Original work published 1926)

Walsh, M. (2020, June 8). The Metaverse Is the Next Universe. [Video File]. YouTube. https://youtu.be/0TCGBhY7qfE

Warwick, K. (1997). In the Mind of the Machine. London: 2nd Story Press. 224 pp.

WCED (World Commission on Environment and Development). (1987). Our Common Future. Oxford University Press.

Yong, G. Y. V. (2022). Climate-resilient urban design from a biomimicry-arcology perspective. *International Journal of Environment, Architecture, and Societies, 2*(1), 1–15. https://doi.org/10.26418/ijeas.2022.2.1.1-15

Yong, G. Y. V., & Hassan, N. H. (2024). Blue urbanism: An insight of navigating climate change in humid tropical cities in Brunei Darussalam. *International Journal of Environment, Architecture, and Societies, 4*(01), 23–36. https://doi.org/10.26418/ijeas.2024.4.01.23-36

Zachos, J. C., Medin, K., & Brown, T. (2010). Geologic history of the Earth's climate and its implications for future climate change. *Global and Planetary Change, 72*(1–2), 4–21. https://doi.org/10.1016/j.gloplacha.2010.01.007

MIX
Papier aus verantwortungsvollen Quellen
Paper from responsible sources
FSC® C105338

If you have any concerns about our products,
you can contact us on
ProductSafety@springernature.com

In case Publisher is established outside the EU,
the EU authorized representative is:
Springer Nature Customer Service Center GmbH
Europaplatz 3, 69115 Heidelberg, Germany

Printed by Libri Plureos GmbH
in Hamburg, Germany